A PRACTICAL GUIDE TO

BUILDING

SMALL GAS

BLIMPS

Copyright 1977
by
Robert J. Recks

1997 Edition; Contains over 200 illustrations in 200 pages

A PRACTICAL GUIDE TO
BUILDING
small gas
BLIMPS

Copyright 1977
Revised 1987, 1997.

All rights to bound volumes of this book are reserved for the author, who is wholly responsible for its content and periodic revisions. Inquiries should be addressed to:

Association of Balloon and Airship Constructors (ABAC)
P.O. Box 3841
City of Industry, Calif. 91744-9991

Library of Congress Cataloging in Publication Data:

A Practical Guide to Building Small Gas Blimps

Bibliography. p. Includes index.

Summary: An introduction to constructing small engine powered gas blimps manned for sport and research as adult reading.

I. Aircraft
I. Airships
I. Author
I. Blimps

ISBN 0-937568-28-7 (hardcover edition)

ISBN 978-1537361451 (paperback edition)

TL685.1 C++ 629.133'+++

To order/reorder the paperback edition:
https://www.createspace.com/6530583

Building Small Gas Blimps

CONTENTS

- Foreword 1
- Preface 2
- Outline 3
- Glossary 4
- Blimp History 6
- Design Criteria 7
  - Ultralights 9
  - Standard Type Certification 10
  - Standard Sizes 11
- Materials 13
- Workmanship 17
- Ground Support 21
- Masts 28
- Hangars 42
- Gondola/Airframe 47
- Instruments 69
  - Instruments & Controls—Suppliers 78
- Engines 80
  - Ducted Fans 84
  - 2-Stroke Engines—Suppliers 86
  - 4-Stroke Engines—Suppliers 87
  - Propellers—Suppliers 88
- Valves and Pressure System 89
- Fins 101
- Envelopes 122
  - Fabric & Webbing—Suppliers 135
  - Assembly Procedure 142
  - Bibliography 152
  - Airship Fin & Rudder Loads (BuAer LTA Design Memorandum No. 169) 154
  - Gas Airship Parts—Suppliers 157
- Catalog Section 158
- Airship Technical Notes 179

Building Small Gas Blimps

RECHS, Robert J. (pronounced Recks) Nickname: Rex. b: 31 Aug. 1935 South Gate, California, USA; U.S. Army 1957-59 (mechanic); e: Minor-Engineering, Major-Aviation Management 1960, University of Miami (FL). p: Airline pilot (PanAm) ret.1988; All aviation ratings. Flew in more than 100 countries (licensed in six). f: First Blimp flight 1960 with R. Widdicomb; First Balloon flight 1969 w/Don Piccard; FAA-CFI, DPE, DER, A&PIA, DAI. Licensed Engineer & Master Rigger; Pioneer of Ultralight aircraft. Engineering Consultant & Inspector on many balloon & blimp projects; Author of many aviation publications. Founder of the Lockheed Balloon Club (over 100 members); Introduced balloon construction, sport ballooning, and B-pilot standards to Brazil in 1971. Member Al Malika Shrine, Amer. Institute of Aeronautics & Astronautics, Balloon Federation of America, BFA Recipient of the first Shields-Trauger award for contributions to the sport of ballooning. r: 1148 Third Ave. Chula Vista, CA. (2005)

Building Small Gas Blimps

FOREWORD
R: 11-02-97

So you saw the *Goodyear* blimp fly over one day and think you could build one better? Let me be the last one to discourage you. It can be done, but NOT without LOTS of money, and a lot of blood, sweat, and tears. Yes, they do look simple; and they are, basically. But all of the historical information you may have read is no primer for that day you turn on that hissing valve that starts the inflation. At that point, any mistakes you have made, are not going to be overcome by several thousand dollars worth of helium. Your creation will be all consuming; you'll **have to live with it,** or leave it.

This book IS directed toward the construction of small blimps. Small meaning less than 60,000 cuft., Helium filled, Engine powered, and manned by a pilot on board. It does have applicability to unmanned, radio controlled, and solar or man powered airships. But has very little relationship to Thermal-air filled.

This book is NOT intended to be a design manual, but rather a guide as to the best of what has been tried on small blimps. At this point, you no doubt have some new ideas to test, and dreams to develop. I say again, do not let me discourage you. But you do owe it to yourself, and the airship community, to visit several successful airship operations and see for yourself what it takes to handle a blimp in high winds & no winds, and in temperature extremes. Talk to the crews, kick the tires, feel the fabric, and by all means, **ask questions.**

The most important thing I want to get across to you is safety. Not just to you, but particularly your crew, and the people you fly over, of course. You won't find much in here on loads, stresses and yields of materials. You, as the pilot are going to have to set the limits. Even the Wright Brothers went through the school of hard knocks (one was killed). Expert advice is available; at least listen and give some thought to all suggestions. **Do not compromise safety,** even if it costs a little more money.

Lastly, the designs and techniques in this book are NOT my ideas. They have been tried and tested over the last 100+ years. The only new innovation in the last 50 years are the raw materials readily available, The biggest, of course is *DuPont* for synthetic fabrics. I **do not warrant anything in this book for your particular application.** So please do not call me at 2 in the morning to ask me what went wrong. Or at any time if you call yourself an *airship* manufacturing company.

The author

PREFACE
R: 11-02-97

My standard question is: **Do you really understand what you are getting into?**

You are contemplating a project that will be all consuming for one to five years. Are you willing to sacrifice family and friends for the possibility of a few moments of pleasure?

Were you born rich, or just win the lottery? What ever amount you have in the bank, it isn't enough. If you are a smooth talker, and plan to get 10 investors, you will never have enough. Each investor will want to tell you how he wants it done, and it never will be. If you think it will pay for itself in advertising, you'll never live long enough. Be realistic, plan on redirecting $100,000 of your net worth to develop even a small blimp. When you start thinking about a second mortgage on your house, you are in over your head.

What do you know about airships in general? What do you know about aerostatics in particular? Are you handy with tools; how about a heliarc welder, or a vertical mill? How about mathematics; ever try a simultaneous equation that your life depended on? Try some simple math, like how much will 100 bottles of helium cost, then 2 per week after inflation.

What are you going to do with it when it is inflated (other than making a million $)? Surely you don't plan to park it in your 2-car garage. It'll probably end up on a "stick" mast in some farmers field, outside of town, where you can't see it when the next hurricane comes through. Forget a hangar, the cost of your airship just doubled.

Know anything about small airplanes? I am just going to assume that you do. If you don't, you are out of your element. Get some training before you turn another page. In 35 years around blimps, I've never heard of an experienced airship pilot wanting to build one. Where do you plan to get your experience? Yes, you can buy a car without the salesman asking if you know how to drive. But how close to the ground are you going to be if you fall out the gondola door? Do you really think you are going to coast to a stop when your engine quits?

Do you have a lot (thousands) of friends? Launching and recovering an airship is definitely NOT something you are going to do by yourself (and friends seldom come out twice). The number in your ground crew will depend on the volume of your ship, AND the extremes of weather you operate in. Ever try to stop a Volkswagen rolling at you at 5mph? Better think about 20mph on a common day. As some famous guy once said, $F=ma$.

So lets do it anyway

Building Small Gas Blimps

A practical guide for small Airships

OUTLINE
R: 12/03/97

Foreword: Limits of what this book is trying to accomplish, staying under 60K, and away from commercial ventures.

Preface: An understanding of the expertise, time, & money involved; personal preparation by you as the pilot.

Design Criteria: Deciding what you want to do with it; How many people onboard, how high, how fast, and how far. What & where the FAA will let you do with it.

Materials: Understanding material strengths and safety factors; What is commonly used & available; where to find it. Buying & Inspecting Fabric.

Workmanship: Understanding your personal limitations, the tools, machines and work area needed, and knowing the limits of those you sub-contract to.

Ground Support: Where are you going to put it once you assemble it? Helium supplies, inflation nets, crew/helpers Hangars, Masts, aux. Generators, Blowers, Vehicles.

Airframe Sub-assemblies: What to expect and what you need
    Gondola: Shape, size, accessibility, structures, maintainability;
        Instruments, mechanical, digital, placement;
        Engines & Propellers: Rules for matching the two;
    Nose reinforcement: Docking cone, battens.
    Fins: Shape, size, structure, controls, rigging, lights.
    Envelope: Shape, Volume, Gore layout, Burnouts,
        Ballonets, valves, access & deflation.

Sources (Proven units, availability off-the-shelf)
    Materials and suppliers
    Airship expertise & Airship Helping organizations

Technical information

Bibliography	History	Glossary
Helium	Mathematics	Strength of materials
Selected excerpts	Licensing	Index

Respect those that ask you seemingly "dumb questions" (you may learn something), And those that want to help (all will be a lasting friends).

Building Small Gas Blimps

GLOSSARY

Actuator; an electro-mechanical device for moving by advantage in a linear or rotational motion.
Aerodynamic; a shape that streamlines or generates lift from the movement of air flow around it.
Aerostat; an aircraft whose buoyancy is derived chiefly from a Lighter-Than-Air gas.
Airship; a powered aerostat with dirigibility.
Ballast; a quantify of weight used to compensate for changes in lift.
Ballonet; a chamber inside the hull used to control the internal pressure.
Battens; envelope nose stiffeners to resisting pressure from forward speed.
Blower; fan used to maintain ballonet pressure.
Breathing; the passage of air through the valves to compensate for pressure changes.
Cables, Suspension; steel cables from the CATENARYS, supporting the gondola weight.
Cables, Control; steel cables from the gondola controlling the elevator & rudder.
Car; a commonly accepted slang term for the gondola.
Catenaries; curtains inside the top of the envelope for attaching cables to support external gondola loads.
Ceiling, ballast; the maximum altitude to which an airship can ascend and return to the surface in equilibrium.
Ceiling, ballonet; the maximum altitude to which a pressure airship can ascend and return to the surface under pressure with the ballonets 100% full of air.
Ceiling, static; the altitude in standard atmosphere at which an aerostat is in static equilibrium after removable of all dischargeable weights.
Center of Gravity; (CG) that point on the longitudinal axis where the hardware weight is equally balanced.
Center of mass; (CM) that point on the longitudinal axis where the aircraft weight (incl. gas & air) is equally balanced.
Cowling; the covering section of the engine and accessories.
Cradle; a wheeled device for moving/storing a gondola when not attached to the envelope.
Damper; a valve to regulate the flow of air to the ballonet(s), commonly referred to in slang as a T-Box .
Deflation; The act of removing the gas and air from an aerostat.
Diffusion; is the ability of an element (gas/air) to pass through a medium (fabric)
Dirigible; an aerostat with the capability of being directionally controled.
Displacement; is a volume of a contained mass. i.e. Gas in the hull, or air in the ballonets.
Drag; the resistance to airflow over a surface.
Drip fence; a strip of fabric glued to the bottom of the envelope to prevent moisture from running off the envelope into critical areas.
Droop; a condition where the airship end(s) sag from lack of pressure, or excessive weight.
Elevator; the control surface(s) affecting pitch (climb & descent).

U.S. Navy Training Manual, 1942

Building Small Gas Blimps

Envelope; is the fabric container of the contained masses, gas and air.
Envelope, gore; the assembled panels that form an envelope section from the nose to the tail apex (lettered clockwise from inside the envelope looking forward).
Envelope, panel; one of many pieces that are attached to form a gore (numbered from the nose to tail).
Equilibrium; an instantaneous condition where the aerostats lift equals its weight.
Fairing; a cover joining two surfaces to improve the air flow over the area.
Full castoring; referring to the gondola wheel that is free to swivel 360 degrees.
Gondola; the enclosed structure housing the occupants and cargo.
Helium; an inert, colorless, odorless gas, toxic in high quantities.
Inspection Window; see View Port.
Manometer; a device for measuring gas pressure.
Mast; a pole device for airship docking.
Mooring; the docking of an airship.
Mooring cup; the stationary receptacle of the mast for receiving the docking cone.
Mooring cone; the mast peg on the airship nose for docking to the mast.
Osmosis; the exchange of gas molecules though a medium (fabric)
Patch, Finger; a reinforced piece of fabric glued to the envelope as a restraint.
Patch, surge, a finger patch to prevent fore & aft movement of a unit.
Patch, brace, a finger patch to prevent side movement of a unit.
Patch, scuff, a strip of fabric to prevent abrasion.
Permeability; is the rate of diffusion of an element (gas/air) through a medium (fabric).
Purging; the process of removing impurities from the gas.
Rigging; is a term given to cable and rope positioning and adjusting of units.
Rigid airship; an airship with a complete internal structure to maintain its shape.
Rip line; a heavy cord found near the pilots exit, used for emergency deflation.
Rip panel; a tear-out panel for emergency airship deflation.
Rudder; the control surface(s) affecting yaw (turn left & right).
Semi-Rigid; an airship with a partial structure to support the external weight.
Stability, inherent; the natural ability to recover from a disturbed attitude.
Stability, static; the natural ability to recover from a roll disturbance.
Static lines; copper wires extending from the metal control surfaces for dissipating static electricity.
Topping up; adding helium to optimum to replace that lost from osmosis.
Turnbuckle; a metal device for adjusting length or tension.
Valve guard; a device used to keep the slipstream from forcing the valves open in flight.
View Port; an inspection window for viewing the inside of gas or air cells.
Volume, refers a size of an airship envelope; is that total quantity of hull displacement, gas and air.
Wheel lock; a device used to keep the gondola wheel perpendicular to the longitudinal axis while the airship is on the mast.
Weight, absolute; the total weight of the airship at equilibrium.
Weight, empty; without crew, fuel, oil, or fixed provisions (**EW**).
Weight, gross; weight at that particular time (**GW**).
Weight, Max.Alowable Gross; structural weight that may not be exceeded (**MAGW**).

U.S. Navy Training Manual, 1942

Building Small Gas Blimps

BLIMP HISTORY

In response to requests, some historical background is presented here to show the progression of blimp technology. Aerodynamic hulls were not understood until 1923. Most library Aeronautics shelves have complete books with details of these events.

year	name	nationality	unique design
1783	MARYTON, Thomas	ENG	Proposed sail & rudder design
1783	MORVEAU, Guyton de	FRA	Used rudder and muscle powered oars
1784	MIOLAN/JANINET	FRA	Proposed thermal propulsion
1784	MEUSNIER, General	FRA	Proposed ellipsoidal shape with ballonet
1784	BLANCHARD, Pierre	FRA	Used muscle powered winged oars
1784	CROSBIE, Richard	ENG	Used fixed & revolving sails
1785	GUYOT,	FRA	Used egg-shaped ellipsoidal hull
1785	PATINHO,	SPA	Used muscle-powered oars with rudder
1785	POTAIN, Dr.	ENG	Used muscle-powered prop
1785	HOOLE, John	ENG	Proposed fish-form, with flapping fins
1785	LUNARDI, Vinc.	ITA	Muscle powered winged oars
1812	BRAMAH, J.	FRA	Proposed trail rope guidance system
1812	DEGEN, Jacob	OUS	Used muscle powered flapping wings
1816	EGG, D./ PAULY,G.	FRA	Experimented with fish-forms
1825	MASON, Monck	ENG	Made many successful models
1825	GENET, Edmond C.	USA	Proposed horse powered paddle wheels
1828	BAYER,	ENG	Proposed multiple balloon hull
1834	LENNOX, Compte de	FRA	Prop driven "Eagle"
1834	DUPIS-Delcourt,	FRA	Used copper hull with prop
1838	ORLANDI, Francisco	ITA	Used thermal balloon with wings
1850	PETIN, Jusque la	FRA	Used four balloon tandem hull
1850	BELL, Hugh	ENG	Used muscle powered cylindrical screw
1850	TAGGART,	USA	Used multiple steering paddles
1850?	LUNTLEY, John	ENG	Proposed coal-gas, engine-driven prop
1851	JULLIEN, Pierre	FRA	Successful working model
1852	GIFFARD, Henri	FRA	Successful steam-engine driven prop
1853?	NYE, James	ENG	Proposed explosive power
1855?	ALDBOROUGH, O'Neal	ENG	Patented dirigible aerostat
1863	ANDREWS, Soloman	USA	Used triple-hulled aerodynamic shape
1865	DELAMARINE,	LUX	Successful helices driven "L'Esperance"
1869	MARRIOT, Frederick	USA	Proposed steam engine driven prop
1872	LOME, Dupuy de	FRA	Successful muscle powered prop & ballonet
1872	HAENLEIN, Paul	OUS	Used spindle-shaped rotating hull
1878	RITCHEL, Charles	USA	Successful muscle-powered prop
1883	TISSANDER, A.& G.	FRA	Successful electric driven prop
1888	WOELFERT, Michael	GER	Proposed gasoline engine driven prop
1897	SCHWARZ, David	GER	Successful gasoline engine driven prop
1900	ZEPPELIN, Ferd. von	GER	Successfully used rigid hull design
1901	SEVERA, Agusto	BRA	Used centerline thrust props

Who's Who of Ballooning, Appendix-E, 1983

Building Small Gas Blimps

DESIGN CRITERIA
R:10-12-97

My standard question is: **What do you really want to do with it?**

With enough money, you can make an airship to fly 200 passengers at 50,000', at 100 mph, but each parameter extreme will be a trade off with something else (money being constant, of course). So lets keep it simple, and in the doable range.

How about a blimp that will fly a pilot and one passenger at 1000' at 30 mph for 2 hours. Does that sound reasonable to you? It does to me, but I need to be more specific about how the numbers are derived for your particular location and operation. Are you Wintering on Everest and Summering in the Gobi Desert? Not likely. Most blimps (commercial) Winter in Florida, and Summer in the North, preferably near water for temperature stability. It is not likely that a seasonal migration is an option for you. But if we have to pick a specific volume for your *Envelope*, lets try to establish the parameters in the order of importance:

KNOWN
Useful Load (varies with desires)
- 2 people @ 170-210 lbs. each (*1) — 380 lbs.
- Fuel, 6 gal. per hour for 2:30 (*2) — 90
- Disposable ballast, 4 bags (*3) — 100

PRELIMINARY CONFIGURATION
- Envelope, 20-40,000 cuft, (*4) — 300
- Gondola, 2-place, dual controls (*5) — 300
- Single engine, ultralight 50-60 hp (*6) — 100
- Removable banners (both sides) (*7) — 50
- Fins & Rigging (cruciform tail) (*8) — 150

Gross Weight (the weight of everything at take-off) (*9) — 1470 lbs.
Volume required on California coast in Summer — 24,000 cuft.

VARIABLES
- Ambient Temperature, 1% for ea. -5 degrees — 180
- Altitude above Mean Sea Level, 1% /375' — 540
- Weight penalty extreme — <u>720</u>
- Total — 2190 lbs.

Volume req. at Minneapolis on a Summer mid-day — 35,900 cuft.

Analysis: On the average, a 30,000 cuft. volume is a realistic number. I have personally flown 2-man airships of 19,000 and 42,000 cuft. limited to one man at the extreme. All of the numbers above are realistic. The trick now is to keep below those numbers or fly only in the most favorable conditions. Keep a running log by setting these numbers in your computer, updating each item with actual weights.

Building Small Gas Blimps

Design Criteria cont.
R: 11/06/97

EXPLANATION of the numbers

(*1)   I recommend 2 seats because an instructor will be needed, a friend impressed, and a reward given to a returning ground crewman.

(*2)   Many smaller, fuel efficient engines are available. But consumption changes significantly when you want more power and reliability. Two and a half hours is plenty before you need a pee-break.

(*3)   Bags are kept in easily remembered weights. A crewmember can carry four 25 bags readily.

(*4)   Decide what volume you can live with **ONLY** after all of the other hardware is constructed and accurately weighed.

(*5)   Have Dual controls, but make one (aft) set removable.

(*6)   Remember to include the Prop. and all accessories in your weights.

(*7)   Definitely a hassle, but easy to see the exposure value.

(*8)   A lot of weight variation is possible, all affecting flutter and controllability.

(*9)   At Standard Temp. & Pressure (STP = Sea level x 59F x 29.92" Hg.).

Variables note: Don't plan on flying in Las Vegas or crossing the Continental Divide.

FAA LIMITS

It is NOT my obligation to quote you the regulations of the Federal Aviation Administration, but I want you aware that this is a powered aircraft that can't help but cross FAA *controlled* airspace. Since you may be required to request an Experimental Airworthiness Certificate, I'd play it safe and invite the FSDO-Airworthiness reps. to visit/inspect your project at critical points in construction. Yes, I've heard it before, he IS there to help you (get you your certificate). He is NOT there to build it for you, so you had better listen to him. If he trusts your judgment, and sees good workmanship, you may be issued Operating Limitations you can live with. By all means, ask him if there is a DAR (Designated Airworthiness Rep.) or DER (Engineering) nearby that may be interested in a low-budget project. If he tells you to call me, be advised that I have done more than my share of charity, and I am not a push-over. Let me do at least 50% of the talking, or I'm gone.

What the FAA does NOT want to hear, is that the extra seat and banners are for hire, or in any way linked to a commercial venture. Crew training seat, Yes. Banners available for a charitable cause, OK. But a slip of the tongue can bring you grief. **I do advise strongly**, to have copies of the applicable Fed. Aviation Regulations (FAR's) and Advisory Circulars available for reference. See **SOURCES** chapter.

DEFINITION

Just in case somebody asks you. An ULTRALIGHT is an *AIRCRAFT* operated for sport or recreational purposes which does not require FAA registration, Airworthiness Certificate, or Pilot Certification. They are primarily single occupant vehicles, but may be authorized 2-place for training purposes. It may **not** be operated for compensation or hire, nor in controlled airspace without prior permission.

Building Small Gas Blimps

ULTRALIGHTS

Background: Just to clarify an important point, the Federal Aviation Administration (FAA) gives the word very specific limitations. If you stay within the guidelines, it can save you some nickels and dimes, but you will still have specific limitations on what you can do and where you can operate. A home builder can NOT claim ignorance if challenged. The complete rules (Federal Aviation Regulation, FAR-103) are available from the Govt. Printing Office, or your local FAA Flight Standards Office. And I suggest you acquire a copy AND keep it available.

Definitions: The specific rules include an aircraft empty weight (E/W) not exceeding 254 pounds. To a blimp, this means the entire aircraft BEFORE any lifting gas is added. The rules also stipulate that this category was designed for the "homebuilder/pilot" as the only "owner/operator" of the aircraft. Yes, some ultralight airplane builders you have seen do push (over) the limits, but one challenge will initiate an investigation. The gray area comes when you put in an extra seat for a passenger and call it "student training". The other gray area is when you see one flying over your house, obviously out of the designated ultralight flying area.

Benefits: For you as a blimp owner, **NOTHING**. Yes, you eliminate a 30 minutes of paperwork and a few dollars. But you are **NOT** required to have a FAA Pilot License to fly your *blimp* regardless. Even if you do not qualify as ultralight, you can still apply for a "Student License" (free) and request a "waiver" to fly your "experimental" aircraft in a mutually agreed upon "flying area". Be sure to mark an X in the block "crew training" of your FAA application Form 8010-6.

Is it Possible to build a blimp with an E/W of less than 254 pounds? Certainly, and I don't mean out of *Saran-Wrap*. I have seen TWO good 2-seaters under 9000 cu.ft. with single Konig engines that worked very well. It is easy to strap a small engine to a lawn chair for a gondola, with a 4 mil. heat sealed *Polyethylene* envelope. One of the two above had the engine gimbaled for directional control, the other had single layer tensioned fins. But then, Santos Dumont & the Wright Brothers had pretty flimsy designs also.

Some Numbers: One *Airship Homebuilder* is presently contemplating the construction of a blimp with the following target weights:

Envelope:	63 # 9000 cu.ft. (3 oz/sy)		
Ballonet:	21 # 10 %	Tail Fins:	35 # 115 sq.ft.
Propulsion	50 # 15 hp.	Framing:	40 # Alum.tubing
Cabin:	15 # enclosure	Nose:	20 # structure

Total equals 244 pounds, leaving only 10 pounds for valves, ducts, control cables, instruments, and comfort (?). I might add that the pilot weighs about 140 # wringing wet.

It CAN be done, but forget it.

Building Small Gas Blimps

STANDARD TYPE CERTIFICATION

Summary:
This status is NOT what this book is intended for, and is definitely not an endeavor for the amateur. It involves a lot of time (one-two years), costs a lot of money ($100,000 up, for a small aircraft), and may require major changes in your design. If you are only going to make one "sport" aircraft for yourself, FORGET Type Certification. Make it as good and as safe as your capabilities.

What does Standard Type Certification (STC) do for you?
This status allows the aircraft to be used the same as a Cessna-150; using the same airspace, AND in commercial ventures. It allows you to go into the manufacturing business, AND export completed aircraft to foreign countries.

What is required for an STC?
It requires additional staffing of permanent employees as "qualified & experienced" managers, technicians, assemblers, AND inspection personnel. These positions are not taken lightly, and constant monitoring by the FAA will preclude declining standards or high turnovers of personnel.

What does STC NOT do for you?
It does not allow you to make design changes without FAA-Engineering approval; which will require additional testing and data. It does NOT allow you to deviate from the regulations, or from the accepted procedures as agreed in your STC. It does not allow you to make or sell "kits".

Supposing I still want to proceed?
Go for it ! It may sound easy for someone that has been around Cessna-150's, but you had better start doing a lot of reading. Sound like you know what you are talking about when you approach your local FAA. Here are a few FAA publications you will need:
    8110-2 Airship Design Criteria
    8100.5 Aircraft Cert. Directorate Procedures
        Design Approval Process
        Certification Basis & Notification
        Standards & Compliance
        Procedures & Inspection
        Testing & Witnessing
        Quality & submission of Data for approval.
    Federal Aviation Regulations Part-1, -21, -23, -27, -29, 33, -36, -39, -43, -45.
        FAR Part-183 Representatives of the Administrator
        (DAR.s ,DER's, DMIR's available for help).
    Advisory Circulars, AC 21-23;
    Airworthiness of Aircraft, ICAO Standards
These are not the sole sources of pertinent information, but it is a starting point.

R.Recks@Juno.com

STANDARD SIZES
11-02-97

Just a brief note to make you aware of International Class Sizes as designated by the Federation Aeronautique International (FAI). The FAI is the World governing body founded (1908) to establish guidelines for In'tl competition and records (Altitude, Distance, & Duration). A few reasons for telling you this are:
1. Very few airship/blimp records have ever been applied for,
    and thus the glory goes to the first that try;
2. There are an increasing number of blimps in the World, and more owners want
    to compete with the latest technology (technology usually beats skill);
3. The French want to start serious (biennial) competitions by the year 2000.
    (Thermal airships have been competing every other year since 1987;
    The first Gas competition is scheduled for 1998).
4. Competitions and records are categorized by CLASS SIZE (volume);
5. The glory usually goes to the designer that builds his envelope just within the
    maximum volume allowed for that class (dumb luck beats technology).

At present (since July 1987), the class sizes for airships are:

Class	Range		Cubic Feet
B?-1 = Less than		400 cu.m.	(14,125 cu.ft.)
B?-2 = 400	cu.m. to	900 cu.m.	(31,783 cu.ft.)
B?-3 = 900	cu.m. to	1600 cu.m.	(56,503 cu.ft.)
B?-4 = 1600	cu.m. to	3000 cu.m.	(105,943 cu.ft.)
B?-5 = 3000	cu.m. to	6000 cu.m.	(211,888 cu.ft.)
B?-6 = 6000	cu.m. to	12000 cu.m.	(423,776 cu.ft.)
B?-7 = 12000	cu.m. to	25000 cu.m.	(882,866 cu.ft.)
B?-8 = 25000	cu.m. to	50000 cu.m.	(1,765,733 cu.ft.)
B?-9 = 50000	cu.m. to	100000 cu.m.	(3,531,467 cu.ft.)
B?10 = 100000 cu.m. and above.			

NOTE: The "B?" prefix is for AIRSHIP. *HOWEVER*, **BA is for GAS blimps**; BX for THERMAL; BR for RIGID's; and BT for OTHER (i.e., R/C, solar or pedal powered).

DISCLAIMER:

The information contained herein, may, or may not fit your particular requirements. It only shows the best of what has been done. Anything *you* fabricate *I* will guarantee 6 feet or 6 seconds. This book *SPECIFICALLY EXCLUDES* building an aircraft for a *Standard* Type Airworthiness Certificate. To meet the requirements, the simplest project can be expected to cost $500,000 and 2-4 years to complete. To meet those numbers, you need a staff of DER's on the payroll. The time and money involved are beyond 99.99% of those expressing an interest, and definitely ***NOT* within the realm of this book**. Trust me, FAA certification is my living.

THINK SMALL, ***THINK SPORT***

Building Small Gas Blimps

Make your own spreadsheet and use actual numbers.

SUMMARY weights	No U/L	W/ U/L	Weight in pounds	Location in feet
			30K AIRSHIP	
Lift POSSIBLE	1830	1380	Sample Spreadsheet	
Weight NOW	1039	589		
Lift AVAILABLE (1)	971	791		
W/ half Ballonets (2)	742	* 562		

ASSEMBLY/Item	Weight	Location	ASSEMBLY / Item	Weight	Location
GONDOLA, Frame	100		**ENVELOPE**, Basic	400	42
Covering	10		Ballonet	25	
Glass	15		Scuff Pads	2	
Seats	60		Reinforce.	3	
Controls	20		Finger Ph.	10	
Instruments	10				
			Bat. Tunnel	10	
PROPULSION			Lace Chan.	5	
As Supplied	60				
Generator	5		**ACCESSORIES**		
Starter	5		Strobe	2	
Carburetor	2		Pos. lights	1	
FUEL SYSTEM (dry)					
Tank	12		Envelope Nose Ring	10	
Aux Pump	1		Ring Cables	5	
Plumbing	6		Battens	15	
Transducers	2		Burn Ring	2	
			Drip fence	5	
PROPELLER			Helium Fill	1	
As supplied	22		He. Window	1	
Hub	3		He. Valve	6	
Bolts	2				
Safety Ring	25				
FINS Frames	90				
Fabric	10		Ballonet View Port	1	
Con. Surface	26		Acc. Port	3	
Guy Cables	30		Air Valve	5	
Cont. Cables	10		T-Box	6	
USEFUL LOAD (U/L)					
Pilot	170		Fuel	60	
Passinger	170		Ballast	50	

(1) Ballonet empty; (2) Assuming ballonets 25% of total volume (12% = 229 lbs.)
NOTE: Try to keep within the weights, GOOD LUCK.

Building Small Gas Blimps

FABRICS

The Yellow Pages are full of fabric suppliers, and the best sources are listed in the Supplier Section of this chapter. You will probably want to contact them all by phone first to understand their "lingo". Yes, the fabric trade has their own language. What they take for granted, you may not understand until too late. Warning, **If** in doubt, ask. **If** still in doubt, get it in writing.

WHAT KIND of fabric are we talking about. Polyester (Dacron), YES. Nylon, NO. Sailcloth MAYBE. The Suppliers listed have all dealt with "blimp fabric", but not with you. Even then, you may get an "eager-beaver" new salesman. Make it plain to the supplier what on, and how it is going to be used. Ask him WHAT other blimp (or inflatables) company he has supplied fabric to (and check it out).

BASE **CLOTH** is *Dacron.* Nylon stretches (10% @ 2" WP*) and does NOT retain the same strength in sunlight (ultraviolet) as Dacron EVEN with inhibitor coatings.
We are usually talking about: High tenacity Polyester (Dacron sailcloth)
        Unpigmented (white, no color dyes added)
        Plain weave cloth (specify min. up to 60" width)
        70 - 200 Denier (yarn size/weave tightness)
        2.5 - 6.5 ounces per square yard (specify).
        100 yard (+/- 10%) error free rolls (specify)
        Polyurethane coated (laminated evenly)
        (one or both sides, but 702. max. finish)
        Helium porosity, Blemish free; Ii-66 (glue) compatible
We could put a hundred more criteria on our "wish-list", but it wouldn't be cost effective. As a point of information, hundreds of coatings have been tried against ultraviolet, but none proven effective yet. Each coating costs extra $ of course.

INSPECTION: Ask for a one yard sample for testing before ANY money changes hands. Before you take any roll out of his place, get a written guarantee that ANY rejected roll (or all) can be returned (for cause). You MUST inspect each foot of each roll BEFORE cutting (into gores).

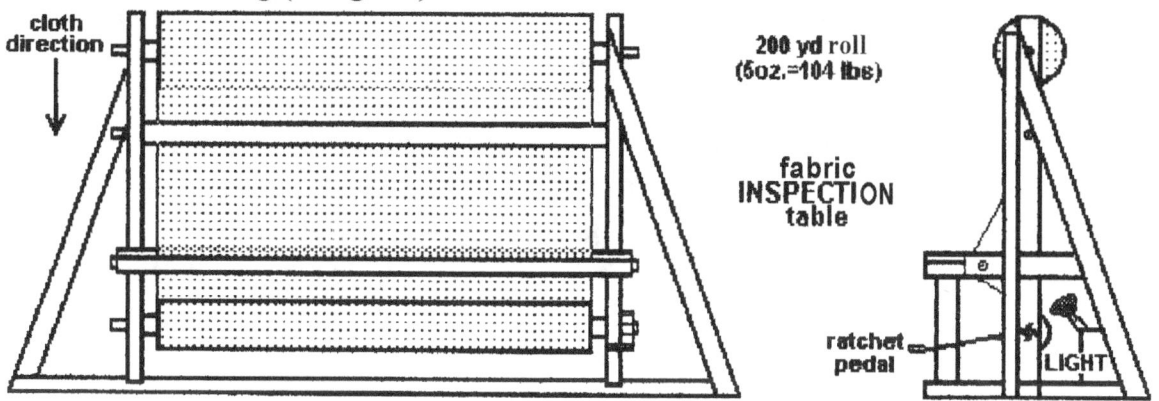

Fabric Trade Rule # 1: **YOU** *CUT IT, YOU BOUGHT IT.*

** Rigging will drive you nuts.

Building Small Gas Blimps

FABRIC ECONOMICS

I don't know what volume you picked, and I don't care. But lets compare some numbers before a lot of your hard earned cash ends up on the cutting room floor.

☞ **FIRST** I'm going to tell you that almost all of the gas blimps made in the last 50 years had an Aspect Ratio (A/R) of 3.8 to one, up to 4.3 to 1. Why? Because that is the best aerodynamic range.

☞ **SECOND** I'm going to tell you that if you settled on 4:1 A/R you will be doing yourself a favor and nobody will ever notice the difference.

☞ **THIRD** I'm going to tell you that the Gore Length **(G/L)** of a 4:1 A/R is always 8.33% longer than the Inflated Length **(I/L)**.

☞ **NOW**, A glance at this simple chart* might affect your approach to how big you build your envelope.

Variations of a 4:1 Aspect Ratio Ellipsoid in 12" increases of Diameter

	Diameter	(I/L)	(G/L)	Volume	Max.Circum.
	19.285	77.14	83.5 feet	15,000 Cu.ft.	727.0 inches
	20 feet	80 feet	86.7	16,755	753.6 "
	21	84	91	19,396	791.7 "
	22	88	95.3	22,301	829.4 "
	23	92	100	25,478	867.6 "
	24	96	104	28,953	904.8 "
✓	24.285' ------	97.14 ------	105.2 ------	30,000 -----	915.5 "
	25	100	108.3	32,725	943.2 "
	26	104	112.6	36,811	980.2 "
	27	108	117	41,224	1018.9"
	28	112	121.3	45,976	1055.6"
	29	116	125.6	51,080	1093.3"
	30	120	130	56,549	1131 "
	31	124	134.3	62,394 out of book limits	

☞ What I'm trying to tell you is, that if you are set on a volume and an A/R, your G/L and circumference will always be the same. Keep the your Circumference and Gore Length numbers on a card in your wallet.

☞ If the salesman says his material is only available in 55' widths, whip out your numbers and a calculator, subtract an inch (cut allowance) from the number he gave you, and keep dividing it by _even_ numbers to get the number of gores and total length of yardage you need. Walla! You can compare the total envelope cost in a minute.

☞ An EVEN number of gores is not a hard-fast requirement, but it makes life a lot easier when marking and locating finger patches and fins.

*****CHART NUMBERS** are accurate within 1/2% for a 4 to 1 aspect ratio. If your A/R is different, or you are looking at the summation of paraboloidal segments, the results will also be different..

Building Small Gas Blimps

Simplified AN Chart
With Superseding MS Numbers

Part	Superseding MS Number
AN 320 NUT – CASTLE, SHEAR	MS 51967 thru MS 51972
USAF 325 NUT – PL. HEX, (NF) (Semi-Fin)	
USAF 330 NUT – CASTLE (Semi-Fin)	
AN 335 NUT – PL. HEX (NC) (Semi-Fin)	MS 51967 thru MS 51972
AN 340 NUT – HEX, MACH. SCREW (NC)	MS 35649
AN 341 NUT – HEX, BRASS (Elec.)	MS 20341
AN 345 NUT – HEX, MACH. SCREW (NF)	MS 35650
AN 350 NUT – WING	MS 35425 / MS 35426
AN 355 NUT – SLOTTED (Engine)	
USAF 356 NUT – PAL	MS 27151
AN 360 NUT – PLAIN (Engine)	
AN 362 NUT – PLATE, SELF-LOCK. (Hi-Temp.)	MS 21047 / MS 21048
AN 363 NUT – HEX, SELF-LOCK. (Hi-Temp.)	MS 21044 / MS 21045 / MS 21046
AN 364 NUT – HEX, SELF-LOCK. (Thin)	MS 21083
AN 365 NUT – HEX, SELF-LOCK.	MS 20365
AN 366 NUT – PLATE, SELF-LOCK.	MS 21047 / MS 21048
AN 373 NUT – PLATE, SELF-LOCK. (100° CTSK)	MS 21049
AN 380 PIN – COTTER	MS 24665
AN 381 PIN – COTTER, STAINLESS	MS 24665
AN 385 PIN – TAPERED, PLAIN	MS 24692
AN 386 PIN – THREADED TAPER	
AN 392 thru AN 406 PIN – CLEVIS	MS 20392
AN 415 PIN – LOCK	
AN 416 PIN – RETAINING, SAFETY	
AN 426 RIVET – 100° FL. HD., ALUM.	MS 20426
AN 427 RIVET – 100° FL. HD., Steel, Monel, & Copper	MS 20427
AN 430 RIVET – RD. HD., ALUM.	MS 20470
AN 435 RIVET – RD. HD., Steel, Monel, & Copper	MS 20435
AN 442 RIVET – FL. HD., ALUM.	
AN 450 RIVET – TUBULAR	MS 20450
AN 470 RIVET – UNIVERSAL HD., ALUM.	MS 20470
AN 481 CLEVIS – ROD END	
AN 486 CLEVIS – ROD END ADJ.	MS 27975
AN 490 ROD END – THREADED	
AN 500 SCREW – FILL. HD. (NC)	MS 35265 / MS 35275 (Dr. Hd.)
AN 501 SCREW – FILL. HD. (NF)	MS 35266 / MS 35276 (Dr. Hd.)
AN 502 SCREW – DR. FILL. HD. (Alloy Stl.) (NF)	
AN 503 SCREW – DR. FILL. HD. (Alloy Stl.) (NC)	
AN 504 SCREW – RD. HD. SELF TAP.	
AN 505 SCREW – FLAT HD., 82° (NC)	MS 35241 SLTD / MS 35190 PHIL
AN 506 SCREW – FLAT HD., 82° SELF TAP.	
AN 507 SCREW – FLAT HD., 100° (NF & NC)	MS 24693 PHIL
AN 508 SCREW – RD. HD. BRASS (Elec.)	
AN 509 SCREW – FL. HD. 100° (Structural) (ALLOY STEEL)	MS 24694 PHIL
AN 510 SCREW – FLAT HD. 82° (NF)	MS 35242 SLTD / MS 35191 PHIL
AN 515 SCREW – RD. HD. (NC)	
AN 520 SCREW – RD. HD. (NF)	
AN 525 SCREW – WASHER HD. (Alloy Stl.)	
AN 526 SCREW – TRUSS HD. (NF & NC)	
AN 530 SCREW – RD. HD., SHEET METAL (TYPE B)	

Lawrence Engineering and Supply, Inc.

PRECISION FASTENERS FOR THE AEROSPACE INDUSTRY
500 SO. FLOWER STREET • P.O. BOX 30 • BURBANK, CALIFORNIA 91503
(213) 845-1763 • (213) 849-1341 • TWX (910) 498-2704

Building Small Gas Blimps

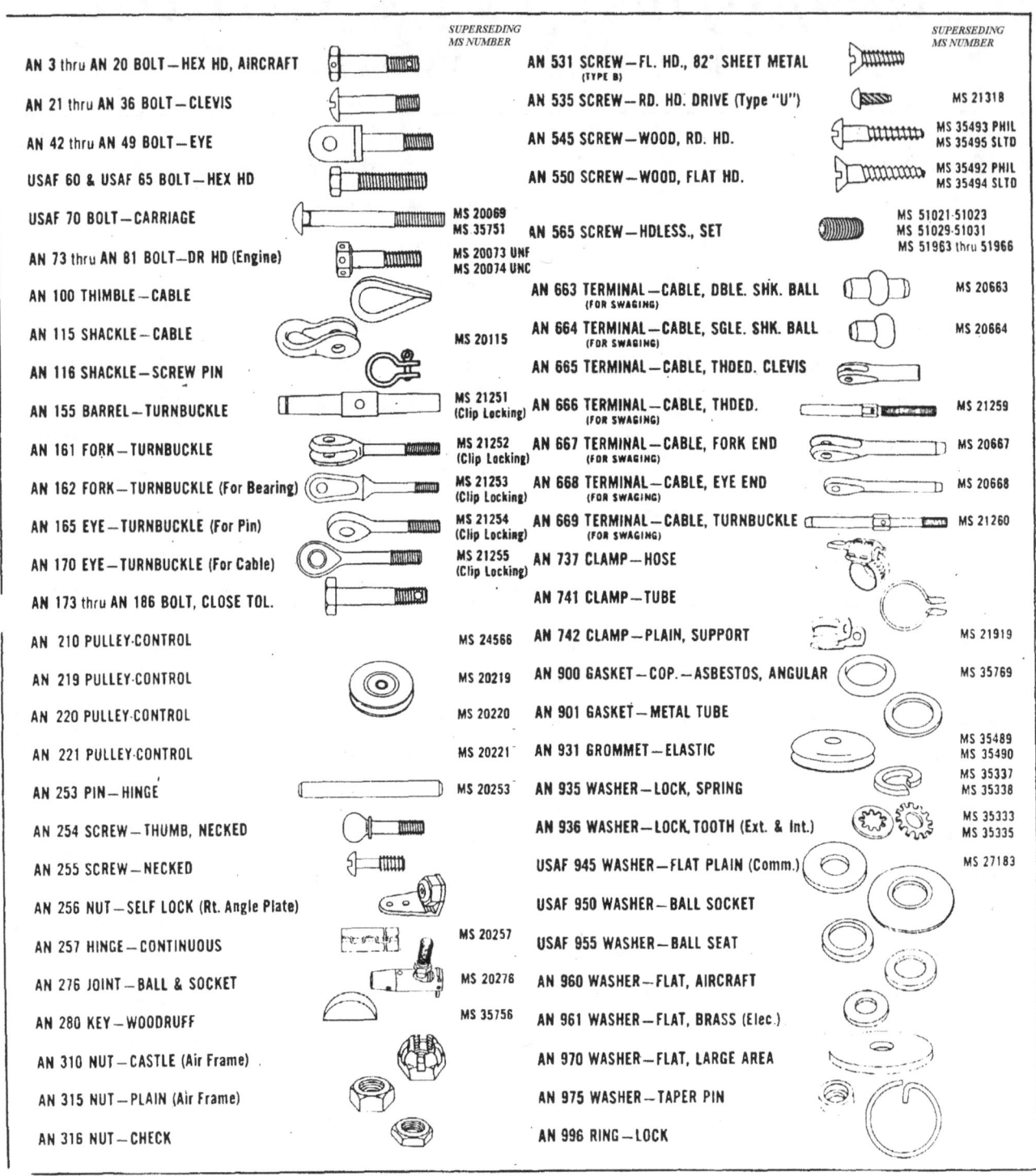

Part		Superseding MS Number
AN 3 thru AN 20 BOLT — HEX HD, AIRCRAFT		
AN 21 thru AN 36 BOLT — CLEVIS		
AN 42 thru AN 49 BOLT — EYE		
USAF 60 & USAF 65 BOLT — HEX HD		
USAF 70 BOLT — CARRIAGE		MS 20069 / MS 35751
AN 73 thru AN 81 BOLT — DR HD (Engine)		MS 20073 UNF / MS 20074 UNC
AN 100 THIMBLE — CABLE		
AN 115 SHACKLE — CABLE		MS 20115
AN 116 SHACKLE — SCREW PIN		
AN 155 BARREL — TURNBUCKLE		MS 21251 (Clip Locking)
AN 161 FORK — TURNBUCKLE		MS 21252 (Clip Locking)
AN 162 FORK — TURNBUCKLE (For Bearing)		MS 21253 (Clip Locking)
AN 165 EYE — TURNBUCKLE (For Pin)		MS 21254 (Clip Locking)
AN 170 EYE — TURNBUCKLE (For Cable)		MS 21255 (Clip Locking)
AN 173 thru AN 186 BOLT, CLOSE TOL.		
AN 210 PULLEY-CONTROL		
AN 219 PULLEY-CONTROL		MS 24566
AN 220 PULLEY-CONTROL		MS 20219
AN 221 PULLEY-CONTROL		MS 20220
AN 253 PIN — HINGE		MS 20221
AN 254 SCREW — THUMB, NECKED		MS 20253
AN 255 SCREW — NECKED		
AN 256 NUT — SELF LOCK (Rt. Angle Plate)		
AN 257 HINGE — CONTINUOUS		MS 20257
AN 276 JOINT — BALL & SOCKET		MS 20276
AN 280 KEY — WOODRUFF		MS 35756
AN 310 NUT — CASTLE (Air Frame)		
AN 315 NUT — PLAIN (Air Frame)		
AN 316 NUT — CHECK		

Part		Superseding MS Number
AN 531 SCREW — FL. HD., 82° SHEET METAL (TYPE B)		
AN 535 SCREW — RD. HD. DRIVE (Type "U")		MS 21318
AN 545 SCREW — WOOD, RD. HD.		MS 35493 PHIL / MS 35495 SLTD
AN 550 SCREW — WOOD, FLAT HD.		MS 35492 PHIL / MS 35494 SLTD
AN 565 SCREW — HDLESS., SET		MS 51021-51023 / MS 51029-51031 / MS 51963 thru 51966
AN 663 TERMINAL — CABLE, DBLE. SHK. BALL (FOR SWAGING)		MS 20663
AN 664 TERMINAL — CABLE, SGLE. SHK. BALL (FOR SWAGING)		MS 20664
AN 665 TERMINAL — CABLE, THDED. CLEVIS		
AN 666 TERMINAL — CABLE, THDED. (FOR SWAGING)		MS 21259
AN 667 TERMINAL — CABLE, FORK END (FOR SWAGING)		MS 20667
AN 668 TERMINAL — CABLE, EYE END (FOR SWAGING)		MS 20668
AN 669 TERMINAL — CABLE, TURNBUCKLE (FOR SWAGING)		MS 21260
AN 737 CLAMP — HOSE		
AN 741 CLAMP — TUBE		
AN 742 CLAMP — PLAIN, SUPPORT		MS 21919
AN 900 GASKET — COP. — ASBESTOS, ANGULAR		MS 35769
AN 901 GASKET — METAL TUBE		
AN 931 GROMMET — ELASTIC		
AN 935 WASHER — LOCK, SPRING		MS 35489 / MS 35490 / MS 35337 / MS 35338
AN 936 WASHER — LOCK, TOOTH (Ext. & Int.)		MS 35333 / MS 35335
USAF 945 WASHER — FLAT PLAIN (Comm.)		MS 27183
USAF 950 WASHER — BALL SOCKET		
USAF 955 WASHER — BALL SEAT		
AN 960 WASHER — FLAT, AIRCRAFT		
AN 961 WASHER — FLAT, BRASS (Elec.)		
AN 970 WASHER — FLAT, LARGE AREA		
AN 975 WASHER — TAPER PIN		
AN 996 RING — LOCK		

SOUTHWEST DIVISION
8656 DENTON DRIVE, P.O. BOX 35067
(AT LOVE FIELD)
DALLAS, TEXAS 75235
214-357-6411

NORTHWEST DIVISION
850 CHERRY STREET
SAN CARLOS, CALIFORNIA 94071
415-591-9473

Building Small Gas Blimps

WORKMANSHIP
R:12-03-97

The WORKPLACE
This is no little project you are about to undertake; your 2-car garage is not going to be adequate. Yes, people have built airplanes in a closet, and boats in the basement. But assuming that you want to enjoy your blimp in your lifetime, you need sufficient space for your tools, machinery & materials, at a work site that is ready (and secure for the next 2 years) with the turn of a key. Definitely, keep the project detached from your living space. I could give a lot of advice on how keep your marriage together, but then quite frankly, I don't give a damn. Airships always come first.

The TOOLS
Just because you own a hammer and crescent wrench, remember: You can't make chicken soup out of chicken shit. If you don't have a roll-a-way tool box, get one. Start with K-Mart wrenches if you want, but expect the tool inventory to increase on a daily basis. The list will be to long to show here.

The MACHINES
You may laugh now, but by the end of the first year, I would expect to see:

a Welding Machine	a Table Saw
an Air Compressor	a Tool Grinder
a Drill Press	2 Work Benches

Start reading the USED TOOLS ads in the classified section of your local newspaper.

WORKMANSHIP
Now that we have covered the preliminaries, lets get into the Nitty-gritty. You may think you know everything about anything, but your education is just beginning. You fix your car and everything around the house? Well your technical expertise is going to be tested to the limit. And how well you do it may determine your <u>afterlife</u>. You may think you have read everything on airships, but it ain't so. You couldn't read 100 years of technical reports in the rest of your life. All you have is your 2 hands and maybe an Internet connection; so lets set some rules.

> If you don't understand something, ask someone that does.
> Go to the source if possible.
> If you have never tried it (i.e. gluing, welding, instruments, composites, valves, etc.), take lessons, or pay a professional to do it.
> If you have any doubt about your ability, DON'T. Practice on a scrap piece until you are comfortable.

SUB CONTRACTING
There are some good blimp people out there that are experts at building critical parts. But make sure you have a meeting of the minds BEFORE you pay anything up front. A good engineering drawing helps. What you want, and what he thinks you need may not be the same.

Building Small Gas Blimps

STANDARD WELDING SYMBOLS[1]

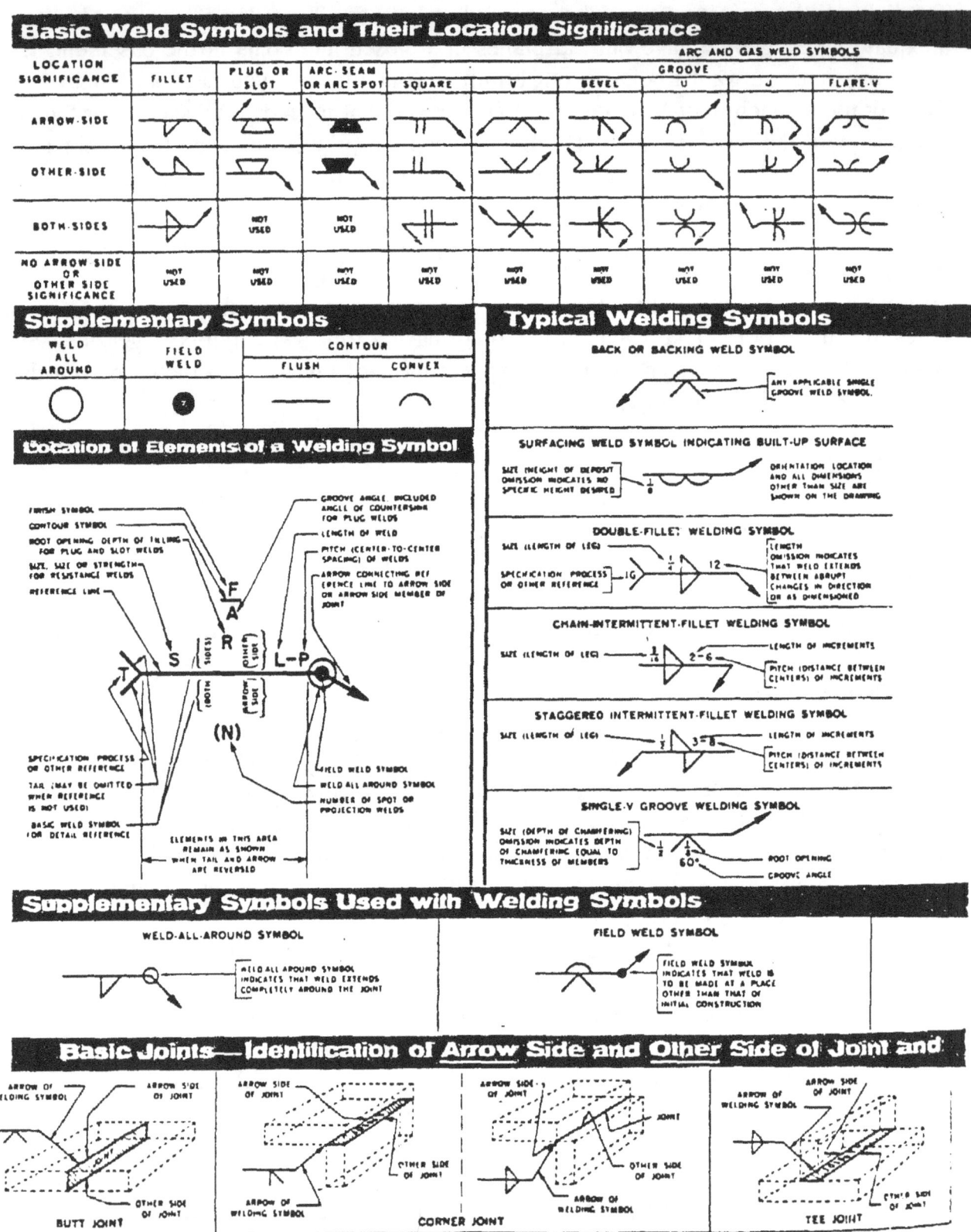

[1] AWS A2.0-58, American Welding Society, 345 East 47th Street, New York, N.Y. 10017.

SPARK PLUGS

The installation torque for spark plugs, recommended by the Society of Automotive Engineers and listed in the SAE Handbook, is as follows

PLUG SIZE	FOOT POUNDS CAST IRON	ALUMINUM
10MM	13–15	9–11
12MM	20–35	15–20
14MM	26–30	18–22
14MM-Tapered	10–15	10–15
18 MM	32–38	28–34
18MM-Tapered	15–25	15–20
7/8" 18	35–43	31–39

TO CONVERT FROM INCH–POUNDS TO FOOT–POUNDS -- USE THE CONVERSION TABLE OUTLINED BELOW

INCH-POUNDS	FOOT-POUNDS
0	0
60	5
120	10
180	15
240	20
300	25
360	30
420	35
480	40
540	45
600	50

SEARS, ROEBUCK AND CO. AND SIMPSONS-SEARS, LTD.
Nut: 150-145 ft-lbs

TORQUE WRENCH CHART BOLTS

GRADE MARKING		0-1-2	5	6	7	8	SUPER
BOLT DIAM.	THREAD PITCH	FOOT POUNDS					
1/4"	20	5.5	9.7	11.0	11.5	13.0	
	28	6.0	11.0	12.0	13.0	15.0	
5/16"	18	10.0	18.0	20.0	21.0	24.0	
	24	11.4	20.0	23.0	24.0	27.5	
3/8"	16	21.7	39.0	43.0	45.0	52.0	
	24	24.5	44.0	49.0	51.0	59.0	
7/16"	14	32.4	58.0	65.0	67.0	78.0	
	20	38.4	69.0	77.0	80.0	92.0	
1/2"	13	43.5	87.0	97.0	102.0	116.0	
	20	54.6	103.0	115.0	121.0	138.0	
9/16"	12	57.0	111.0	123.0	129.0	147.0	
	18	68.0	131.0	146.0	153.0	175.0	
5/8"	11	86.0	173.0	192.0	201.0	230.0	
	18	102.0	200.0	224.0	235.0	269.0	
3/4"	10	152.0	290.0	324.0	336.0	389.0	
	16	182.0	345.0	384.0	403.0	461.0	
7/8"	9	222.0	500.0	555.0	583.0	666.0	
	14	261.0	585.0	653.0	685.0	784.0	
1"	8	307.0	690.0	769.0	807.0	923.0	
	14	370.0	830.0	925.0	967.0	1111.0	

Courtesy Premier Industrial Corp.
Cleveland, Ohio

1. Specifications contained in this chart are correct to the best of our knowledge. Consult the exact specifications as given by original equipment manufacturer.
2. Consult your service manual before adjusting or repairing any automatic transmission.
3. Any torque wrench is a delicate measuring instrument and should be treated with care. Should the pointer become bent by accident reset to zero when tool lies in rest position and accurate readings will result. Make certain pointer floats.
4. Craftsman repair and service tools available for all normal servicing requirements.

ALWAYS USE A TORQUE WRENCH WHEN SERVICING

- Wheel Bearings
  - Oil Filters
    - Cylinders Heads
      - Motor Bearings
        - Manifolds
          - Valve Springs
            - Fly Wheels
              - Automatic Transmissions

Consult exact specifications as given by original equipment manufacturer.

Building Small Gas Blimps

R: 10/21/97

HARDWARE & TOOLS
SUPPLIERS

AGM Container Controls 3526 E.Lowell (POB-40020) Tucson, AZ 85717	602/881-2130 manufacturer	HARDWARE, binding Cam-Lok tensioners Tie-Down load binders
ALL AIRCRAFT Parts 16673 Roscoe Blvd. Van Nuys, CA 91406	818/894-9115 retail sales good walk-in stock	HARDWARE, aircraft Bolts, fittings, instruments pilot supplies
AVIBANK Manufacturing 210-T S.Victory Blvd. Burbank, CA 91503	818/843-4330 moved?	HARDWARE, specialty Ball-Lock pins All sizes & kinds
BARON Manufacturing 1200 Capitol Drive Addison, IL 60101	708/628-9110 some retail	HARDWARE, mfg. Rings, snaps, carribeners load bearing hardware
BELT Makers #304 1615 W.205th St.(POB 630) Torrance, CA 90508	310/618-8868 AmSafe rep. mfg. outlet	HARDWARE, aircraft Rigging & Webbing Seat belts/harnesses
BUNGEE International 207-40 Plumber Street Chatsworth, CA 91311	818/998-6601 RTS-115	HARDWARE Restraint Ratchet Tie-downs Bungee products
COAST Marine Supply San Jose, CA moving Hollister, CA	408451-9620 retail/wholesale major supplier	HARDWARE, marine Rigging & tools Cable & connectors
CONSOLIDATED A/C Supply 55 Raynor Ave. Ronkonkoma, NY 11779	516/981-7700 Fax:981-7706 major supplier	HARDWARE, aircraft Great catalog
HALKEY-Roberts 11600A Ninth Street Petersburg, FL 33716	813/577-2511 Fax/578-0450	HARDWARE, safety Air valves Relief Valves
Rutland Tool Co. Inc. 20845 Prairie Street Chatsworth, CA 91311-6013	800 / 825-5393 major supplier	HARDWARE, tools Great catalog for machining
UNITED Textile (Unitex) 5175 Commerce Drive. Baldwin Park, CA 90012	818/962-6281 Retail/wholesale good source	HARDWARE, for cloth Webbing, canvas H-66 fabric glue

Building Small Gas Blimps

GROUND SUPPORT
R: 12-02-97

Why would I want to start talking about Ground Support equipment before the big items, like Envelopes and Gondolas? Because your finished blimp needs these things designed into it. It is hard for neophytes to understand that an airship is a beast to handle once outside its cage (hangar). After you roll it out, it is to late too start building a mast; don't think you are going to tie it to the bumper of a car. Believe me, I've seen it (the car was saved). The following items are of great importance, but not necessarily in order of priority.

VEHICLES: Do yourself a favor. If you don't have a pick-up truck, get one. If you do have one, fix it up first so you have something reliable. You need one for the duration of this project. Make sure it has a decent trailer hitch AND a secure point on the front to install a shackle for lifting a mast. If you plan to take your blimp "on the the road", you are going to need a covered utility trailer also.

HANGAR: OK, so its too rich for your budget. At least I told you. But are you going to stay out there 24 hours a day 8 days a week, to keep the locals from punching holes in it, or worse? Nice to be able to close the doors and go home for a hot meal. Many kinds are available, and can be ordered to fit. Yes, $10,000 will only be a down payment.

MAST: You can kid yourself, but you can't kid me. You *GOTTA* have one, even if you have a hangar. Sooner or later you are going to get caught out in the weather, away from your base. A simple stick mast will save your butt in an emergency. *BEFORE* you ever think about flying, make sure your crew knows how to use it. Several styles are shown in this chapter. For less than 20,000 cuft. (20K) use a simple lever lock type. Over 20K, I recomend useing a standard docking cone.

GENERATOR: A small Honda (sic) is always needed out in the field. Remember, you need a blower for ballonet pressure any time your engine isn't running. Just about everything you want to fix needs an electric tool to do it. And even light at night helps in making inspections. Which brings to mind:

BLOWERS: Weed blowers are NOT reliable for the continuous duty you require. Make one with a 1/4 hp. motor that you can plug into your Honda. Keep it simple and easy to get on, and stay on, until you want it off.

GROUND CREW: You've got to trust someone on the ground to get you launched and recovered safely. And you can't fly the blimp and direct a ground crew at the same time. The key is in developing a crew "chief" who not only understands airships, but has your safety and interests at heart. Get him involved in the construction early, and keep him involved on how things in the airship work. Then let him select and train his own helpers. But keep an eye on them; They can make you, or break you. When someone gets hurt, its too late. Face the fact airship operations are inherently unsafe, not just from ropes and props, but just plain "Murphy's Law".

PLAN AHEAD: Look thru the yellow pages for an equipment rental place that carries self propelled "Man Lifts". Sooner or later you'll need one to inspect an upper fin.

Ground Support cont.
R:12/01/97

Inflation Crew:
You gotta have people, good people. A standard inflation can easily turn into a Chinese Fire Drill. It seldom goes as planned, due to the large number last minute items that need installation. Before you call the Helium supplier you had better have a check-list going, and helpers assigned with specific duties. I am going to assume you are smart enough to order 30,000 cuft. of gas from a bulk (tank) truck. Filling from 100 steel bottles (even 3) is no fun. When you pick the date and time for the truck to arrive, be there early, and plan to complete it before anyone leaves. Remember, the truck driver gets paid by the hour, and the trailer $100 per day if he has to leave it.

I know that money is tight by this stage of the project, but please plan on making your first inflation INSIDE a suitable building, AND using a net. The hangar eliminates the wind from over stressing the fabric; the net to keep the gas bubble from forming an unmanageable pocket of lift. As the envelope rises, a 4-man crew can easily adjust the tension by repositioning the sand bags (1500 lbs. worth) down the net diamonds. This gives you plenty of time to see that the tail fins don't hang up, and that nothing snags the envelope fabric. Then just wheel the gondola under, and attach it at a leisurely pace.

Now if you are really brave, and plan on a "Shotgun Inflation" (nose attached to an outside mast, and without a net), you had better have all your ducks in a row. Pick a leader that has done it before (no sane person will admit it), and train your helpers carefully. I don't have to tell you that the risk of damage, or just forgetting critical items in the panic, increases greatly.

What Now?
Now that you have it inflated, where are you going to put it. You are not going to empty out $4000 worth of helium and put that 300 pound (now 500) envelope back in the box each night.

*Ultra-light engines aren't exactly rocket science.
Woofo: A person overwhelmed by what he sees

Building Small Gas Blimps

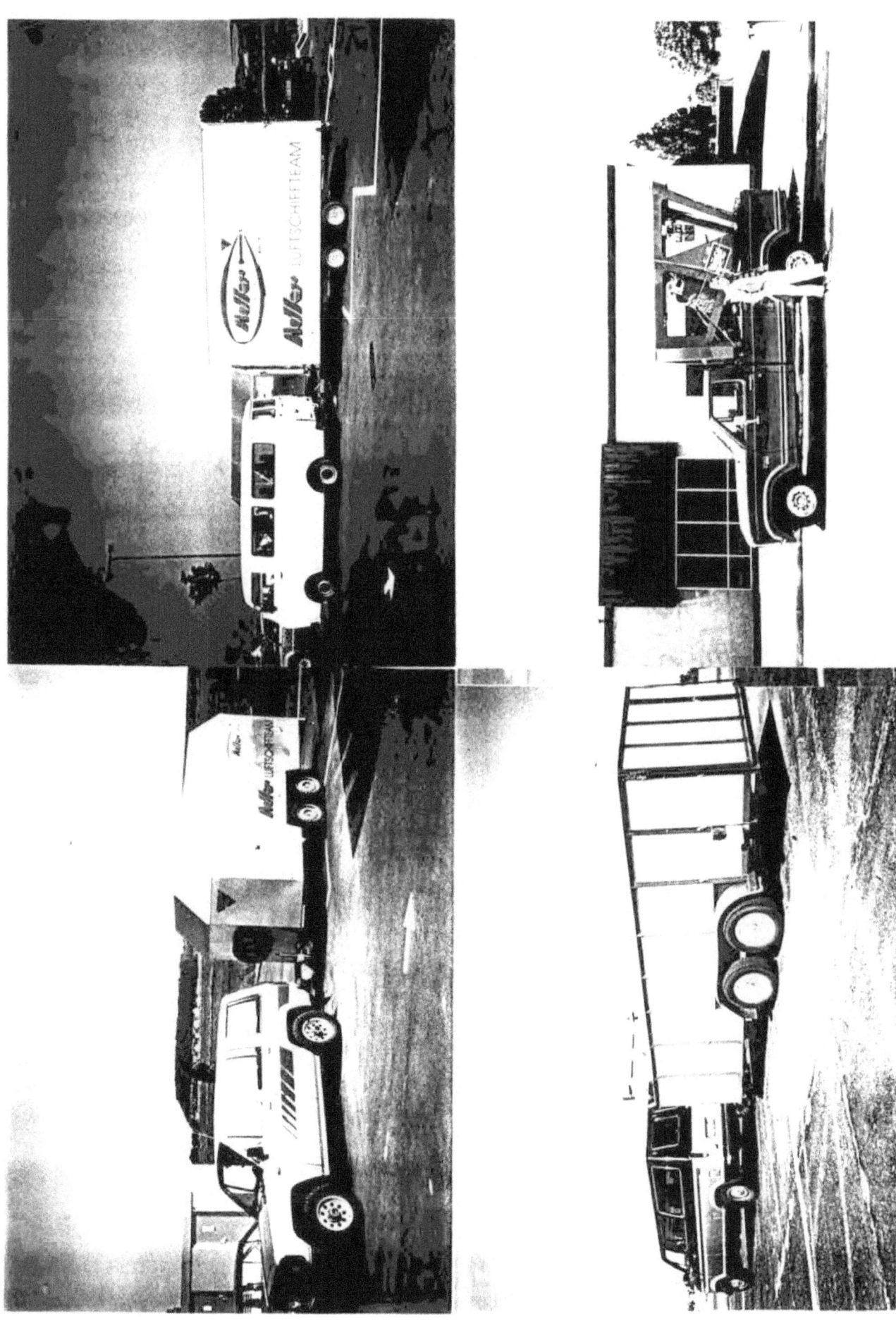

23

Building Small Gas Blimps

1.8 PresCont
1.3 R3

PRESSURE CONTROLLER
(Blower Actuator)

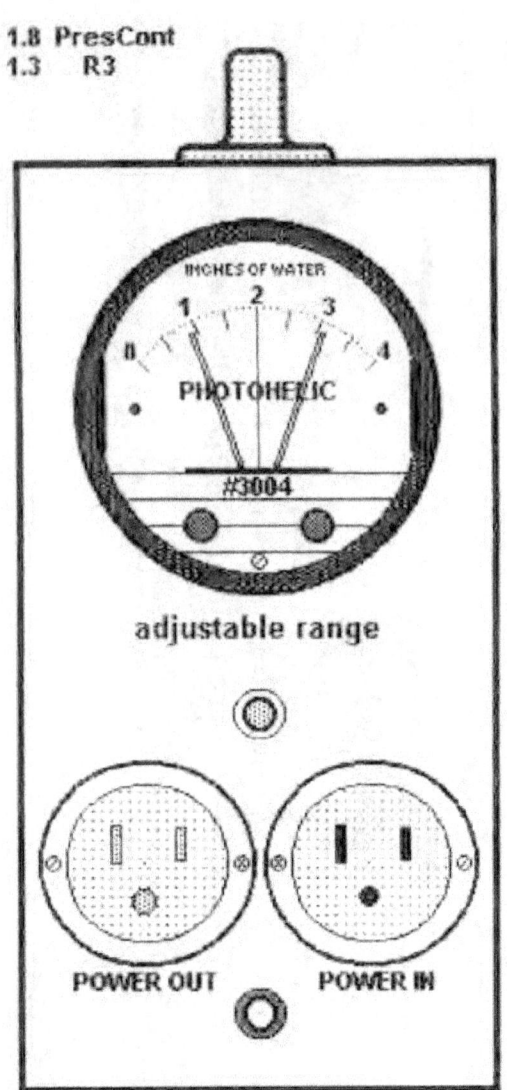

adjustable range

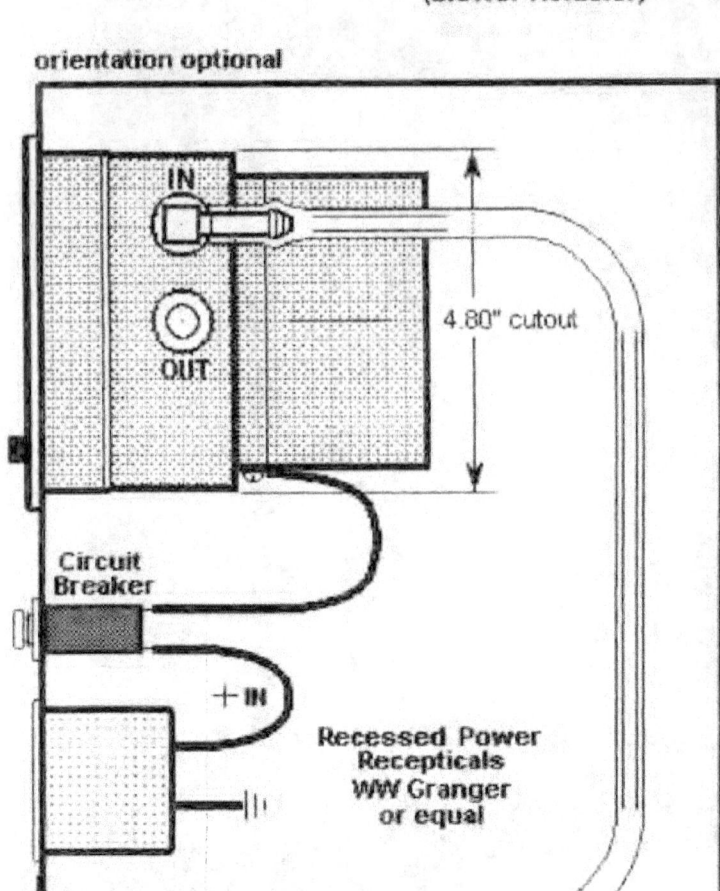

Manufactured by:
Dwyer Instruments Inc.
P.O. Box 373
Michigan City, IN 46360
219 / 879-8000
fax: 872-9057

Available from:
Dwyer Retail Outlets
Anaheim, CA 714 / 630-6424
Marietta, GA 770 / 427-9406
Ivyland, PA 215 / 957-0355
Houston, TX 218 / 446-1146

Also in Australia and U.K.

NOTE: 3" gage = #5004

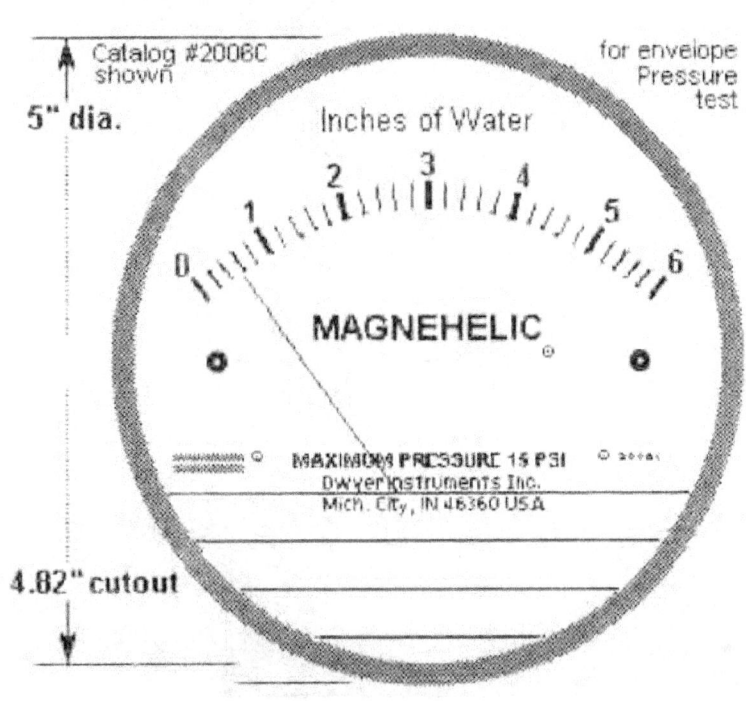

24

Building Small Gas Blimps

1.3 BALLAST
1.0

SAND BALLAST BAGS
(& Lead Shot Bags)

NOTES:
    All bags must be tied securely to prevent contents from leaking
        & spilling, and be made of waterproof material;
      Inventory should be enough to equal your useful load + 10%
Sand bags can be packed to 35-50 pounds, and should
        be yellow or neutral in color;
Shot bags should be packed with the lead shot secure in a
        separate container before putting inside Vinyl bag;
    can be 2" shorter and 1" smaller in diameter than sand bags;
    should be Red or a contrasting color to your sand bags;
    should be of a known weight, usually 25 lbs.;
    Lead is hazardous material, the corrosion dust will
        permeiate your clothes and cockpit;
Treat bags with care.

Materials required:
    Fabric, Vinyl coated 3 oz. Dacron
    Cord, 3/8" braided Nylon
    Grommets, 3/8" brass (#2)
    Clips, 3/8" plastic

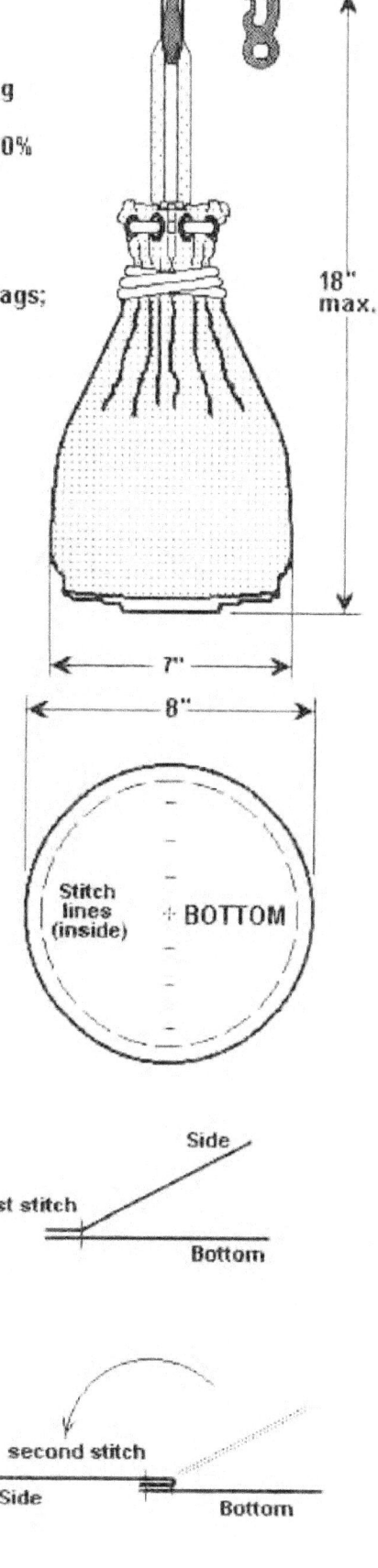

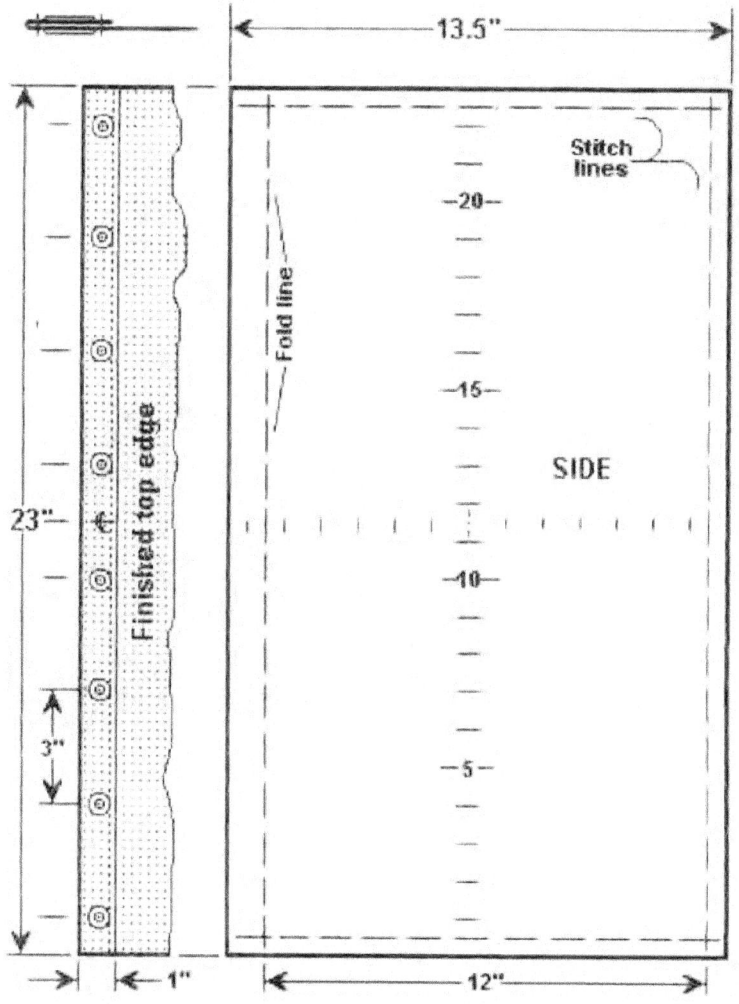

Building Small Gas Blimps

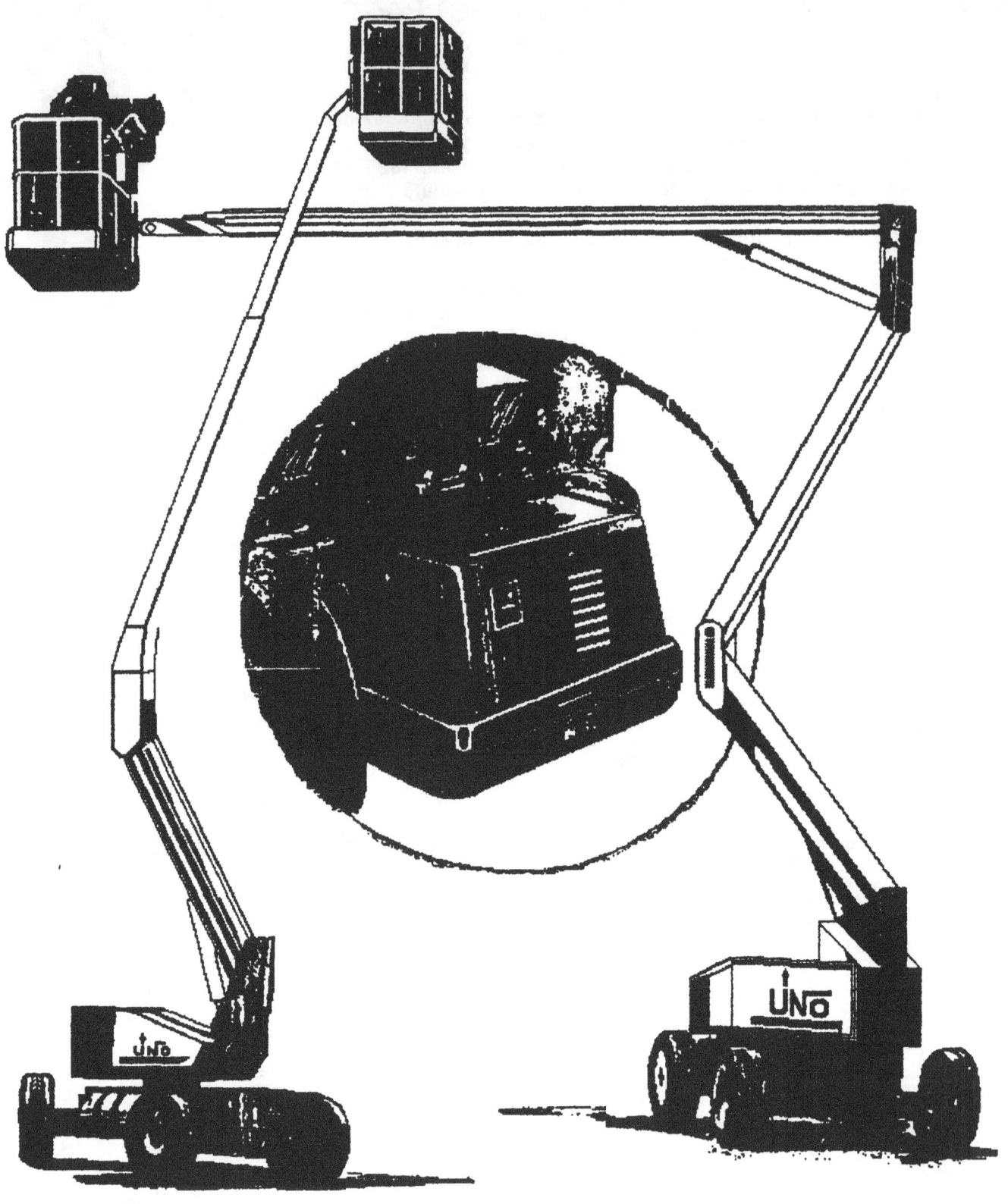

Engine Powered Work Platforms
MODELS: UNO 33 & UNO 41 G and D

R: 12/02/97

SUPPORT EQUIPMENT
SUPPLIERS

AGM Container Controls 3526 E.Lowell (POB-40020) Tucson, AZ 85717	602/881-2130 manufacturer	HARDWARE, binding Cam-Lok tensioners Tie-Down load binders
BARON Manufacturing 1200 Capitol Drive Addison, IL 60101	708/628-9110 some retail	HARDWARE, mfg. Rings, snaps, carabiners load bearing hardware
BUNGEE International 207-40 Plumber Street Chatsworth, CA 91311	818/998-6601 RTS-115	HARDWARE Restraint Ratchet Tie-downs Bungee products
HICKLIN Power Systems 5303 N.W. 111th Drive Grimes, IA 50111	800/234-8989	GENERATORS Honda dealer EX-1000 = $634
W.W. Grainger 1401 E. Third Street Los Angeles, CA 90001	nationwide great source ck yellow pages	HARDWARE, general Blowers, Generators support equipment

Building Small Gas Blimps

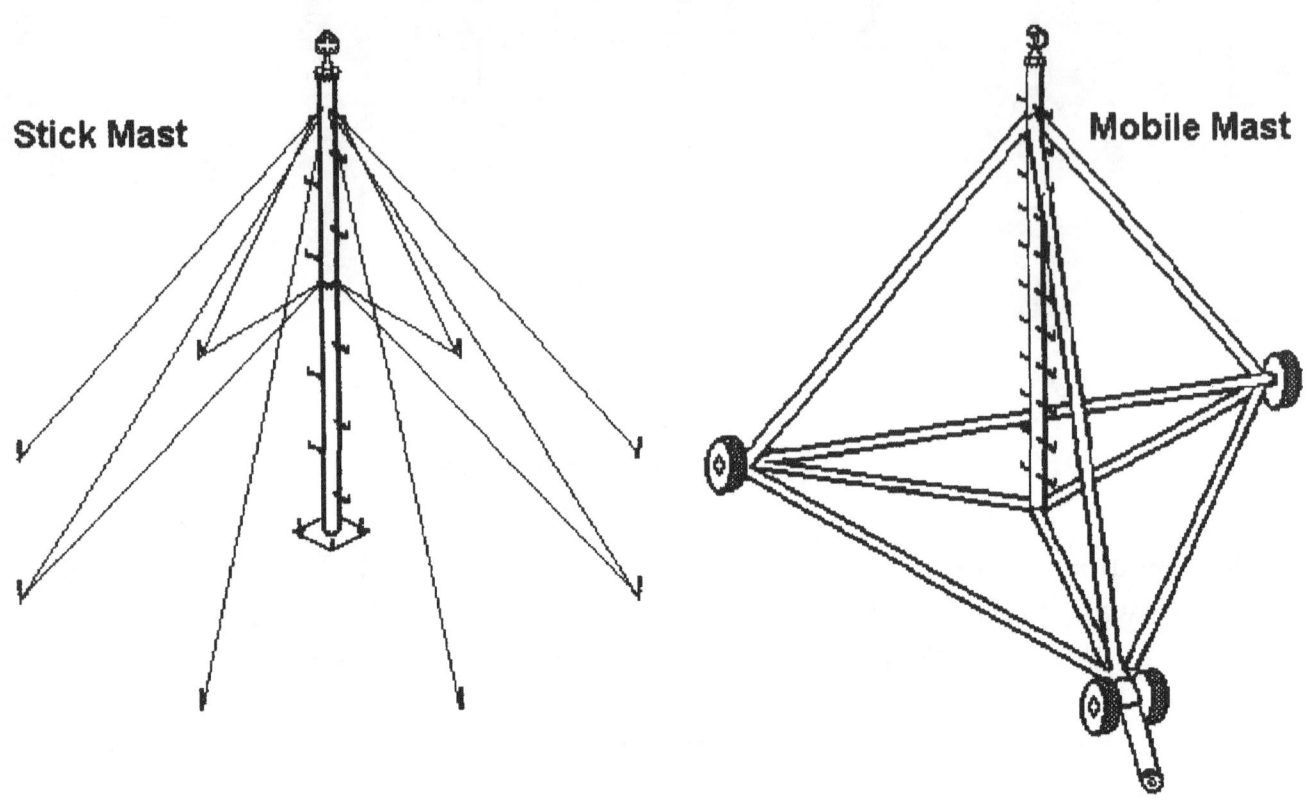

Stick Mast

Mobile Mast

MASTS

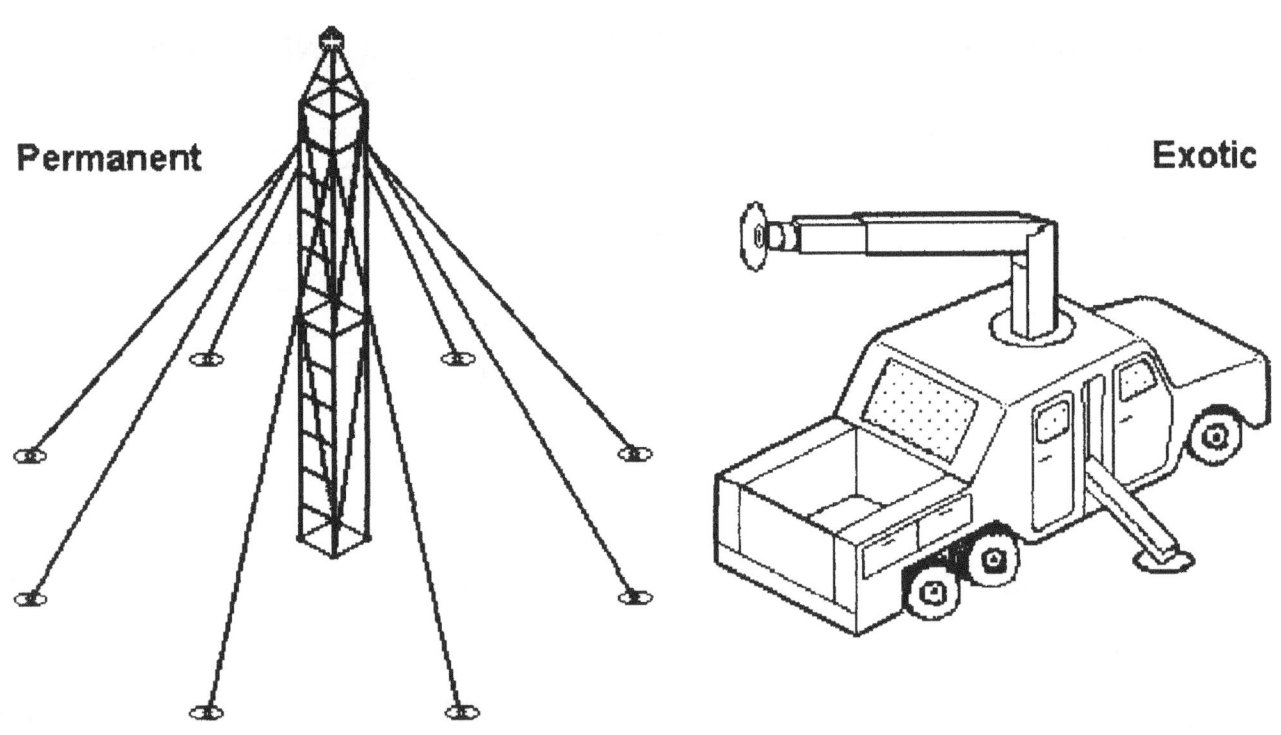

Permanent

Exotic

Building Small Gas Blimps

MAST ANCHORING
GA-42 layout

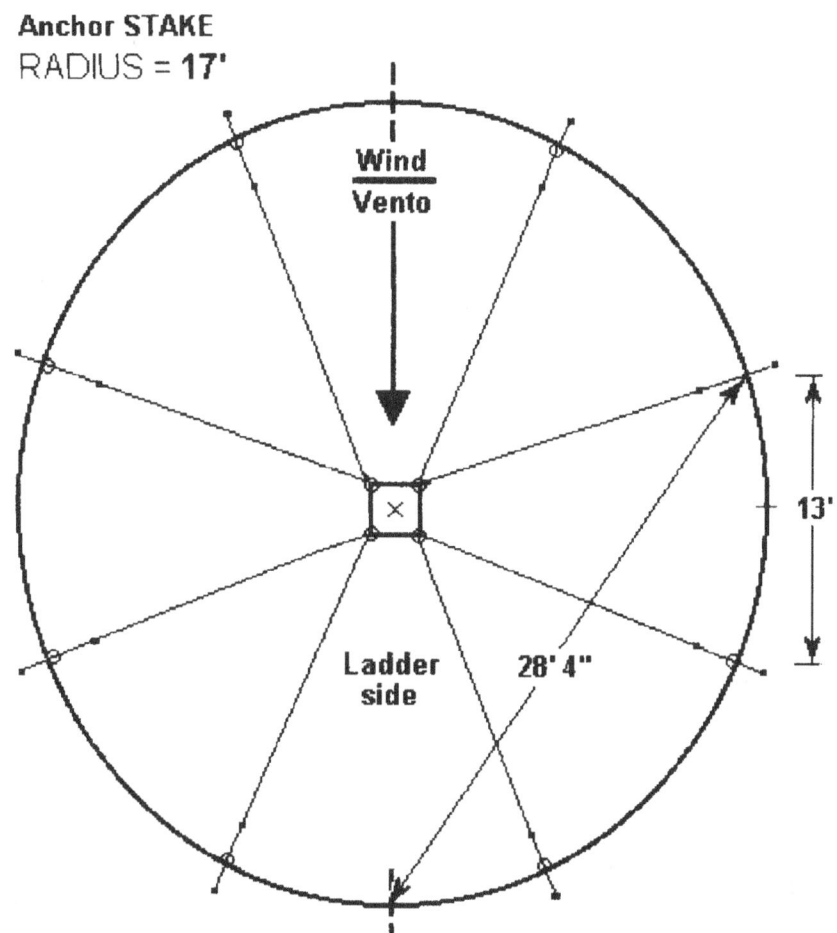

Anchor STAKE
RADIUS = **17'**

Wind / Vento

Ladder side

28' 4"

13'

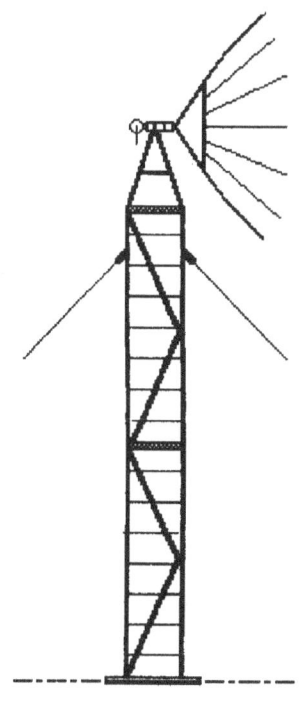

8-CABLE pie section equals 45° each, or **13'** between Anchor Stakes

NOTE: Mounting BASE-PLATE fits 24" x 24" leg frame.

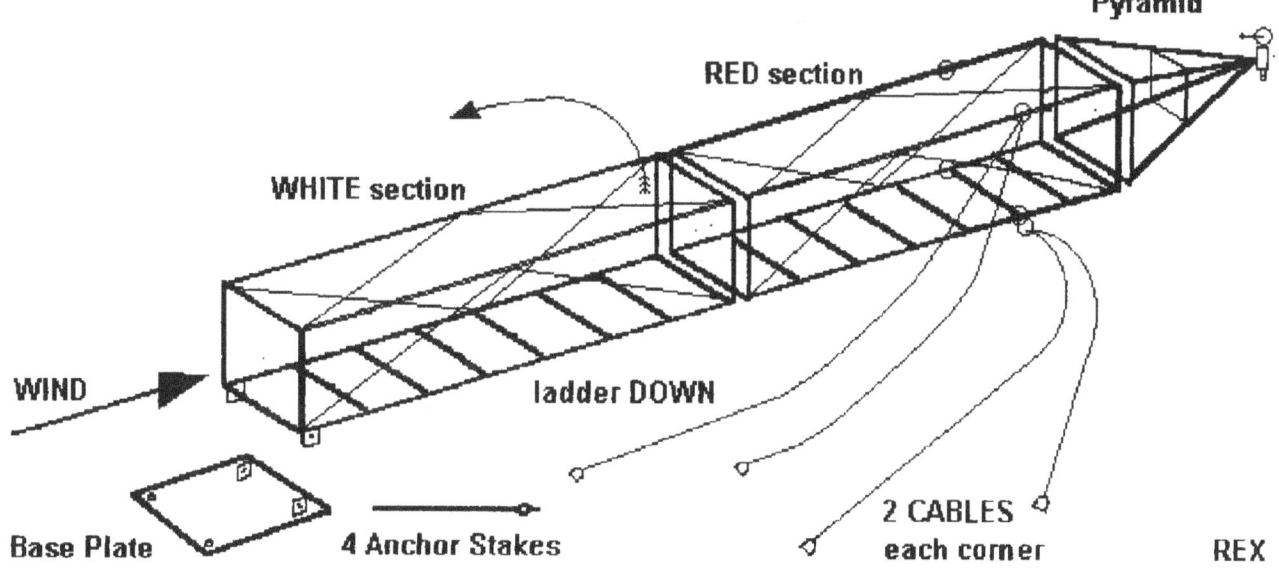

Pyramid
RED section
WHITE section
ladder DOWN
WIND
Base Plate
4 Anchor Stakes
2 CABLES each corner
REX

Building Small Gas Blimps

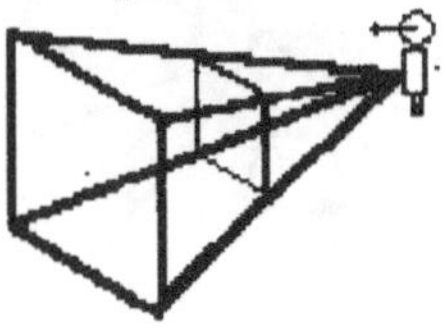

GA-42 MAST MOORING LOCK details

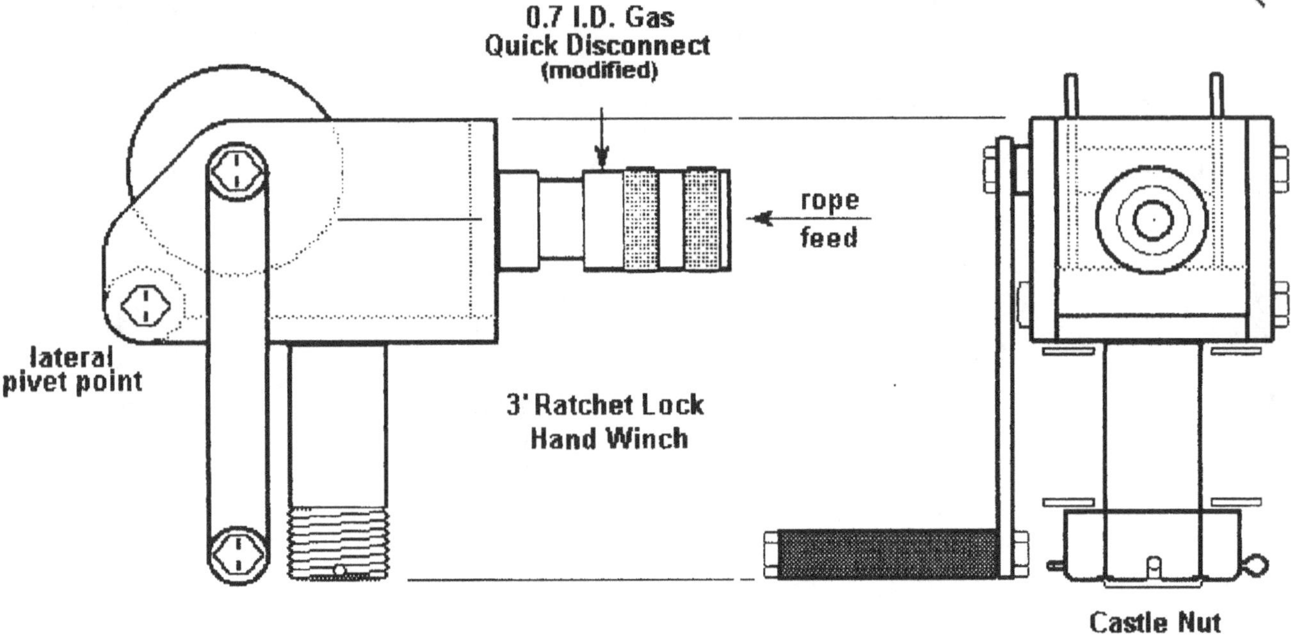

1.3 G42 WINCH
1.0
Pyramid

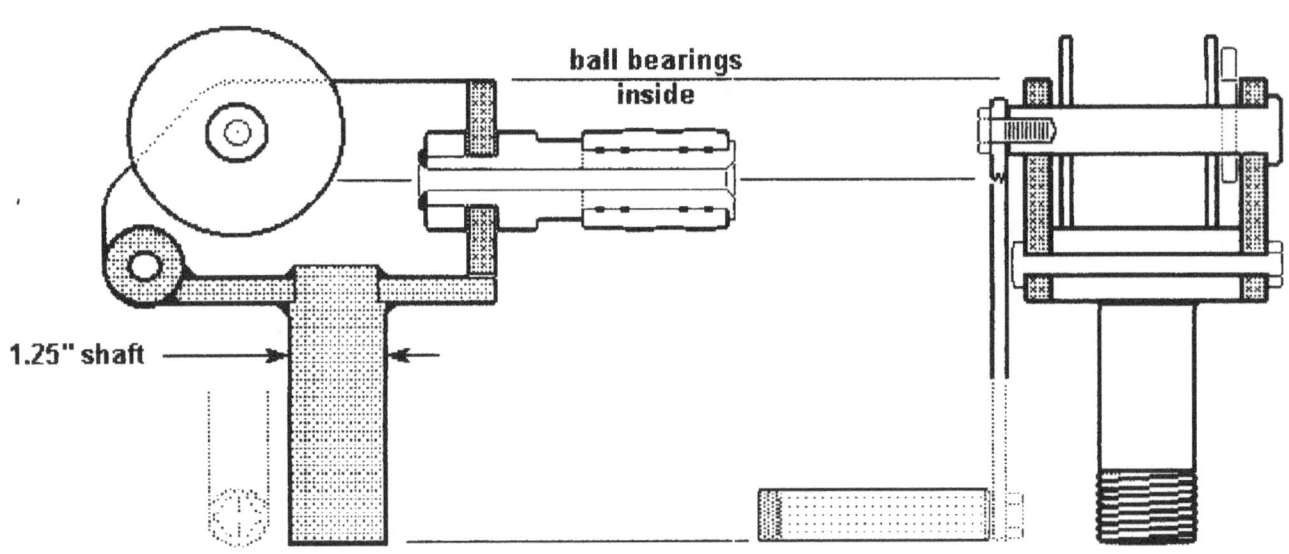

0.7 I.D. Gas Quick Disconnect (modified)

rope feed

lateral pivet point

3' Ratchet Lock Hand Winch

Castle Nut

ball bearings inside

1.25" shaft

Raw Material:
Steel plate, .375 x 3" x 24"
Steel bar, 1.25 x 6"
Steel bar, 1.0 x 3"

Purchased Parts:
Quick Disconnect
Castle Nut, 1.25 - 10 NC
Cotter Pin
3" Hand Winch

REX

Building Small Gas Blimps

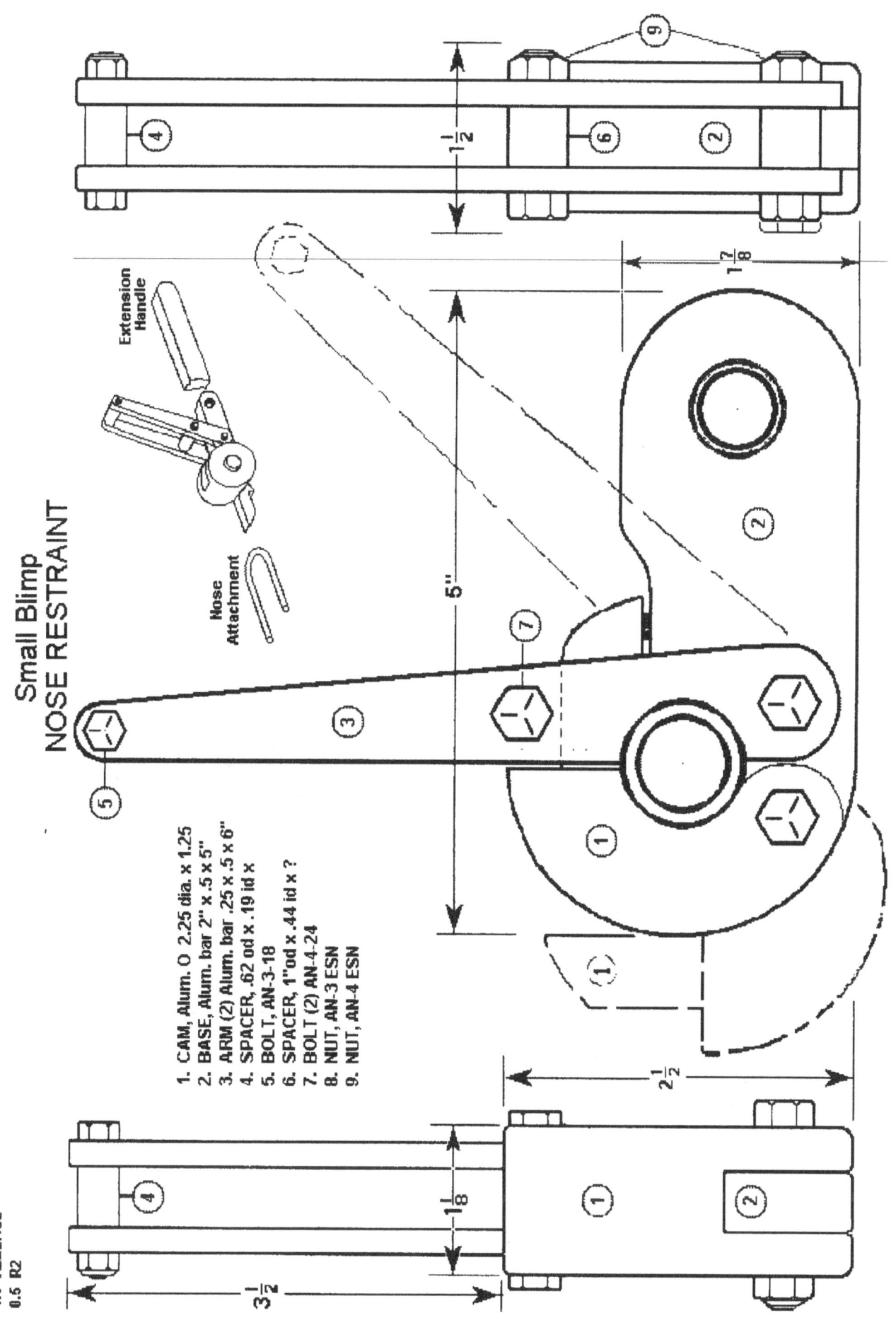

Building Small Gas Blimps

Standard
DOCKING CONE
(Nose Peg)

Use on volumes 30K to 100K

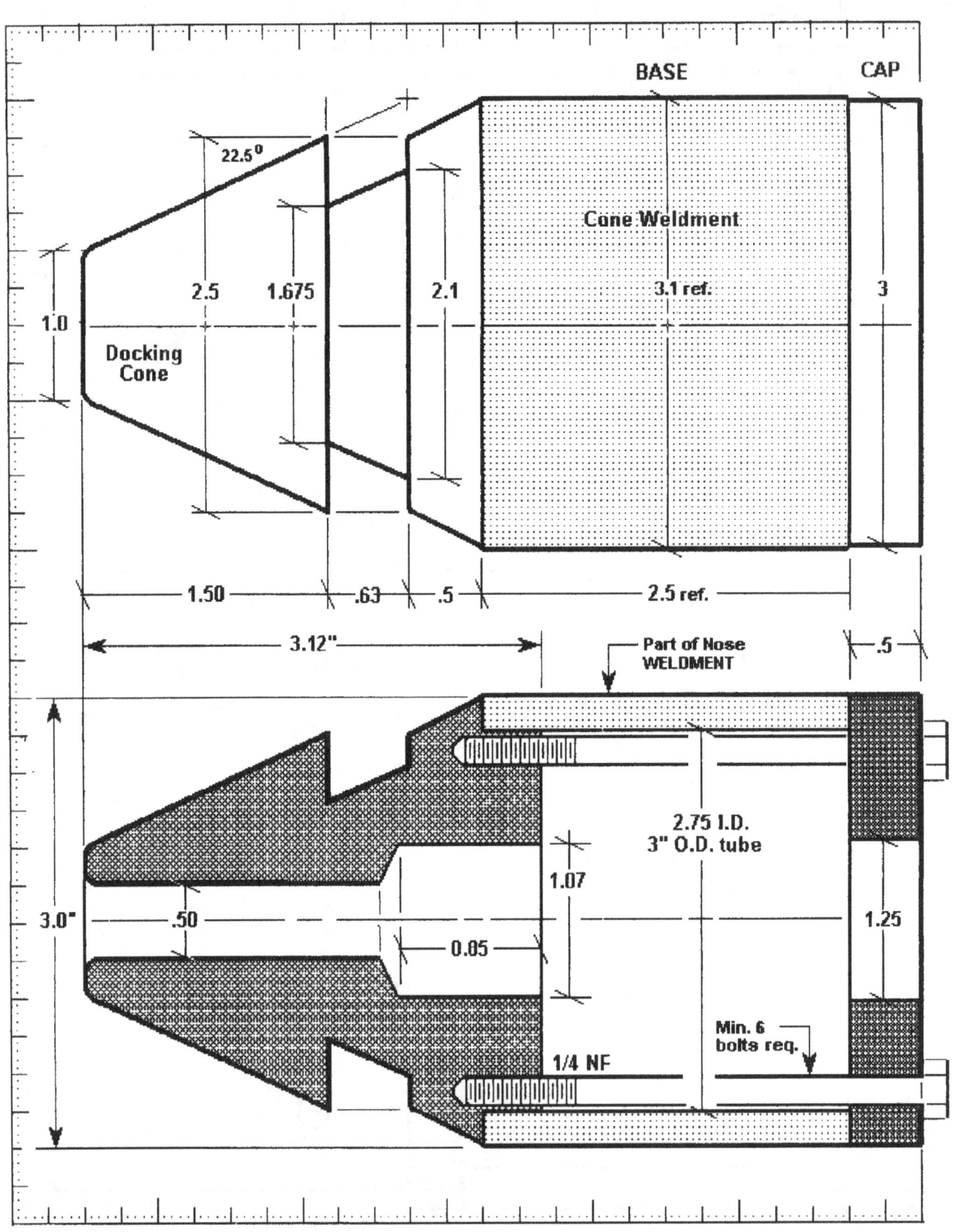

Building Small Gas Blimps

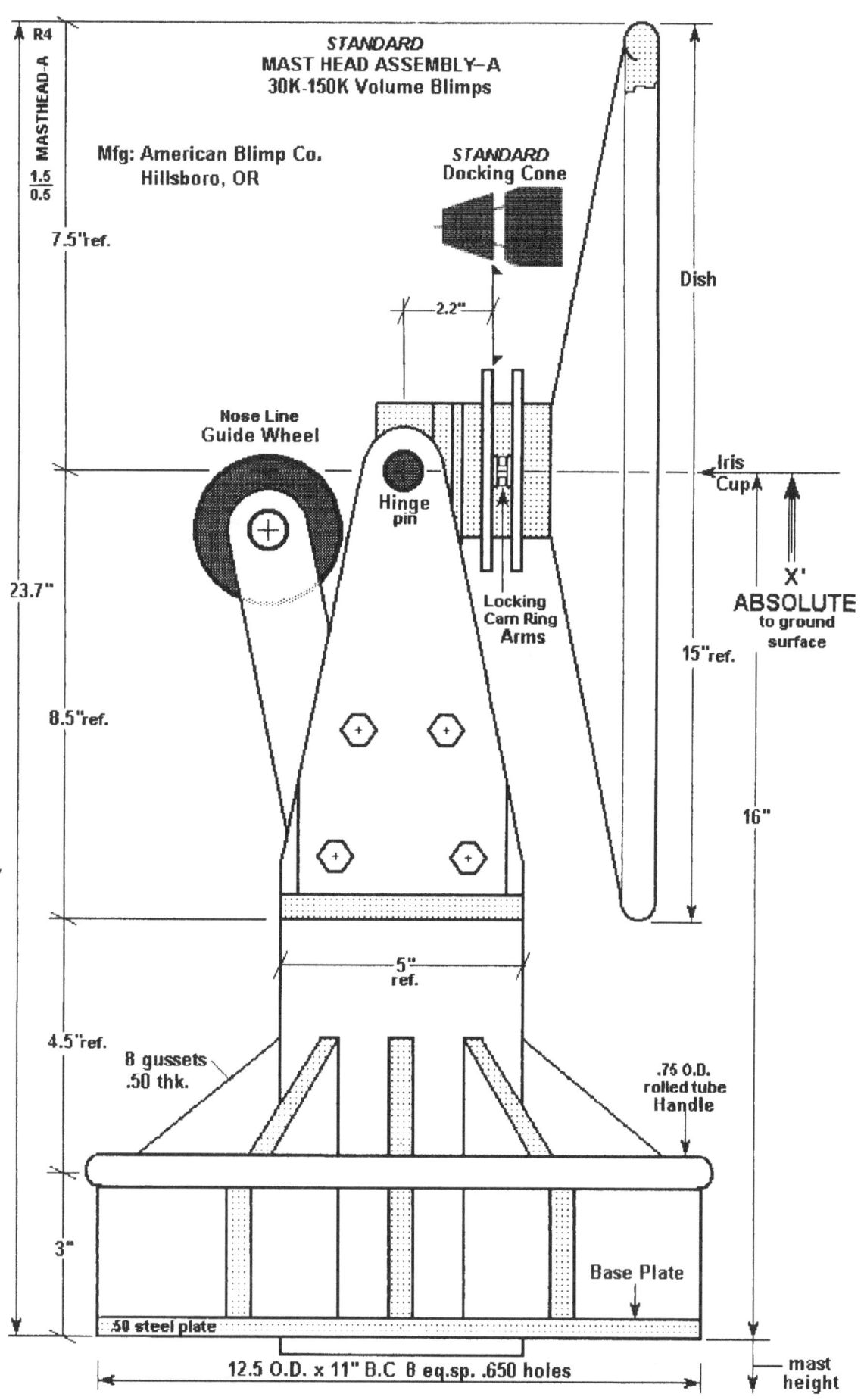

Building Small Gas Blimps

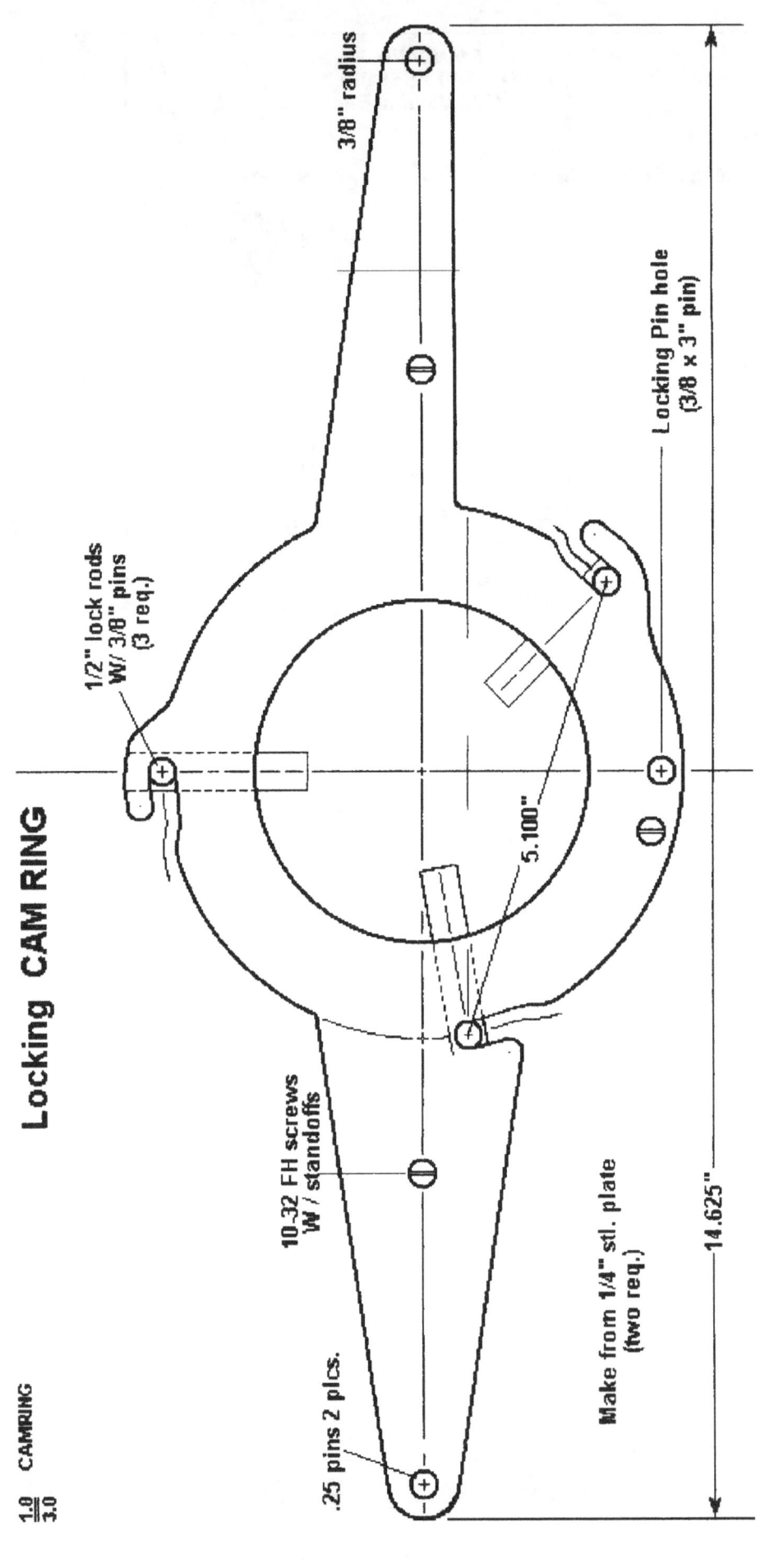

Building Small Gas Blimps

MASTHEAD Asm. B

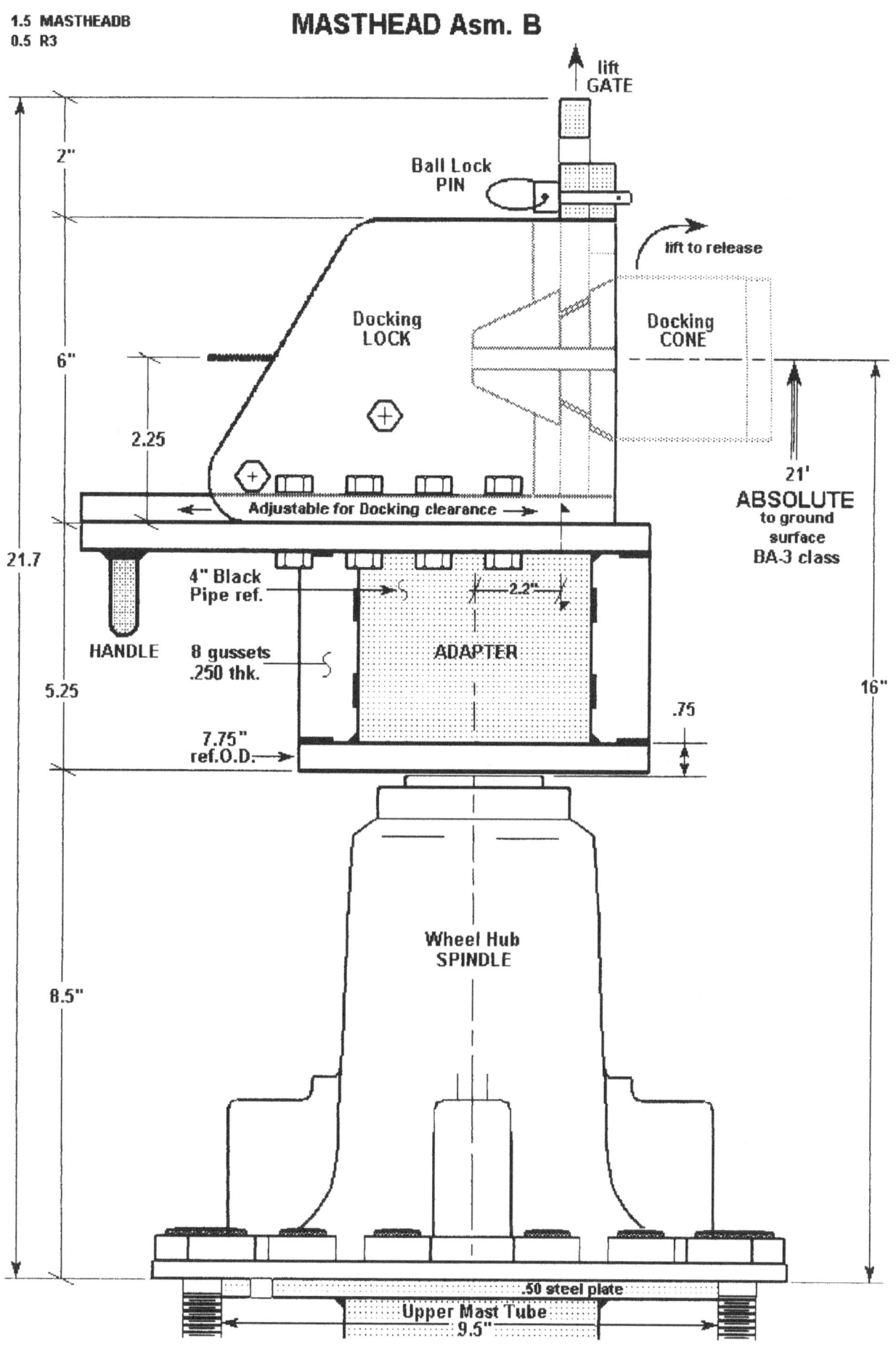

35

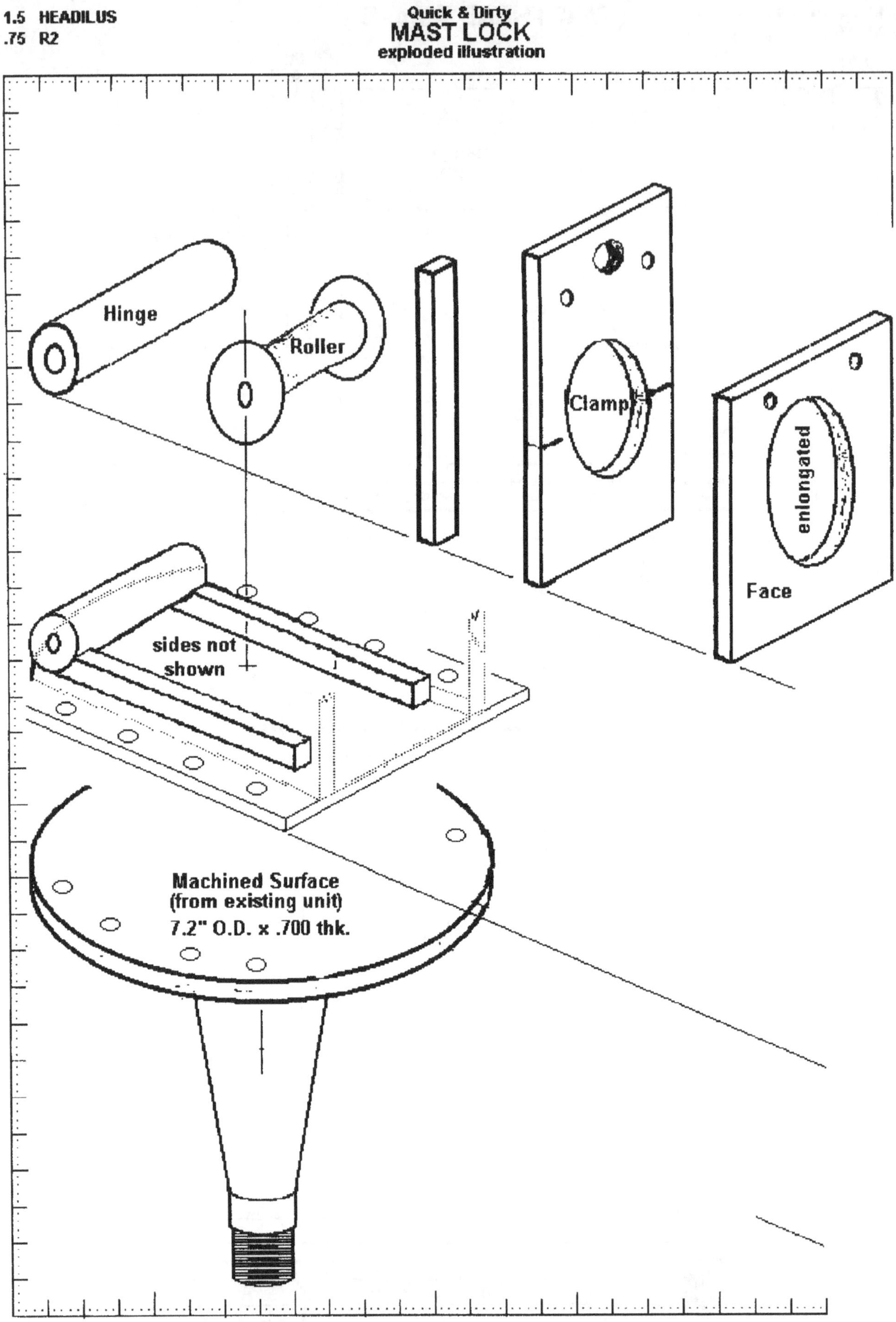

Building Small Gas Blimps

1.5 HEADILUS
.75 R2

Quick & Dirty
MAST LOCK
exploded illustration

Hinge

Roller

Clamp

enlongated

Face

sides not shown

Machined Surface
(from existing unit)
7.2" O.D. x .700 thk.

36

Building Small Gas Blimps

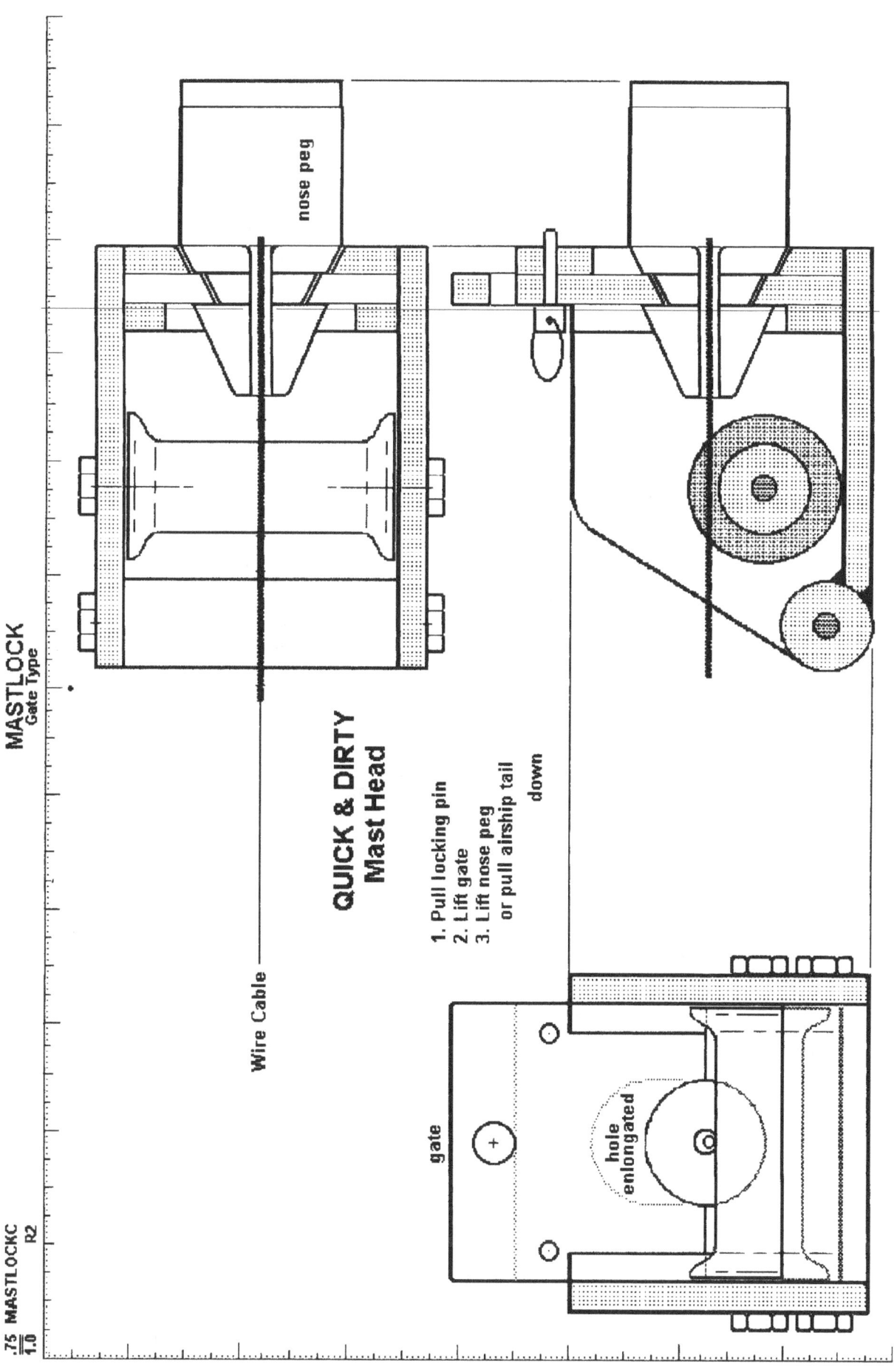

37

Building Small Gas Blimps

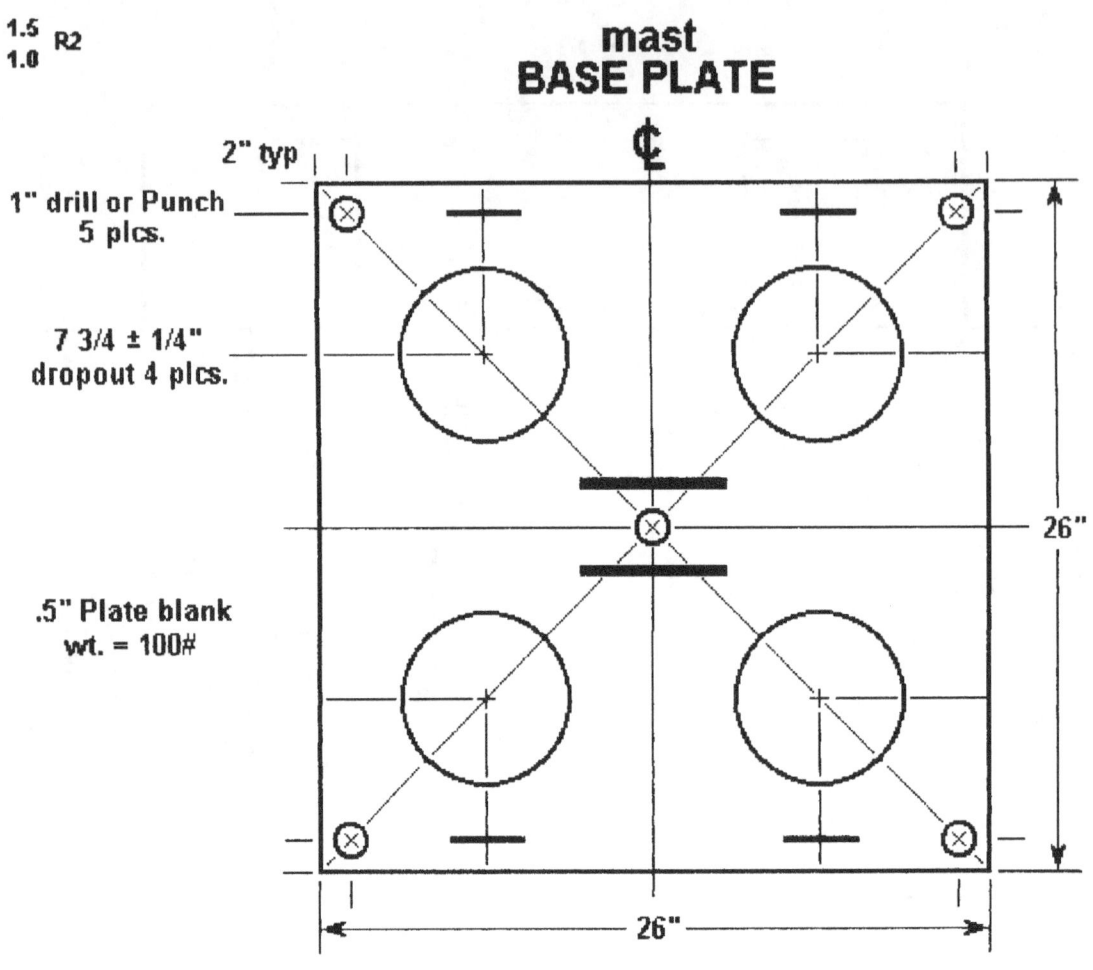

mast BASE PLATE

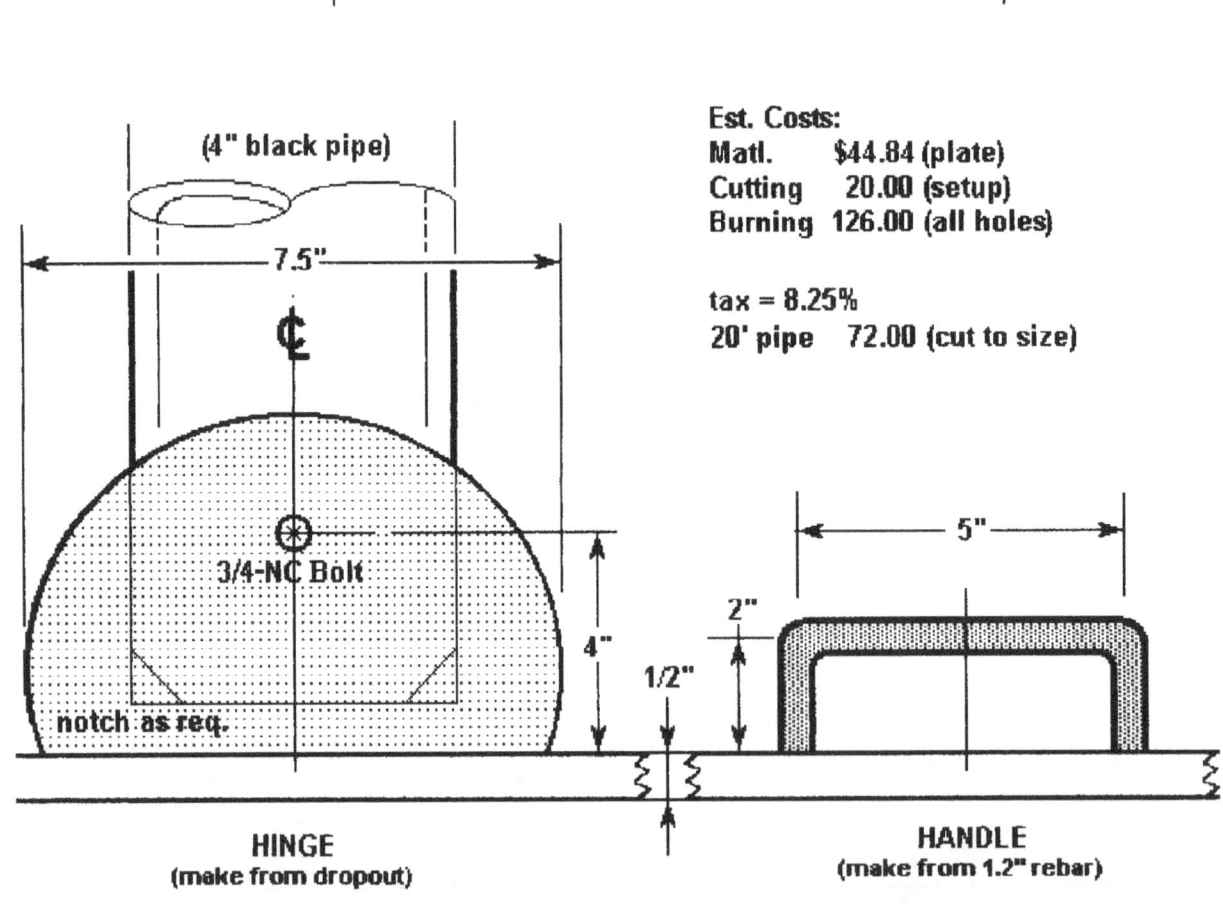

Est. Costs:
Matl. $44.84 (plate)
Cutting   20.00 (setup)
Burning 126.00 (all holes)

tax = 8.25%
20' pipe   72.00 (cut to size)

38

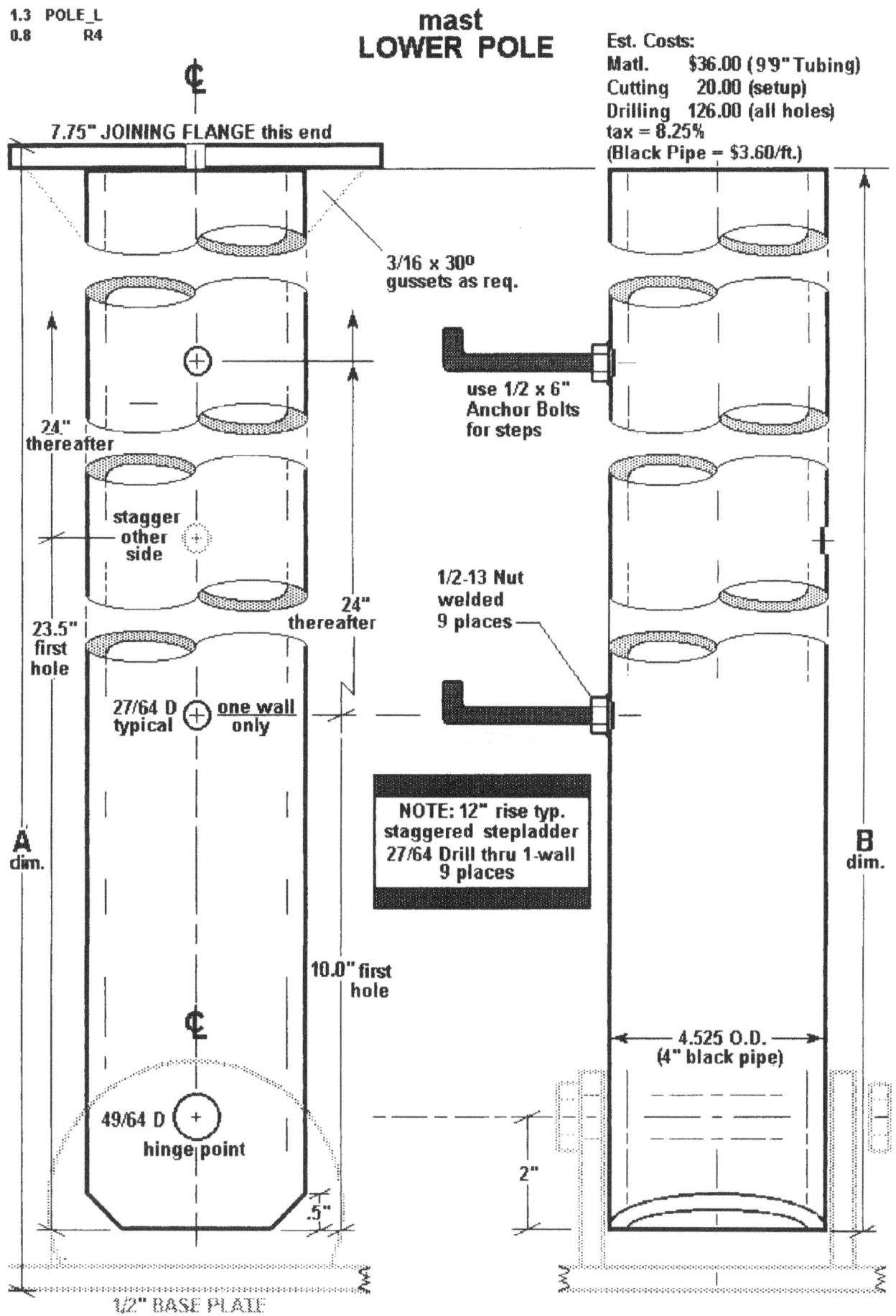

Building Small Gas Blimps

1.5 FLANGES
1.0 R4

MAST FLANGES

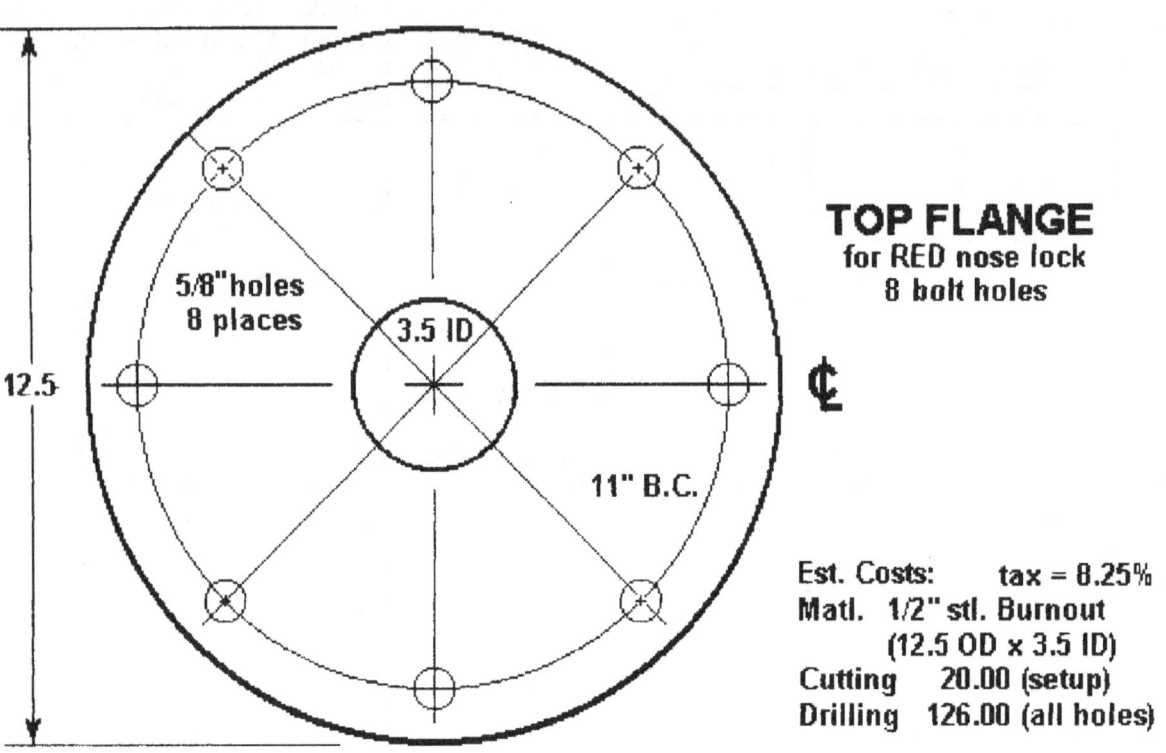

TOP FLANGE
for RED nose lock
8 bolt holes

Est. Costs: tax = 8.25%
Matl. 1/2" stl. Burnout
        (12.5 OD x 3.5 ID)
Cutting   20.00 (setup)
Drilling   126.00 (all holes)

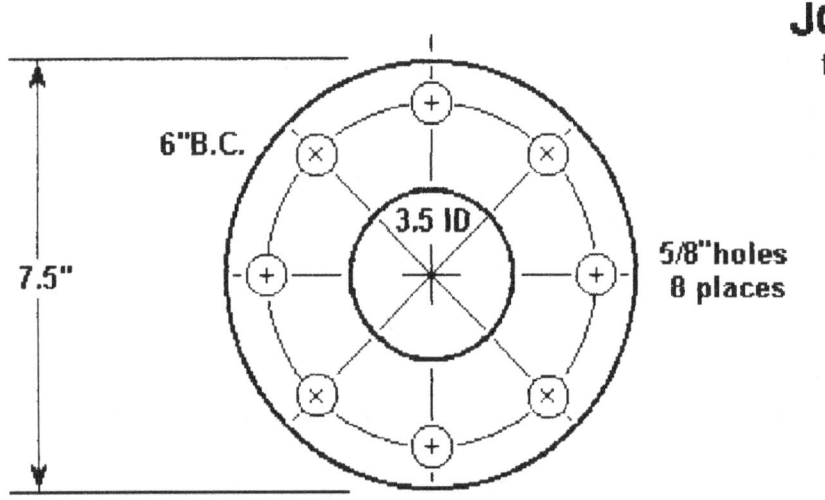

JOINING FLANGES
for the two MAST POLES

NOTE: Make from
BURN-OUTS from
the mast BASE PLATE

Building Small Gas Blimps

1.5 FOOTINGS
0.5 R3

ANCHORING SYSTEM
Mast Footings

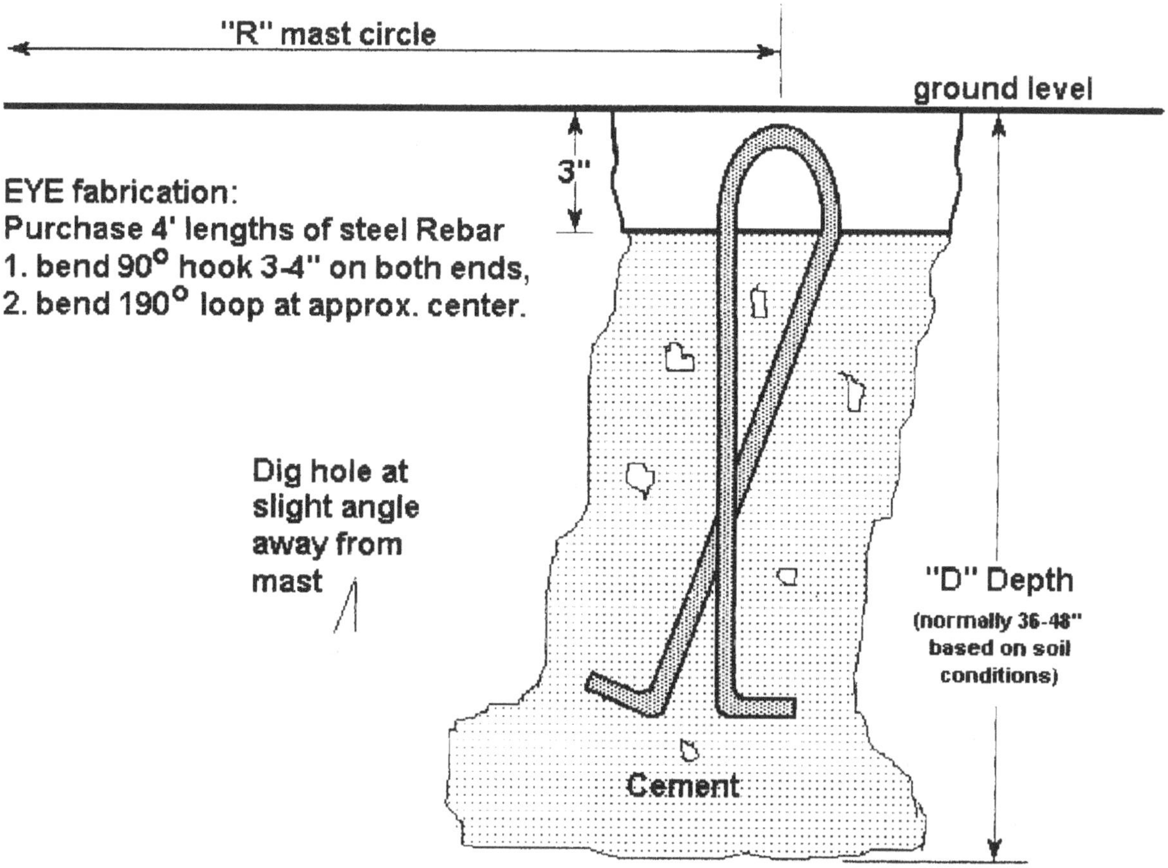

MASTS

"R" mast circle

ground level

3"

EYE fabrication:
Purchase 4' lengths of steel Rebar
1. bend 90° hook 3-4" on both ends,
2. bend 190° loop at approx. center.

Dig hole at slight angle away from mast

"D" Depth
(normally 36-48" based on soil conditions)

Cement

ALTERNATE ANCHORING
in soft ground

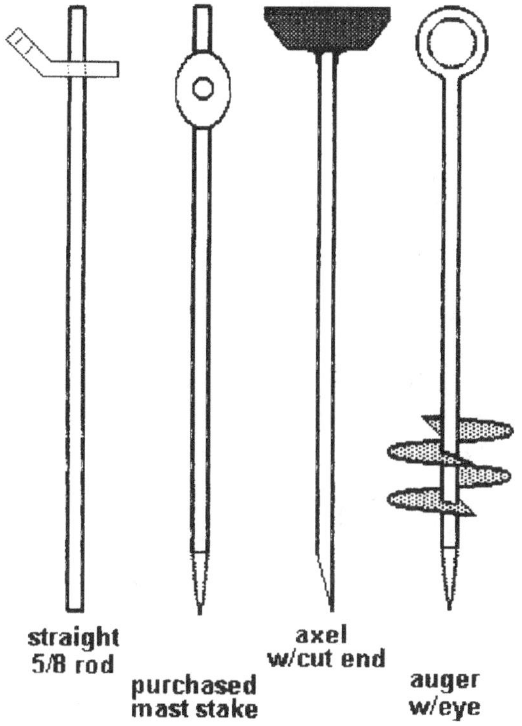

- straight 5/8 rod
- purchased mast stake
- axel w/cut end
- auger w/eye

TAKE-UP ADJUSTERS
guy tensioners

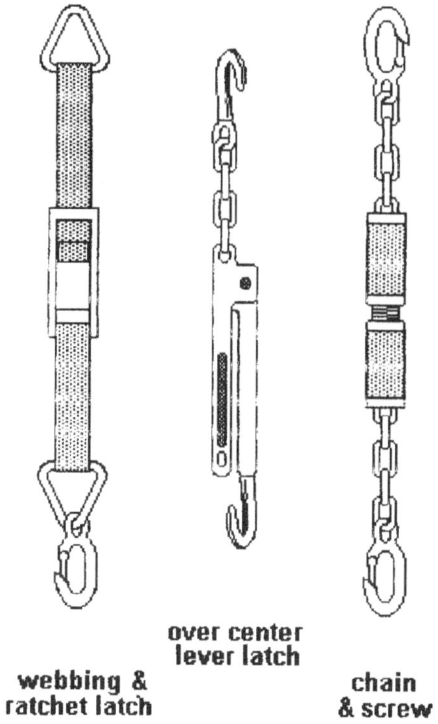

- webbing & ratchet latch
- over center lever latch
- chain & screw

41

Building Small Gas Blimps

HANGARS

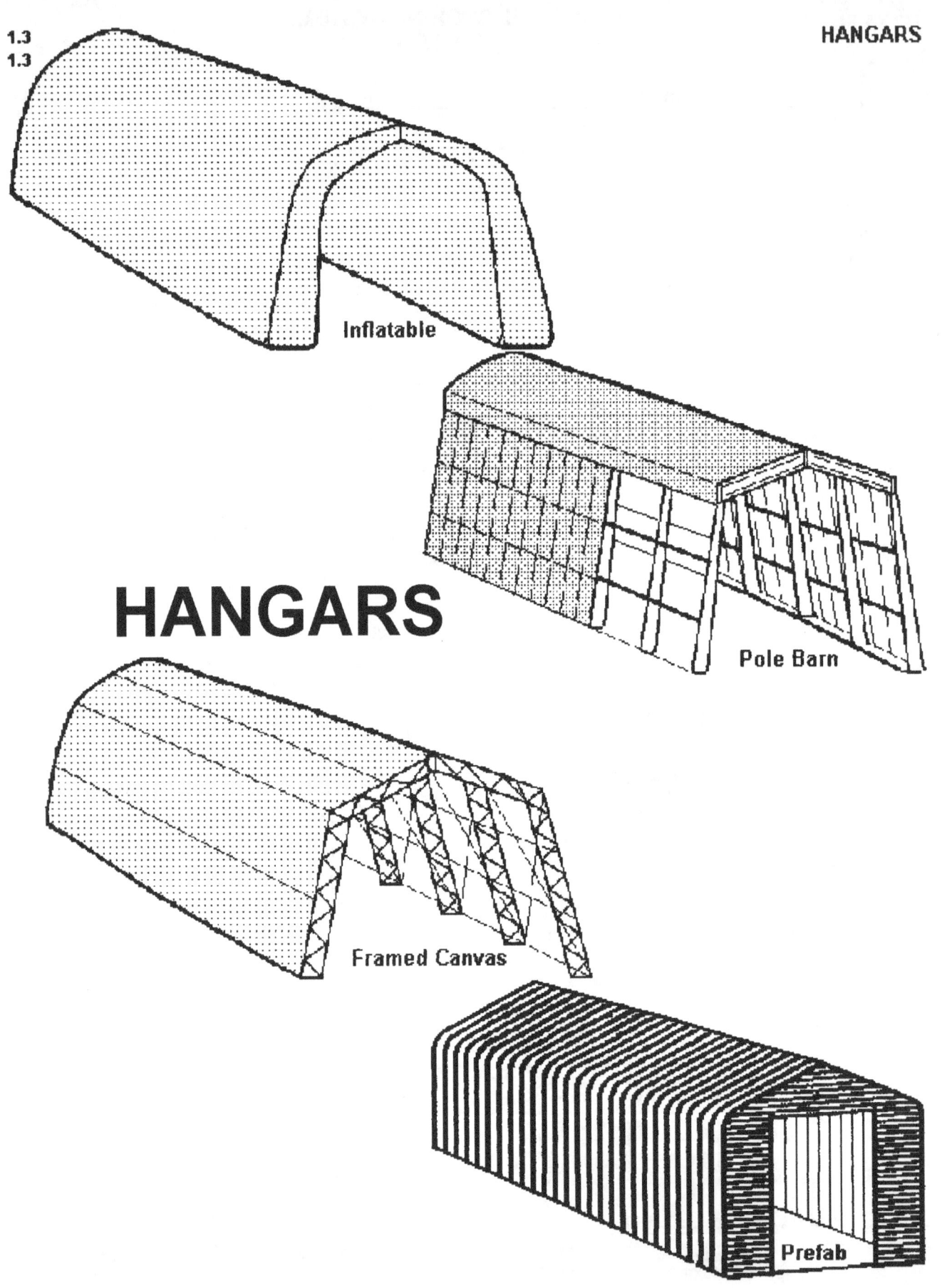

HANGARS

Inflatable

Pole Barn

Framed Canvas

Prefab

Building Small Gas Blimps

1.0 HANGAR3
0.3 R3

BLIMP HANGAR
Fabric covered, modular sections, open web truss

1992 prices

WIDTH x HEIGHT *	LENGTHS						END PANEL EACH
	24'	32'	40'	48'	56'	96'	
36' x 11'	$3689	$4919	$6149	$7379	$8608	$14757	$430
40' x 12'	4099	5466	6832	8198	9565	16397	522
44' x 13'	4509	6012	7515	9018	10521	18036	622
48' x 14'	4919	6559	8198	9838	11478	19676	730
52' x 14'	5329	7105	8882	10658	12434	21315	791
56' x 15'	5739	7652	9565	11478	13391	22956	913

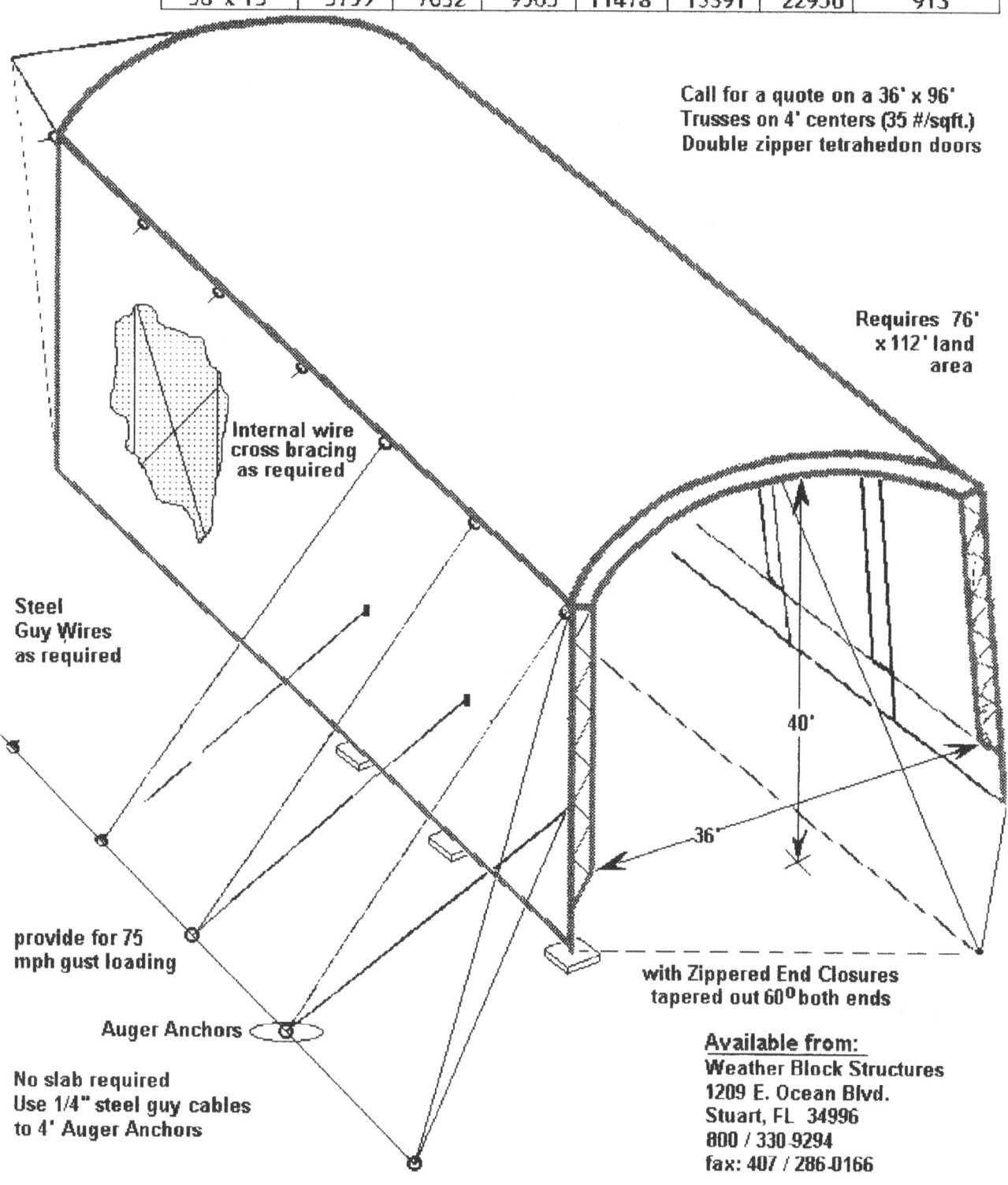

Call for a quote on a 36' x 96'
Trusses on 4' centers (35 #/sqft.)
Double zipper tetrahedon doors

Requires 76' x 112' land area

Internal wire cross bracing as required

Steel Guy Wires as required

40'

36'

provide for 75 mph gust loading

with Zippered End Closures tapered out 60° both ends

Auger Anchors

Available from:
Weather Block Structures
1209 E. Ocean Blvd.
Stuart, FL 34996
800 / 330-9294
fax: 407 / 286-0166

No slab required
Use 1/4" steel guy cables
to 4' Auger Anchors

Building Small Gas Blimps

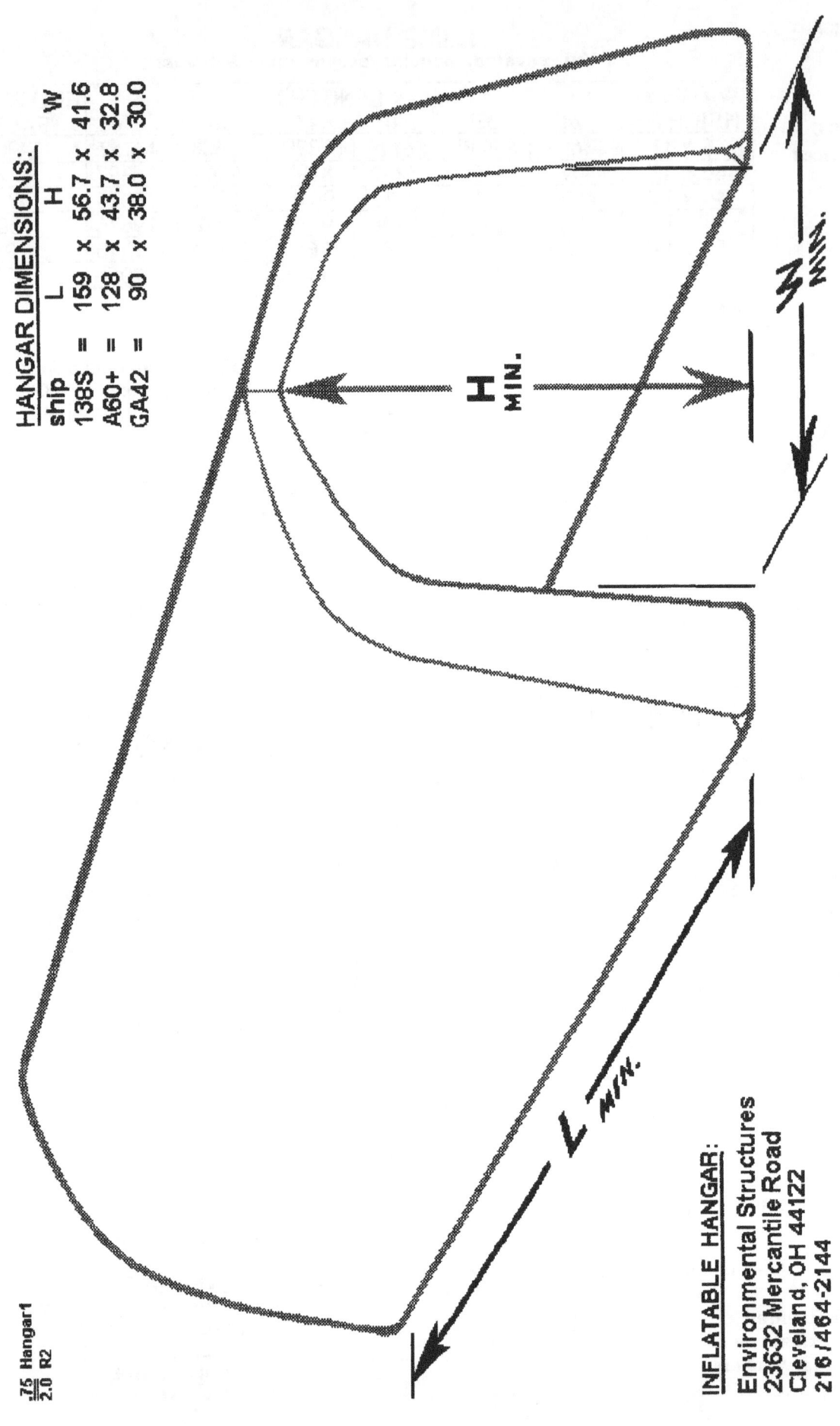

HANGAR DIMENSIONS:
ship	L	H	W
138S	159	56.7	41.6
A60+	128	43.7	32.8
GA42	90	38.0	30.0

INFLATABLE HANGAR:
Environmental Structures
23632 Mercantile Road
Cleveland, OH 44122
216/464-2144

44

Building Small Gas Blimps

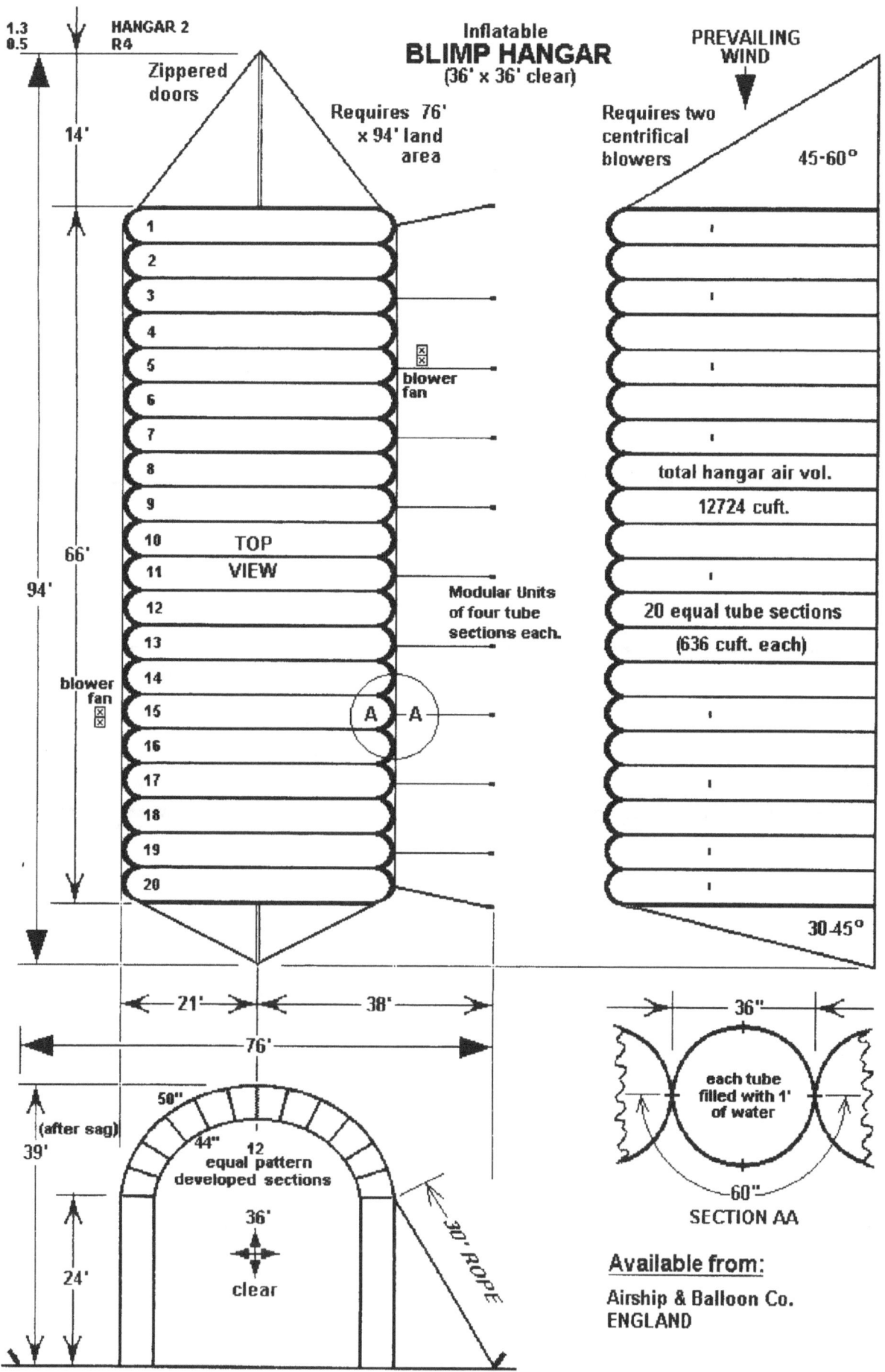

45

Building Small Gas Blimps

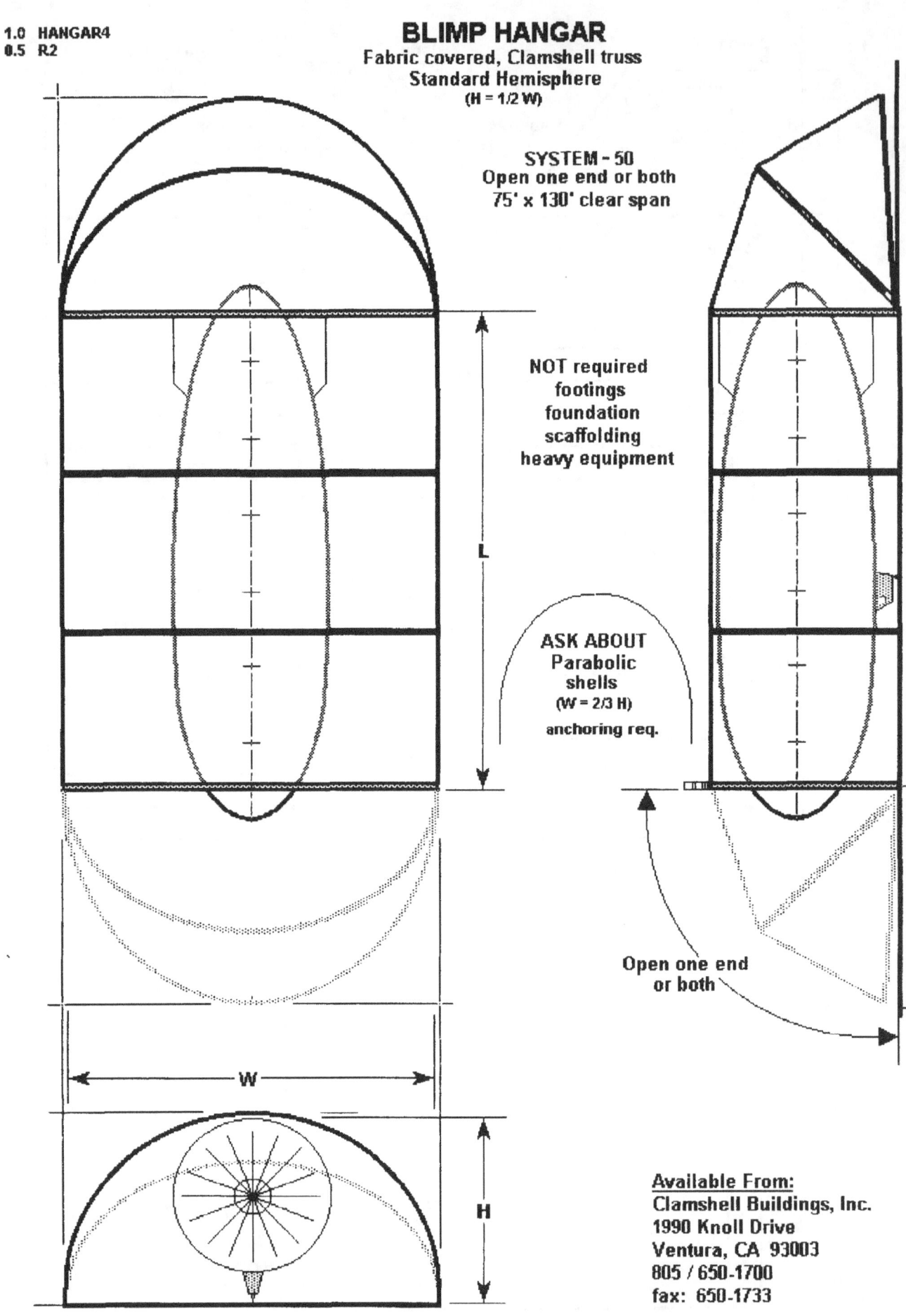

BLIMP HANGAR
Fabric covered, Clamshell truss
Standard Hemisphere
(H = 1/2 W)

SYSTEM - 50
Open one end or both
75' x 130' clear span

NOT required
footings
foundation
scaffolding
heavy equipment

ASK ABOUT
Parabolic
shells
(W = 2/3 H)
anchoring req.

Open one end or both

Available From:
Clamshell Buildings, Inc.
1990 Knoll Drive
Ventura, CA 93003
805 / 650-1700
fax: 650-1733

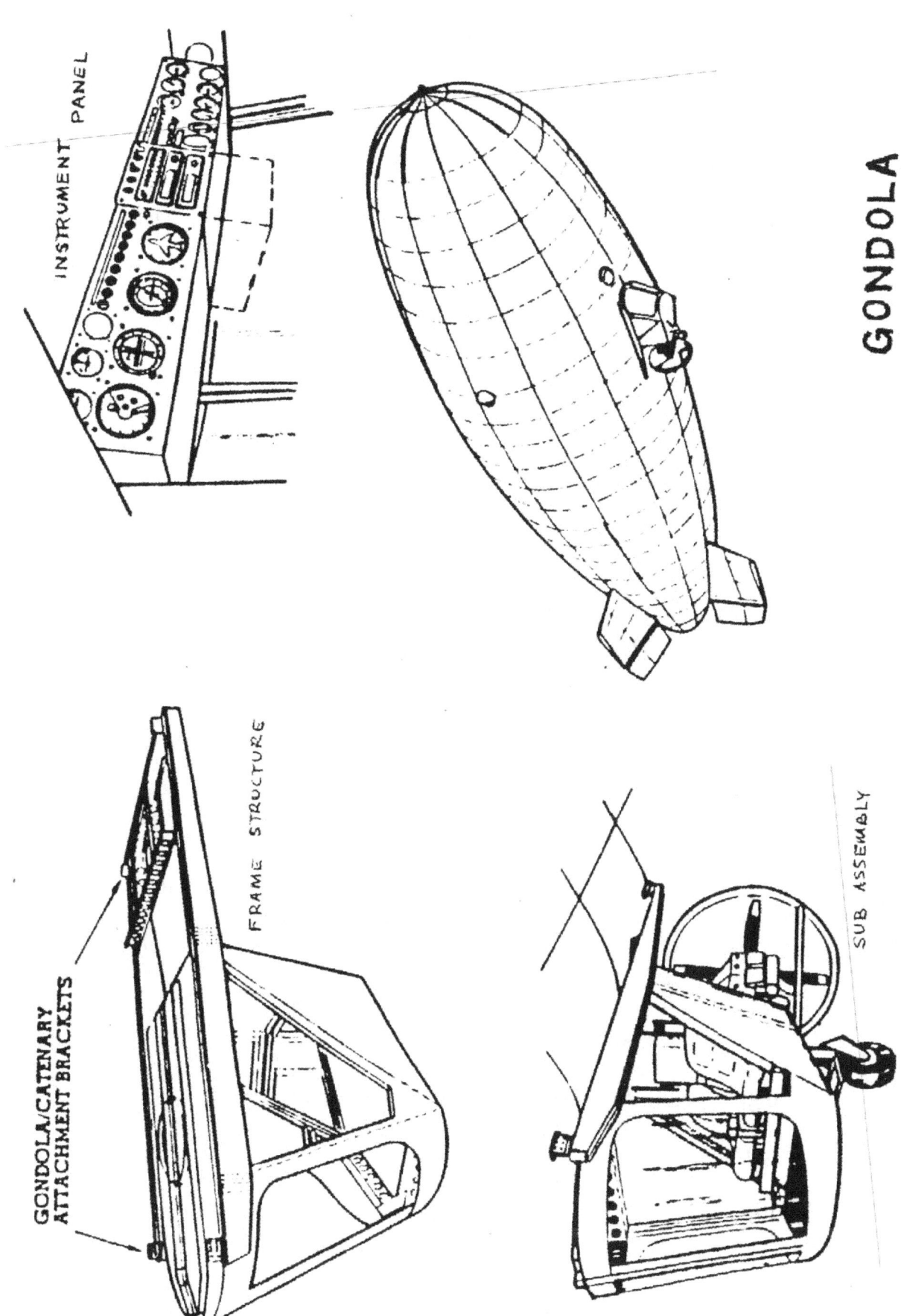

Building Small Gas Blimps

AIRFRAME
R:12-04-97

In the great scheme of things in aviation, aircraft assemblies are divided into airframes, powerplants, and accessories. A blimp is an aircraft, but the FAA has a problem deciding if the envelope or the gondola is the airframe. Since it has been a dilemma for 50 years, let's move on to something more realistic. In this book the major

Assemblies are: (Sub-Assemblies are:)
- Envelope (Gores, Ballonets, reinforcements, finger patches)
- Nose (docking cone, pendant, battens)
- Fins (stabilizers, control surfaces, guy wires)
- Gondola (structure, seats, instruments, controls)
- Powerplant (engine, mount, prop, fuel tank, carb.)

Accessories are:

Valves, ballonet & helium	actuators, linear & rotating
Instruments	Avionics, nav & com
Handling lines	heaters
Masts	blowers
Generator/Alternator	Elect. pumps

I am going to assume you know the definition of each of those words, and what each is expected to do. Hopefully, if you don't, you'll pick it up as I explain the criteria of each item. Again, I am going to leave the envelope for last, as the weight of every thing else determines its size. If you do it the other way around, you'll be lucky to get off the ground.

GONDOLA

Criteria
1. keep it light but strong
2. keep the controls visible & within reach
3. keep it livable in all seasons
4. keep it isolated from the engine noise
5. make the seats comfortable
6. optimize ground visibility
7. incorporate hand rails outside for the crew & ballast.
8. make it secure to the envelope, but easy to detach for overhaul/modification.
9. use a full swivel, but lockable landing gear
10. Design for a minimum of a 2G landing.
11. keep it compact enough to fit on a 4' x 8' pallet.

Styles

Single Seat; Easy to meet all of the criteria. But is it really what you want?
5% more money gives you twice the utility.

Dual seat; Author recommended. But item #2 becomes a challenge.
Side seating; helps overcome #2, but increases aerodynamic drag, and restricts opposite side visibility.
Tandem seating; works well if the gondola CG range is acceptable with an unoccupied seat. **Recommended**

Multi-seating; Don't do it. Licensing requirements are beyond a homebuilders means.

GONDOLAS cont.

Passing thoughts
How about buying one ready made? If you have any doubt about your ability, check around. Several good models have been designed, flown, and proven functional and reliable. Check the SOURCES section.

Getting started
Start with a kitchen chair, a foot stool, and a cardboard box the size of a small engine. Mark the floor area you think you would be comfortable in for 2 hours at a time. Next, start shopping for automotive bucket seating, airplane/helicopter foot pedals, and a reliable Ultra-Light engine. Start rearranging your floor marks accordingly, allowing a liberal amount for a fuel tank. When you feel comfortable, make a wood frame to support all this about 2 feet off the floor. Try getting into it several times. Get your wife in on it too, before she thinks you are completely nuts. If it is hard for her, it will be hard for a neophyte passenger. That is called planning ahead.

Getting serious
Start looking for some flight instruments for a dummy panel. Make an adjustable dolly (with casters) capable of supporting the gondola you envisioned. Start welding up a rough frame to support your sub assemblies. Change it 100 times if needed, before you settle on a final design. Pay particular attention to the landing gear and how it will transfer shock loads through the structure. Give great consideration to how the gondola will attach easily to the envelope. Then draw it on paper to an accurate scale, with dimensions, for future reference.

Look to the future
Some things that will also need some thought are:
1. a "skidding nose" landing (The envelope nose actually touches the ground while landing). Never intentional, but it happens. Any <u>badly positioned</u> lights or antennas on the front of the gondola will be destroyed. If you plan on doing a lot of training, a leaf-spring skid may be in order.
2. a "Tail drag" takeoff (very common) will destroy your prop if the tip touches the ground. (It can even rip the bottom tail fin off).
3. either 1 OR 2 can dislocate any rigid ducting to the ballonet. Control cables need bottoming springs as the impact makes rapid changes to the envelope shape. These are best positioned at/in the gondola for ease of replacement.
4. Keep in mind that a view window or dome will be installed in the envelope over the pilots head (to see the ballonet). Keep that area free of obstructions.

Building Small Gas Blimps

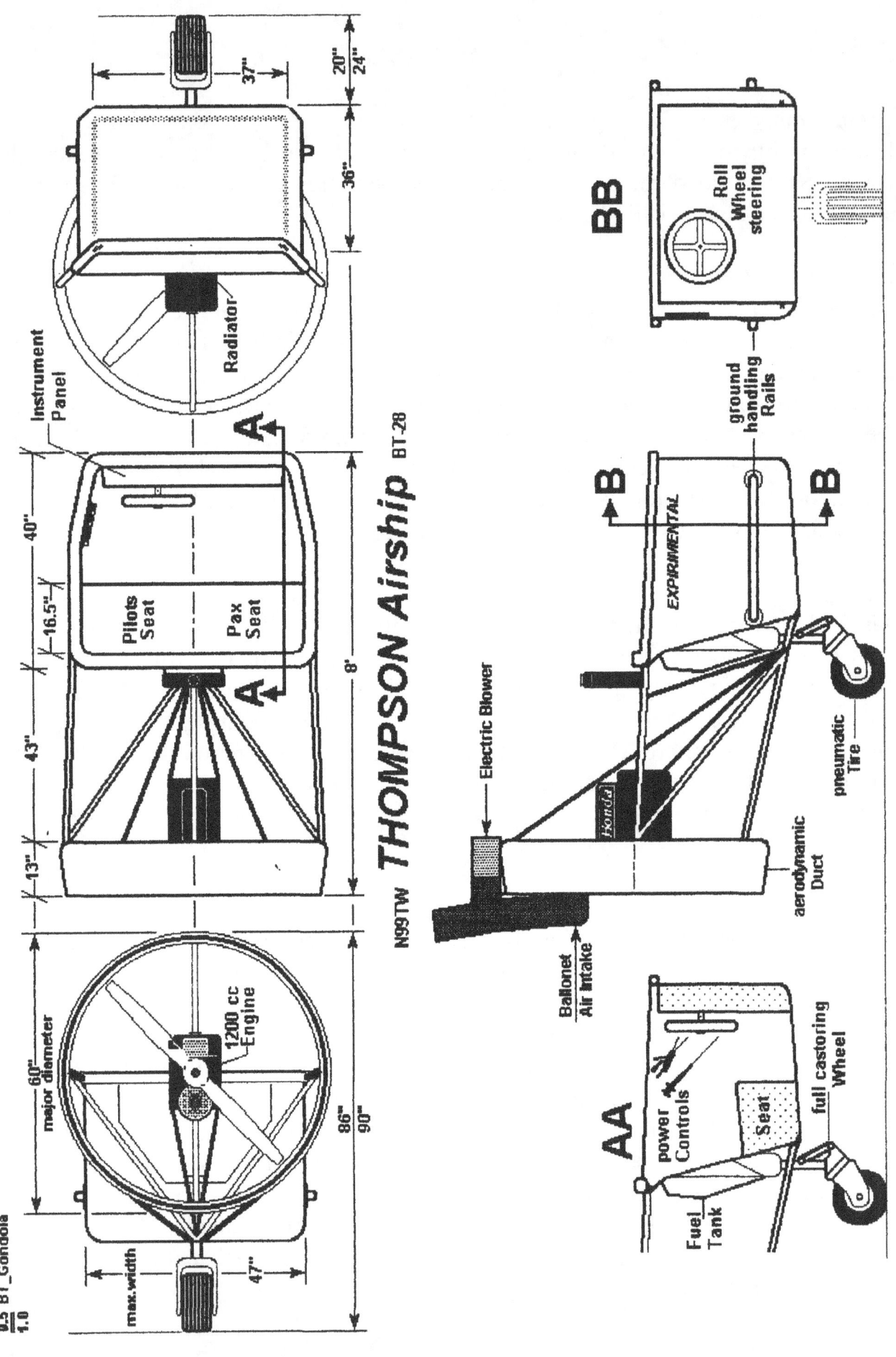

50

Building Small Gas Blimps

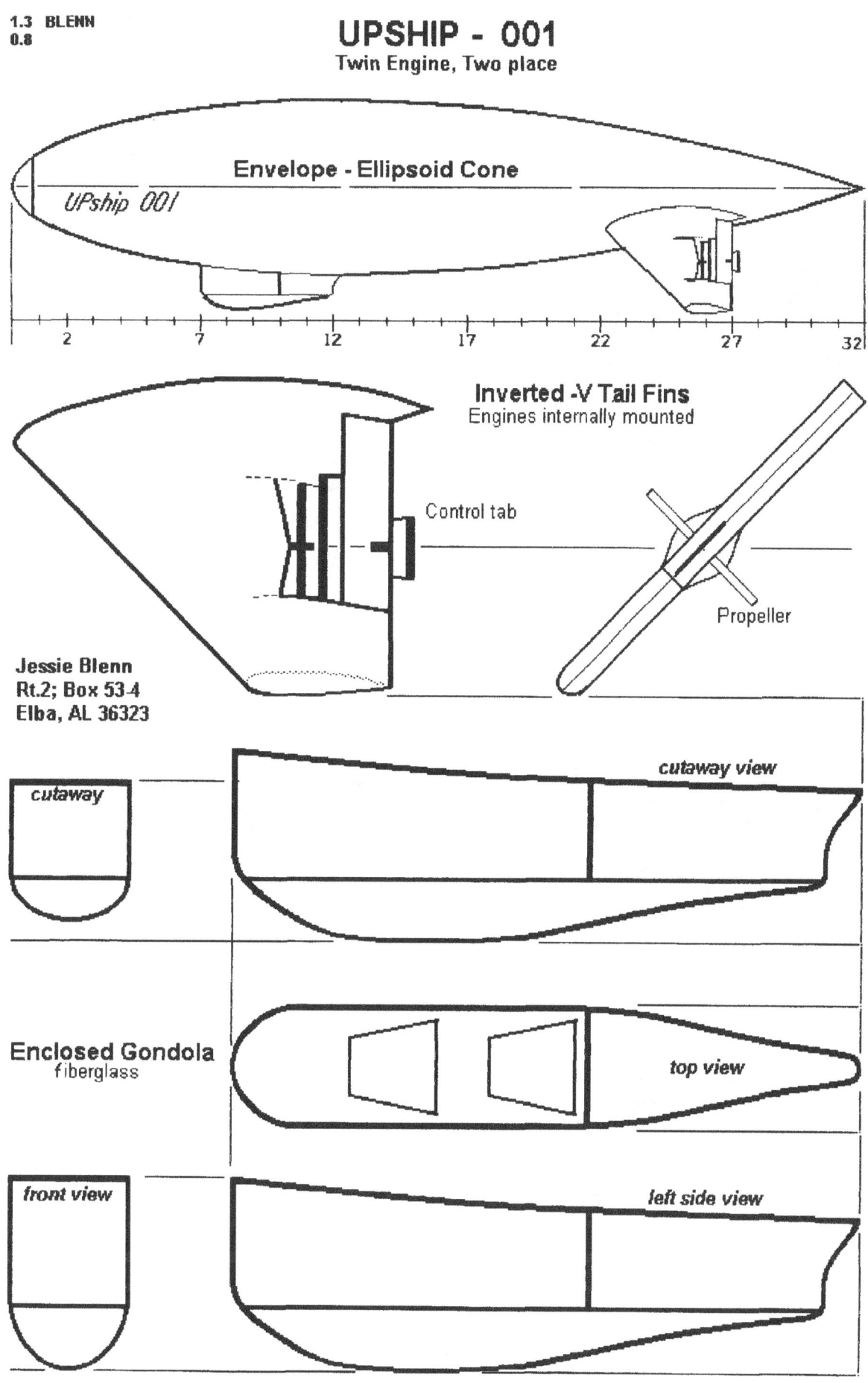

UPSHIP - 001
Twin Engine, Two place

1.3 BLENN
0.8

Envelope - Ellipsoid Cone
UPship 001

Inverted -V Tail Fins
Engines internally mounted

Control tab

Propeller

Jessie Blenn
Rt.2; Box 53-4
Elba, AL 36323

cutaway
cutaway view

Enclosed Gondola
fiberglass

top view

front view
left side view

Building Small Gas Blimps

1.3
0.8

CAMERON DG-19
With two Gondola Options

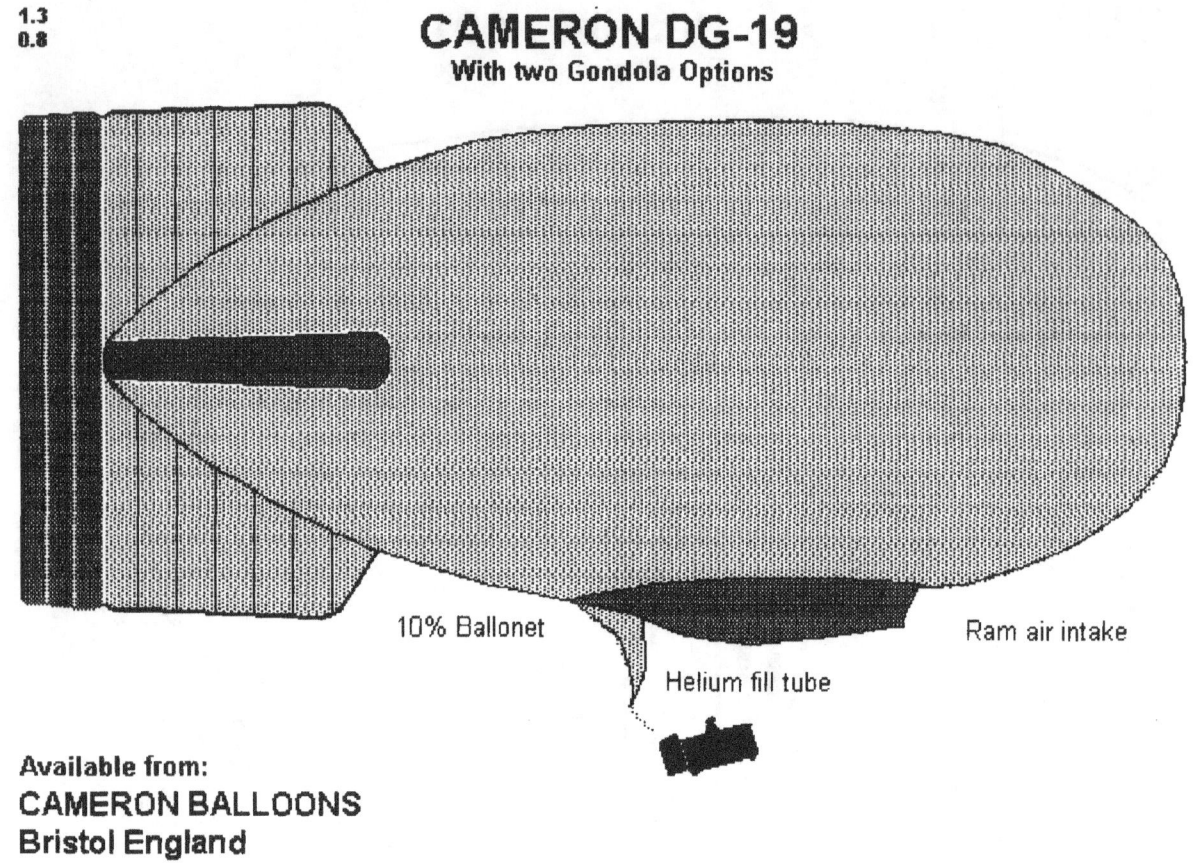

Available from:
CAMERON BALLOONS
Bristol England

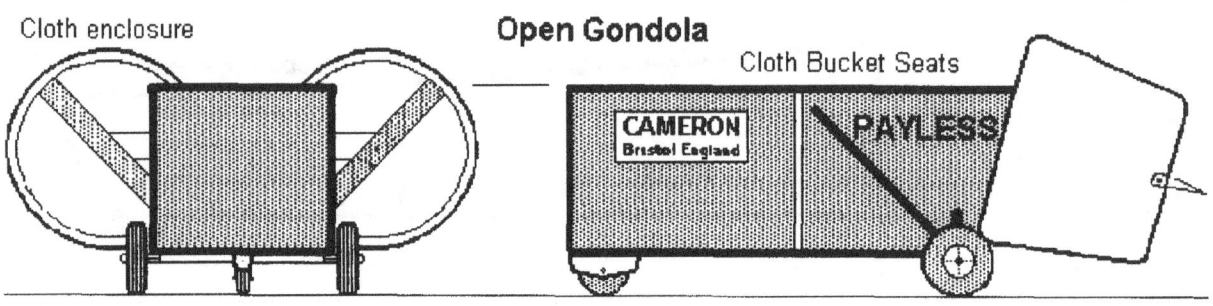

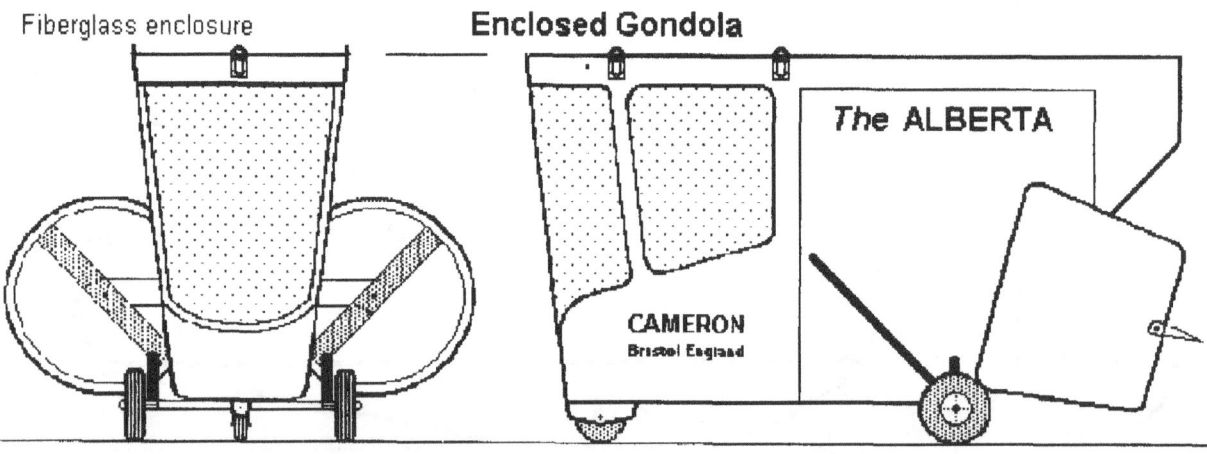

Building Small Gas Blimps

Pre-made MK-3 GONDOLA

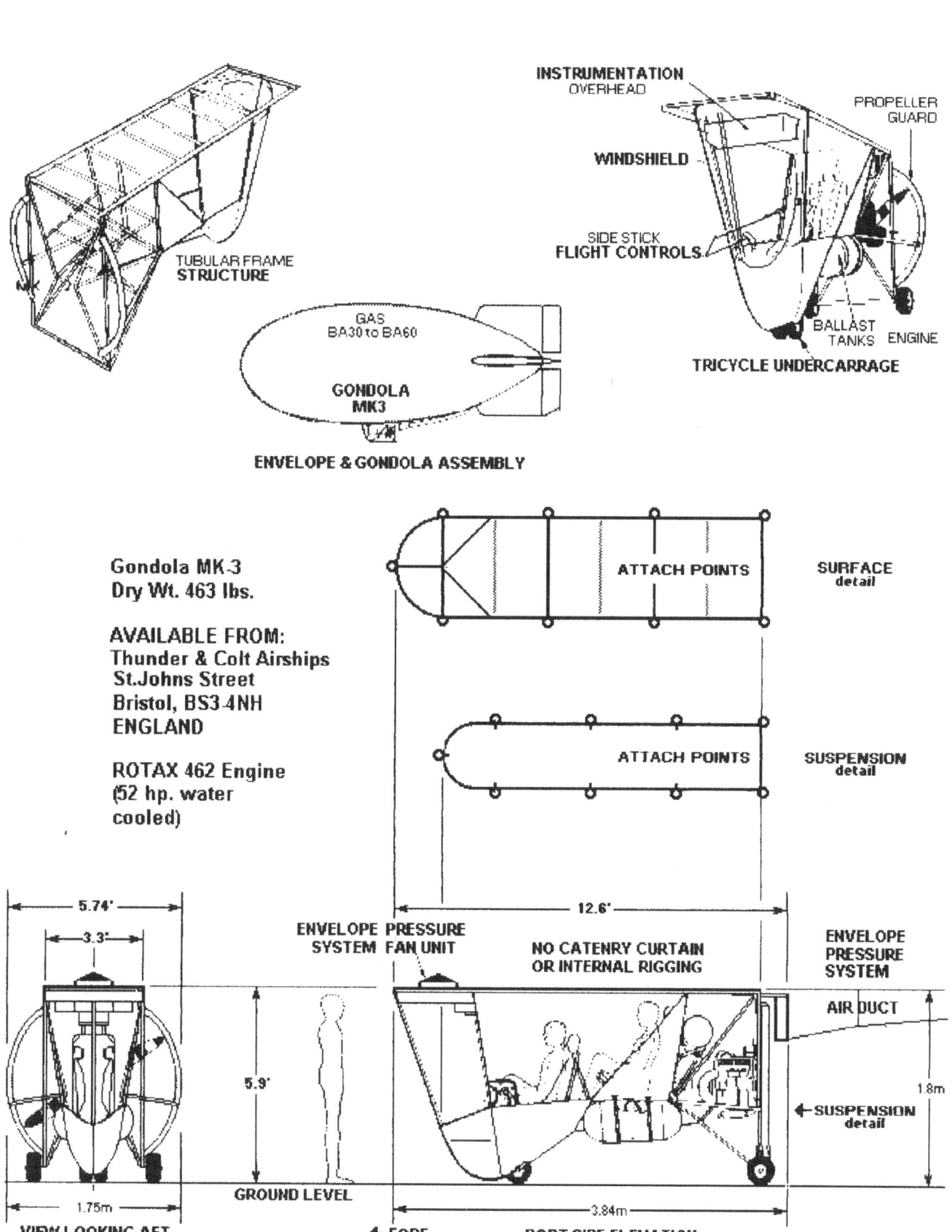

Gondola MK-3
Dry Wt. 463 lbs.

AVAILABLE FROM:
Thunder & Colt Airships
St. Johns Street
Bristol, BS3 4NH
ENGLAND

ROTAX 462 Engine
(52 hp. water cooled)

Building Small Gas Blimps

1.3 KIT1
1.0 R5

PREFAB KIT

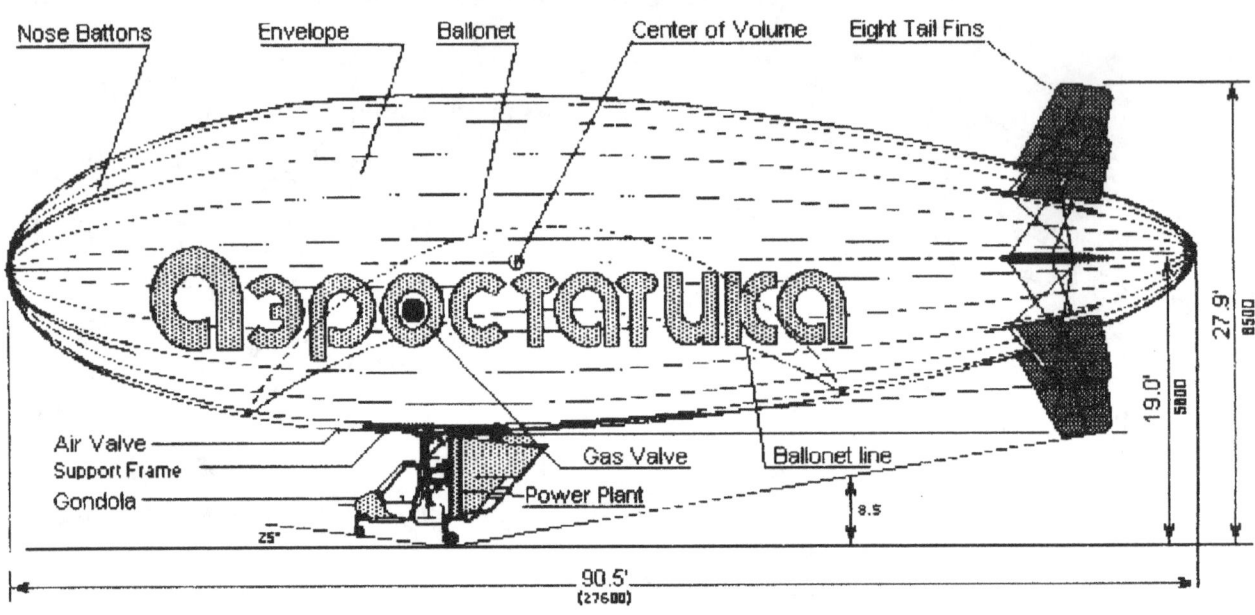

Available from:
Firma AEROSTATICA
31-1-315 Krylatskaya str.
Moscow, 121614 RUSSIA
Fax: +7 095 414 26

NOT TO BE CONFUSED WITH:
World Wide Aeros
of California
(who have NO affiliation)

GONDOLA

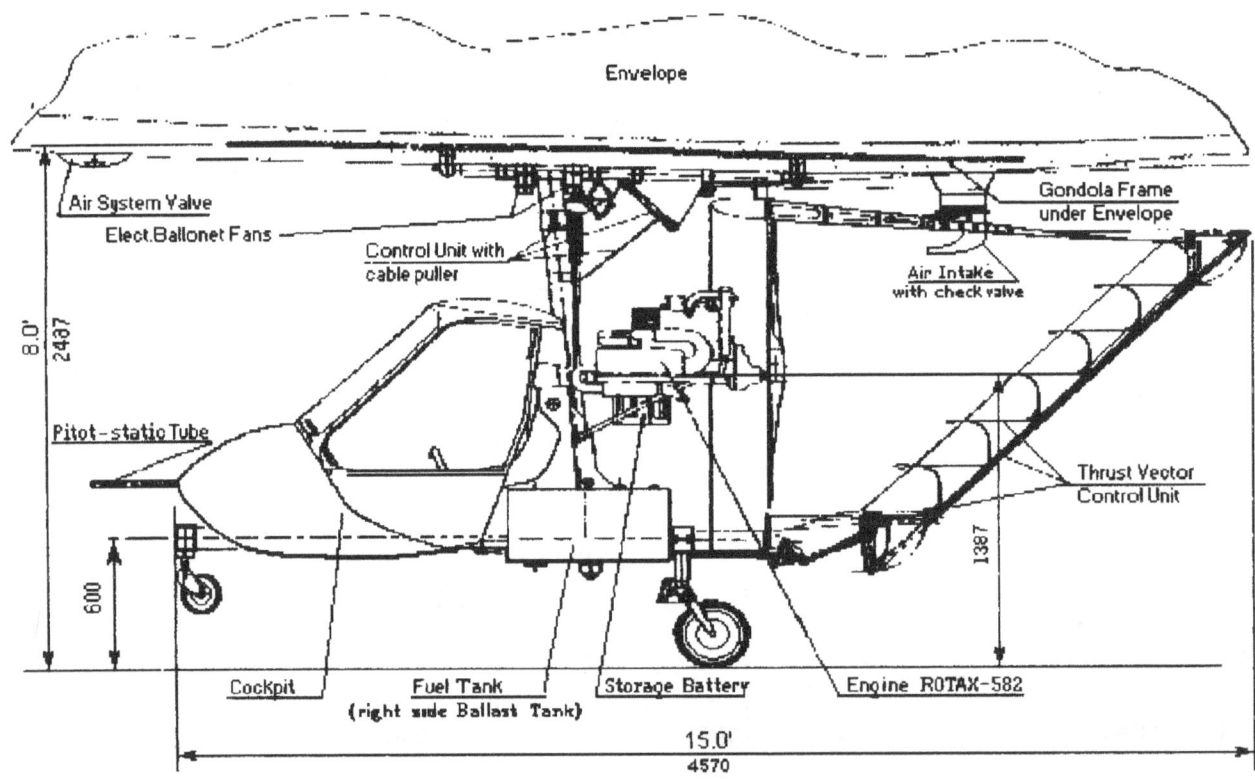

54

Building Small Gas Blimps

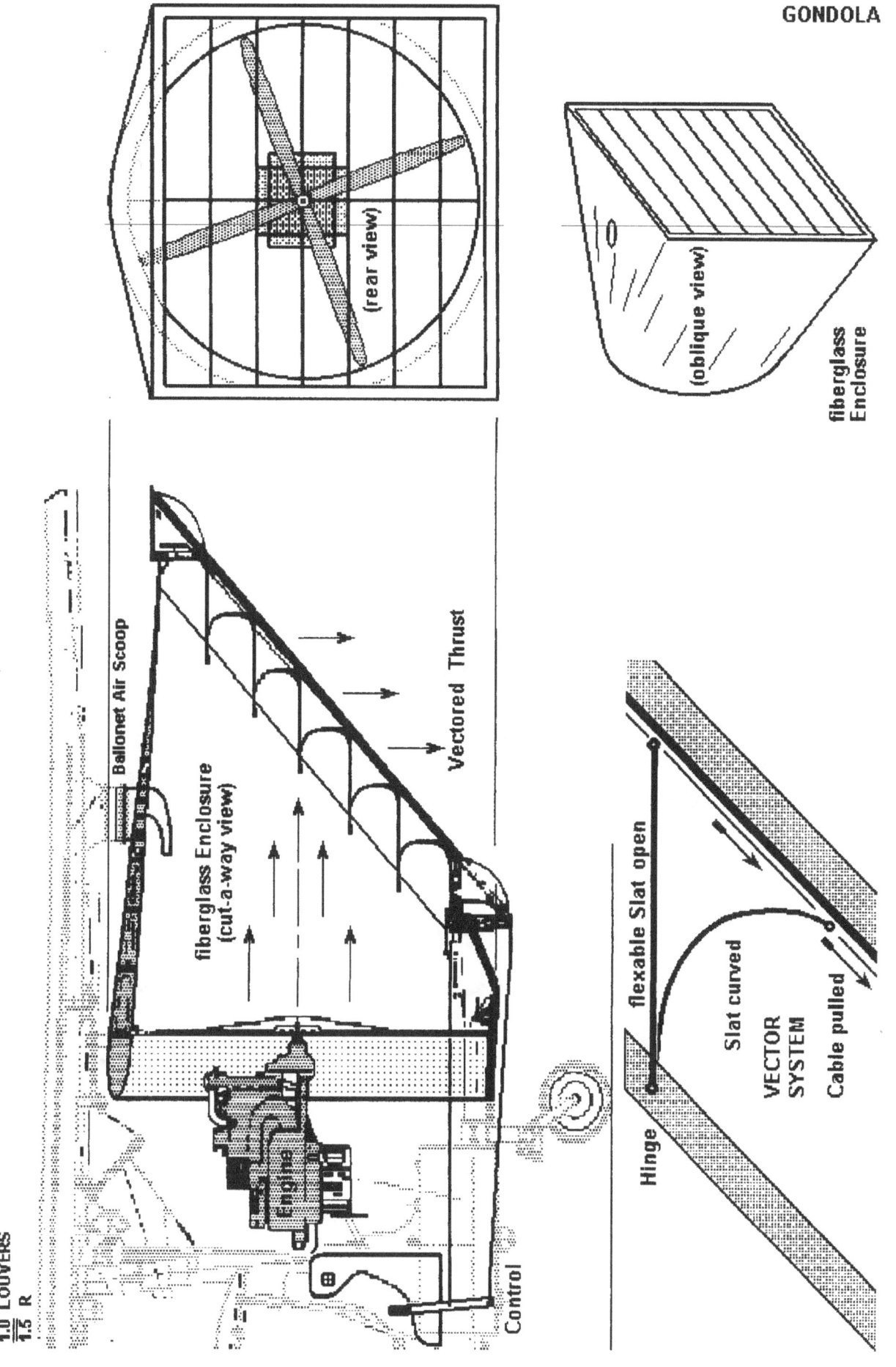

Building Small Gas Blimps

1.3 KIT_LV
1.0 R2

PREFAB KIT
or Ultralight

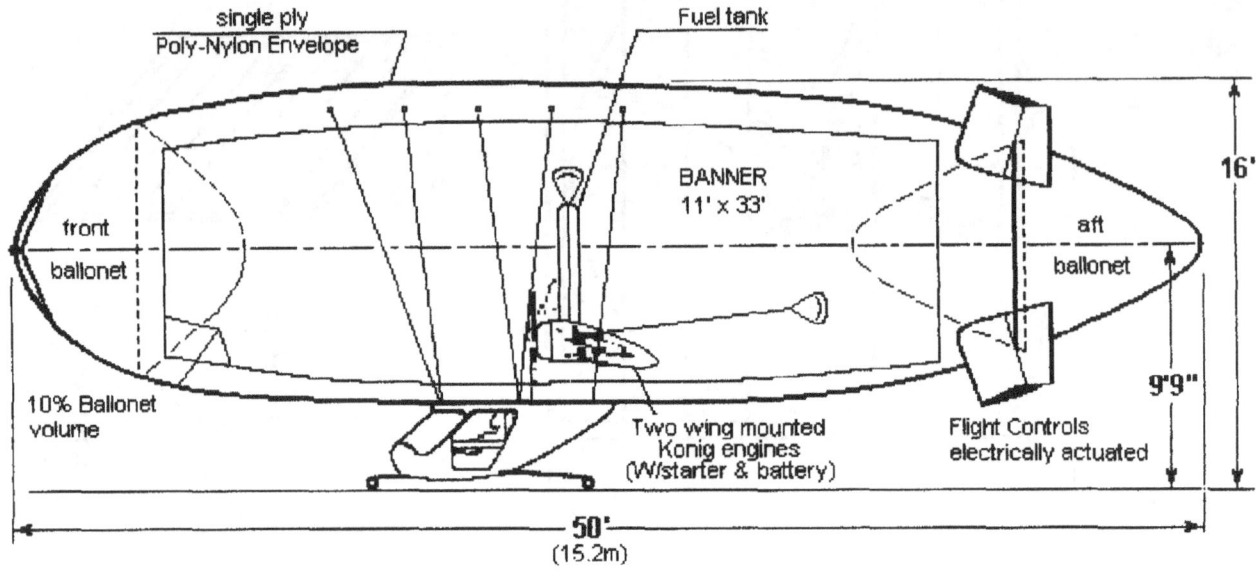

Available from:
Advanced HYBRID Aircraft
P.O. Box 144
Eugene, OR 97440

HORNET LV
(LV2 = 2-place)

http:www.ahausa.com

CONTACT MANUFACTURER
for options and
latest prices.
541/344-5323

specifications:

Volume: 4300 cuft.   (122 cu.m.)
Envelope Diameter: 12.5'   (3.8m)
Ceiling: 3500'   (1070m)
Useful Load: 204 lbs.   (92.5kg.)

Wing span: 22.5'   (6.9m)
Engines: Two @ 24hp ea.
Vector angle: 0-30 up
Speed: 51 kts.   (94kph.)

GONDOLA
cut-a-way view

Building Small Gas Blimps

The KERMIKRAFT Blimp

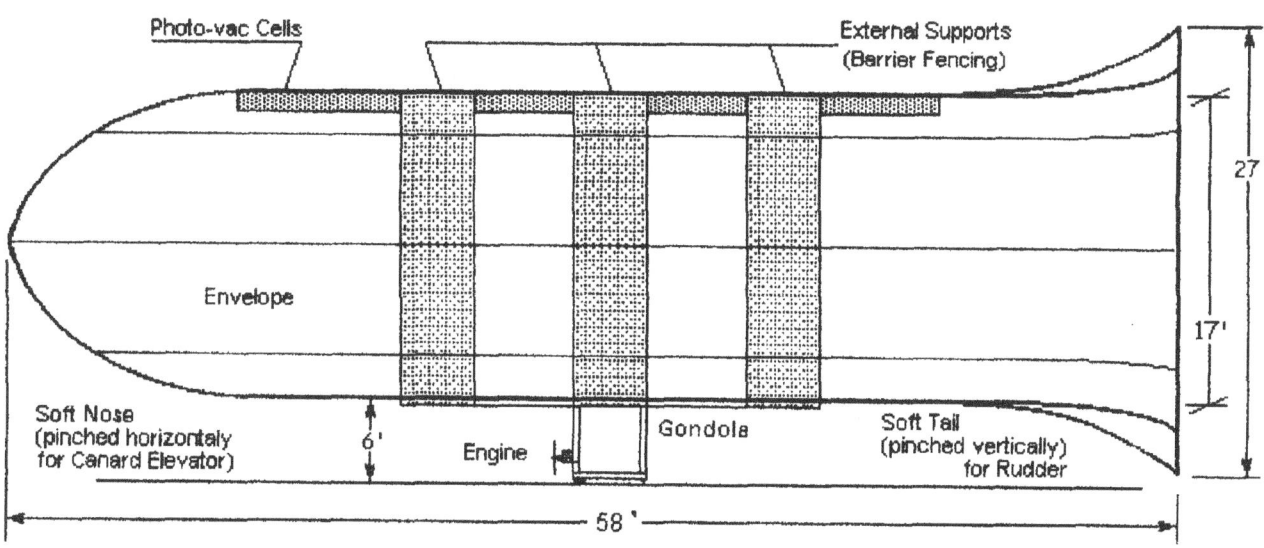

Designed – Builder:
KERMIT WIESELQUIST
174 Depot Road
Westford, MA 01886

ULTRA-LIGHT
Personal Aircraft

OPTIONS:	ENGINE	VOLUME	WEIGHT
Dash-1 =	Twin Electric	7,700 cuft.	264 #
Dash-2 =	Single Rotax-277	13,400 cuft.	496 #

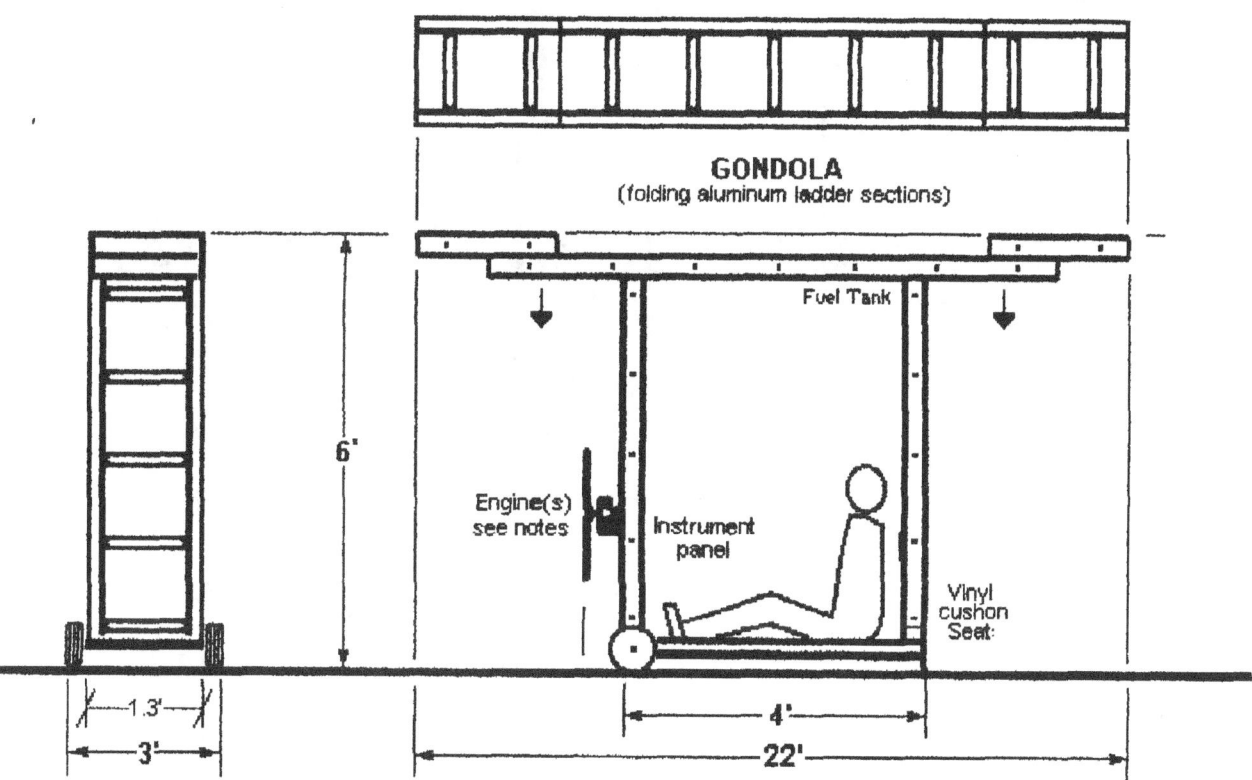

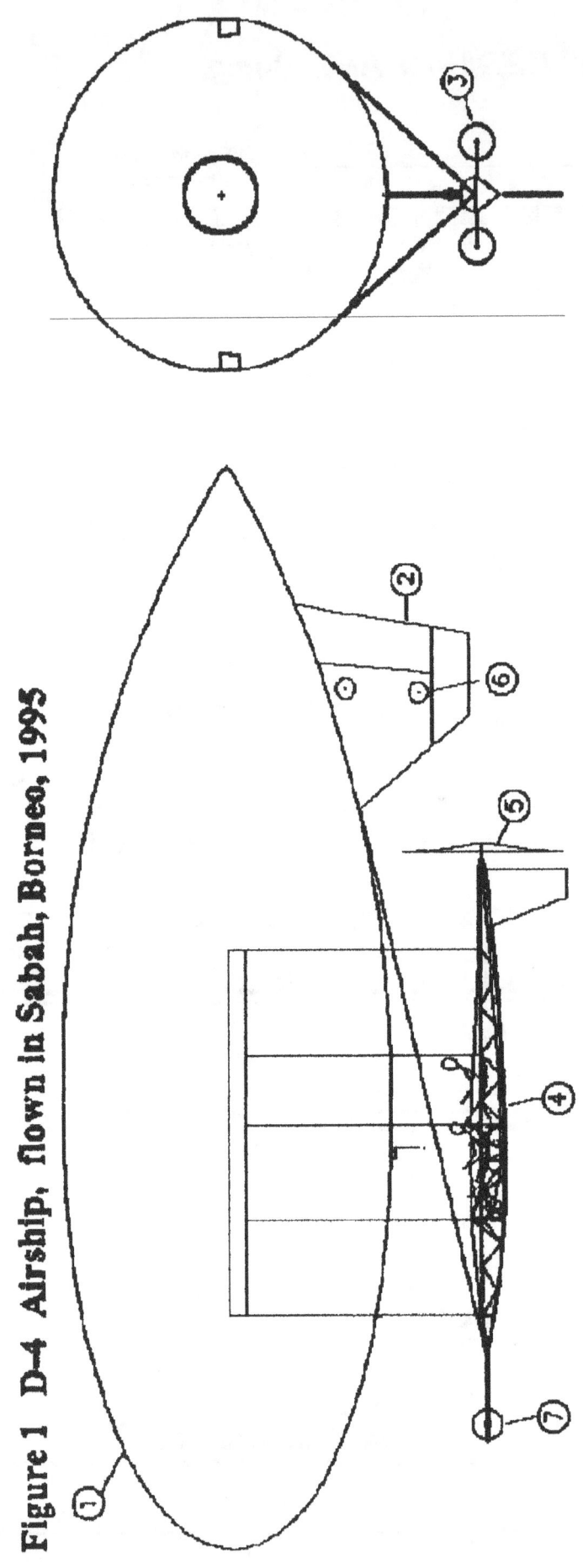

Figure 1 D-4 Airship, flown in Sabah, Borneo, 1995

Key: 1. Envelope (400m^3); 2. Fin (9.25m^2); 3. Pitch vectored 24" ducted propeller (2 off); 4. al.alloy keel structure (9m, 35kg) ; 5. Rear Propeller (diameter 2.6m); 6. Tail thruster (2 off) 7. Bow thruster.

Copyright © 1999, by G.E. Dorrington
Published by the American Institute of Aeronautics and Astronautics Inc., with permission.

Building Small Gas Blimps

The BLIMP WORKS BW-3
(13,000 cuft)

1.3 BARNES
1.0 R3

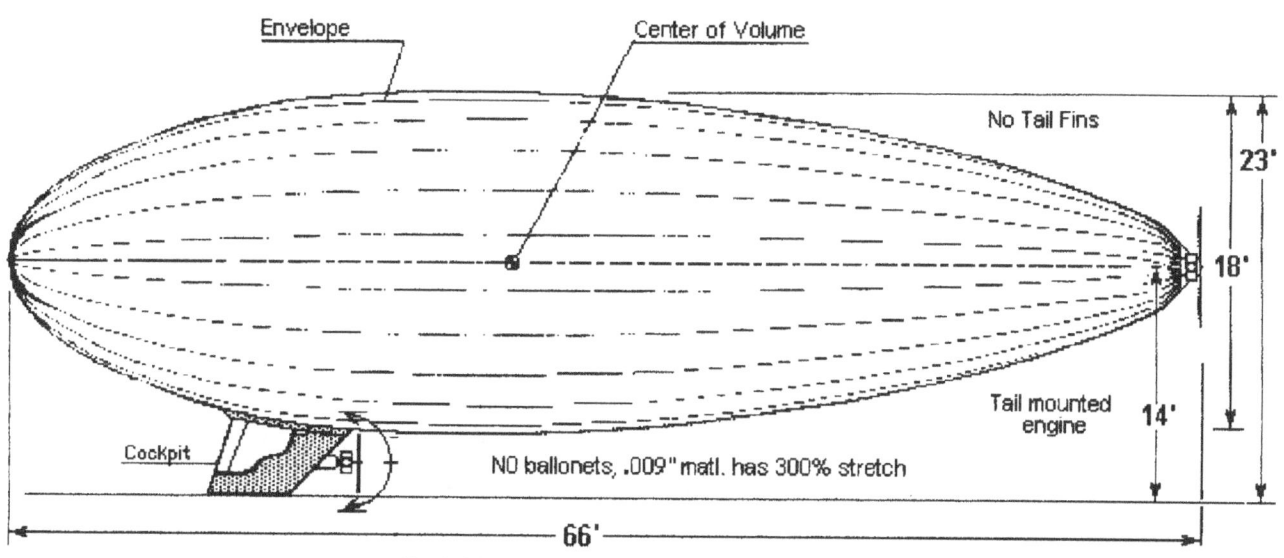

Empty Weight approx. 496 Pounds

Fabricated by:
TRACY BARNES
Rt. 2; Box 86
Statesville, NC 28677
704/876-6848

Specifications & Details
UNKNOWN as of 2/98
No production units ever produced
Proof of concept only
No parts available

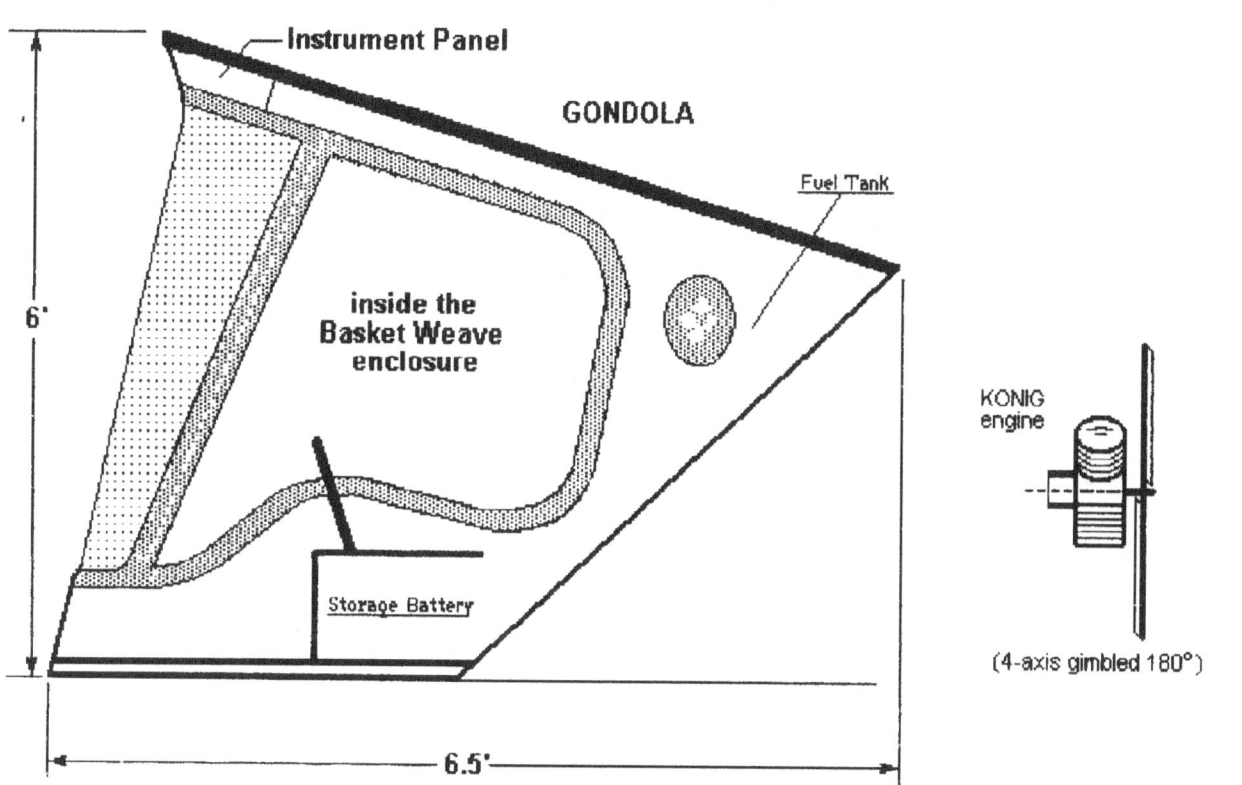

(4-axis gimbled 180°)

Building Small Gas Blimps

BW-1   GONDOLAS

BLIMP WORKS BW-1

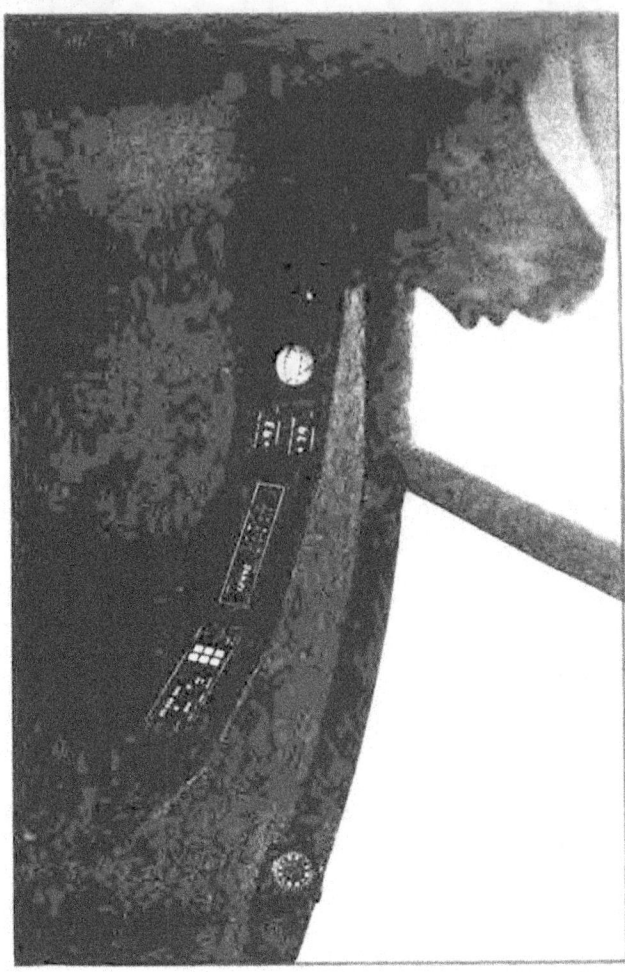

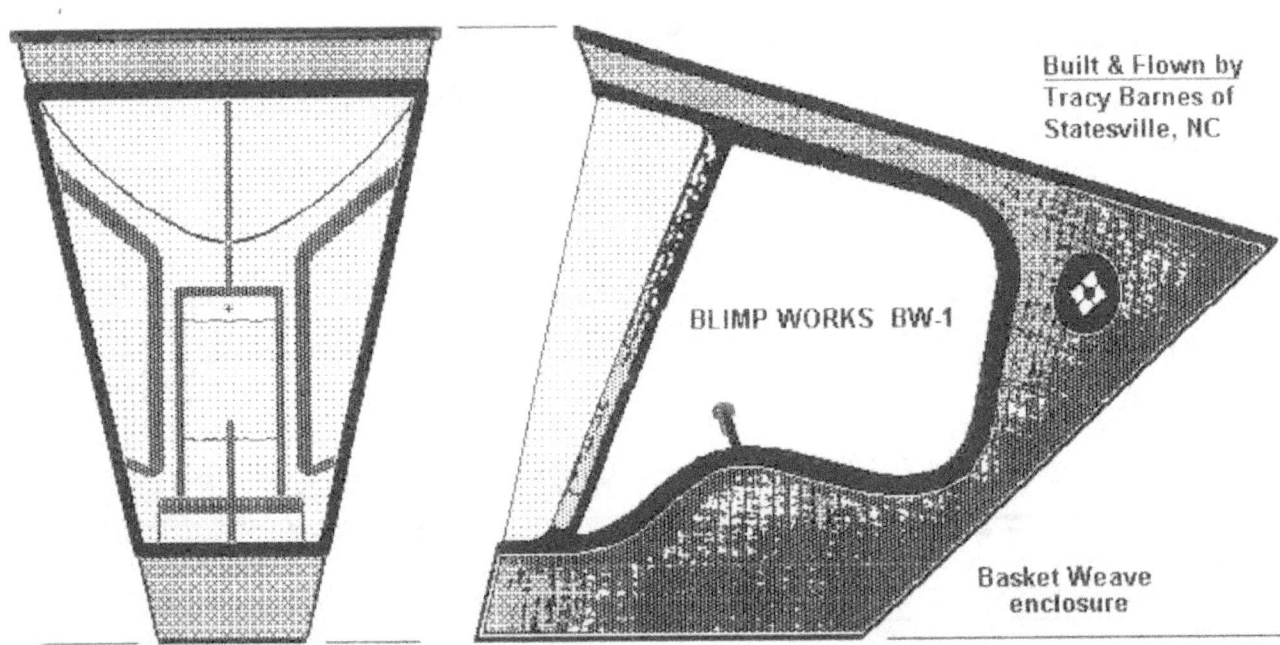

BLIMP WORKS BW-1

Built & Flown by Tracy Barnes of Statesville, NC

Basket Weave enclosure

Building Small Gas Blimps

SPARKS BS-14
(14,000 cubic feet)

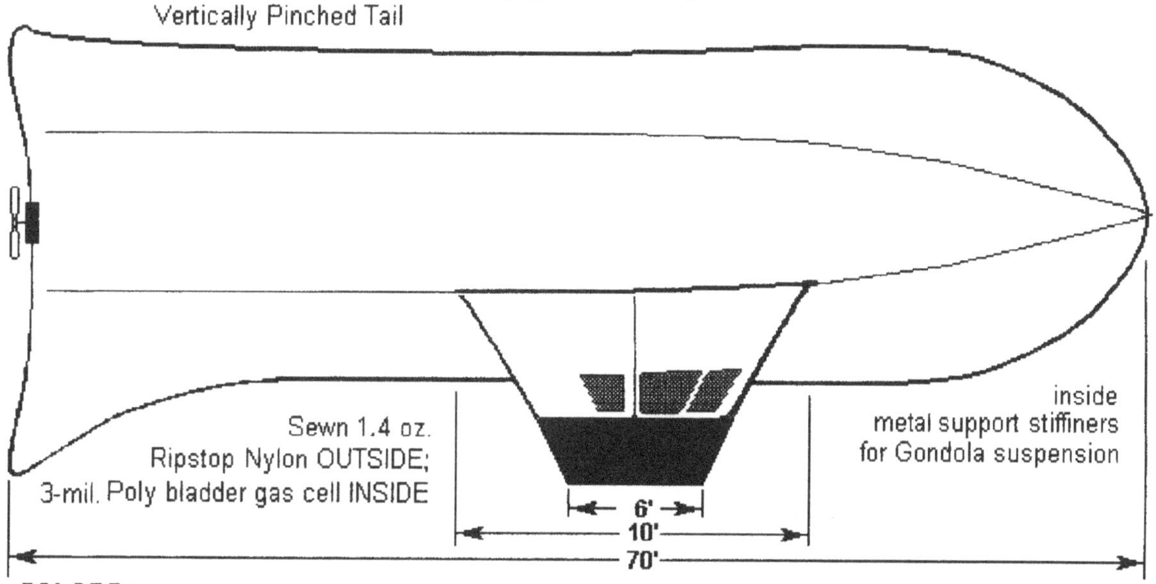

1.3/0.8 R1

- Vertically Pinched Tail
- Sewn 1.4 oz. Ripstop Nylon OUTSIDE; 3-mil. Poly bladder gas cell INSIDE
- inside metal support stiffeners for Gondola suspension
- 6'
- 10'
- 70'

COLORS:
Envelope & upper portion of Gondola ALL YELLOW
Bottom portion of Gondola, BLUE.

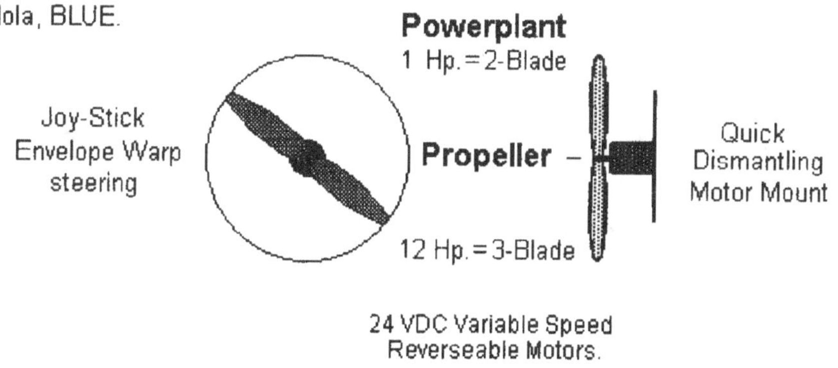

Joy-Stick Envelope Warp steering

Target Speed=15 kts.

Powerplant
1 Hp. = 2-Blade
Propeller
12 Hp. = 3-Blade

Quick Dismantling Motor Mount

24 VDC Variable Speed Reverseable Motors.

Available from:
Bob Sparks
300 E. Franklin St.
Slatington, PA 18080
1800/411-5676

Enclosed Gondola
Cloth enclosure

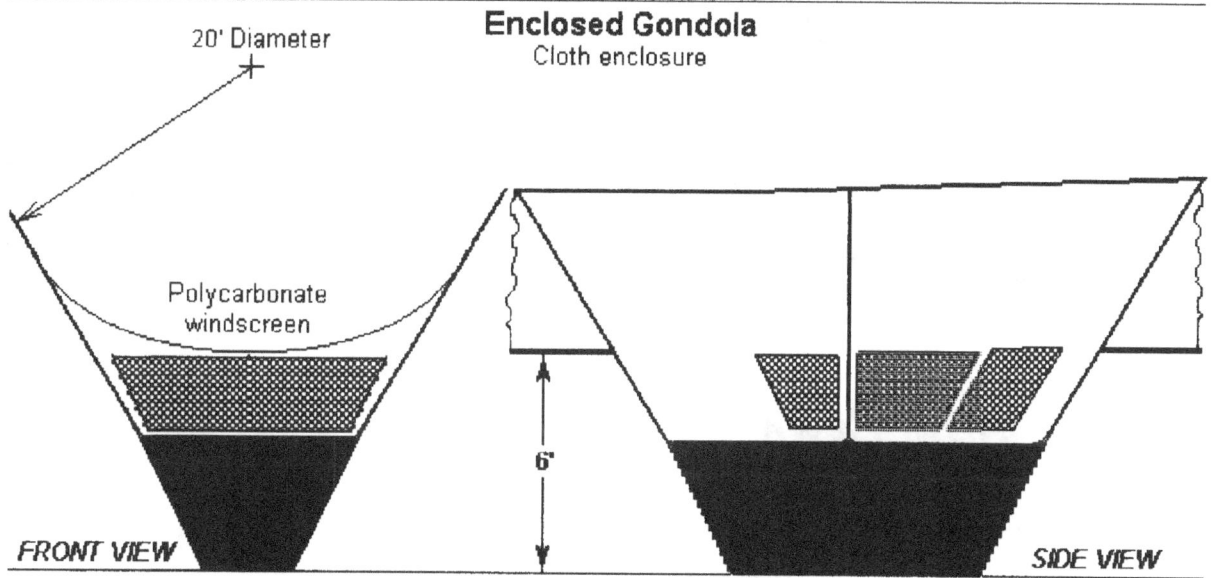

- 20' Diameter
- Polycarbonate windscreen
- 6'

FRONT VIEW *SIDE VIEW*

Building Small Gas Blimps

The TUCKER AIRSHIP TX-1
(14,000 cuft)

1.3 R2
1.0

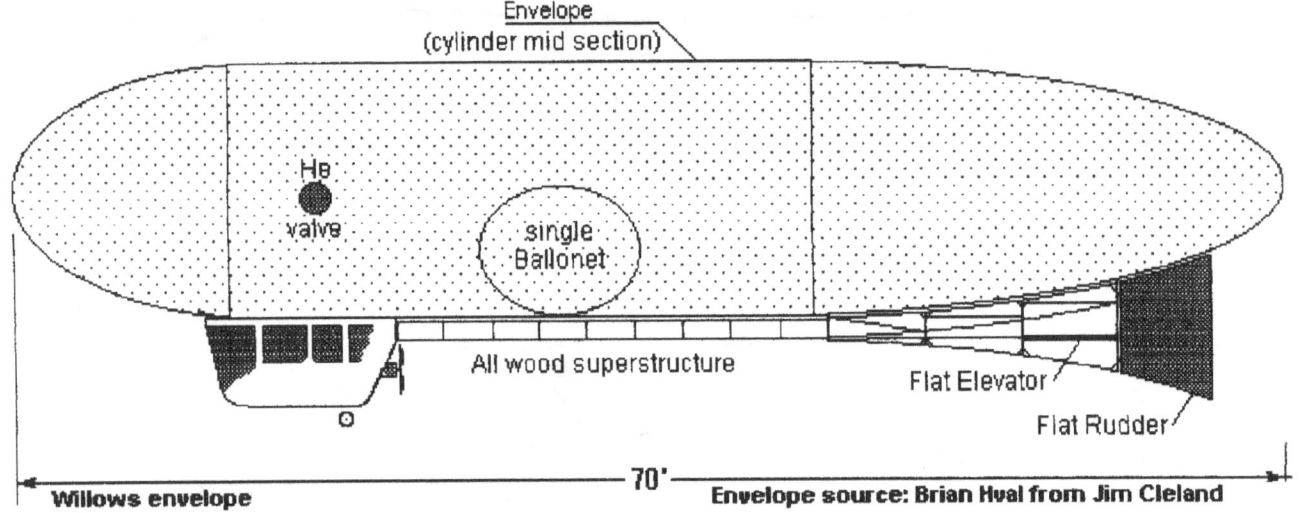

Fabricated by: Curtis E. Tucker Jr.
13218 Lake Street
Los Angeles, CA 90066-2207
310 / 398 - 6907

No production units ever produced
Source: JANES - AIRSHIPS pg.180-1.
AEROSTATION Magazine Vol.4 #1.

Specifications & Details

Length = 65'; Max.Dia.= 20'
Vol.= 14,000 cuft.

Empty Weight: 650 lbs.
Max.Gross Weight: 854 lbs.
Performance target:
12 hrs. flight endurance at 50 mph
to 3000' msl.

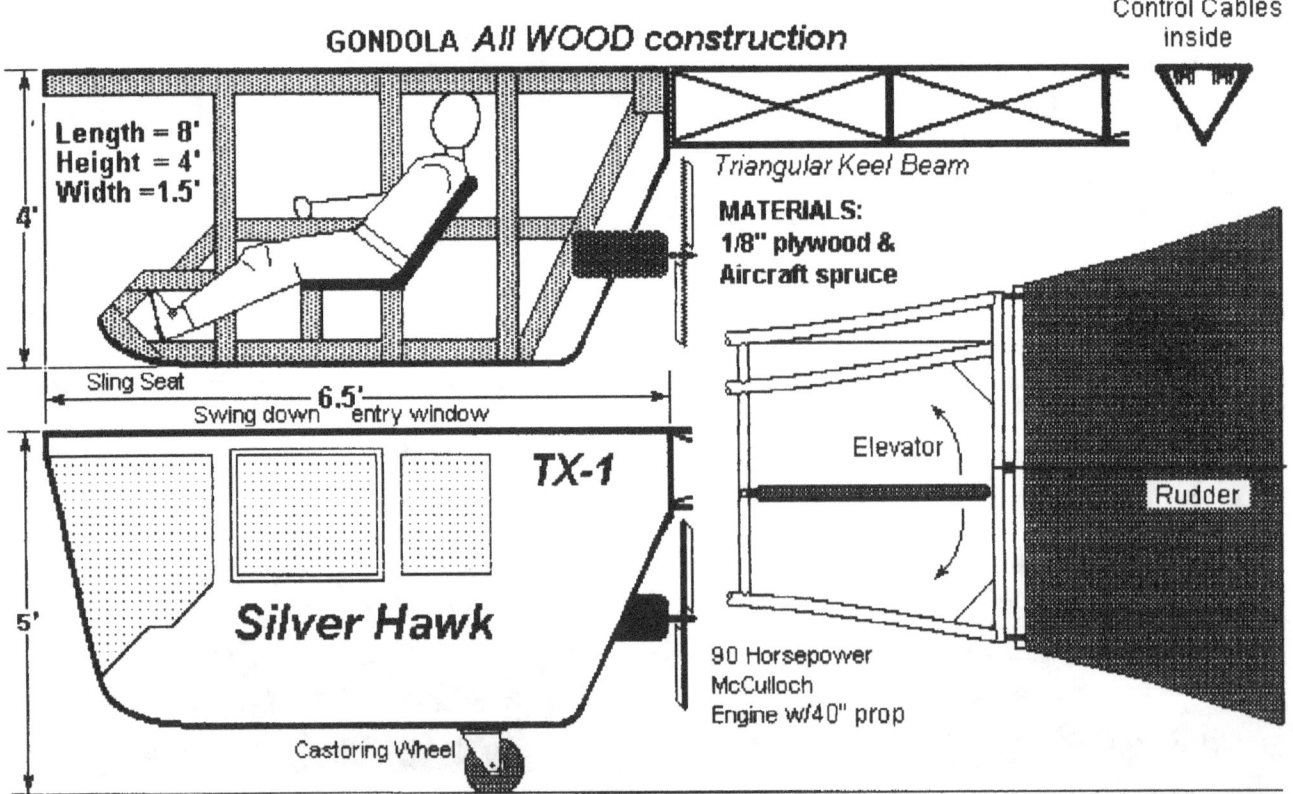

62

Building Small Gas Blimps

U.S. Airships International
(18,000 cu.ft.)

1.3 Meadow
1.0 R6

Ultralight

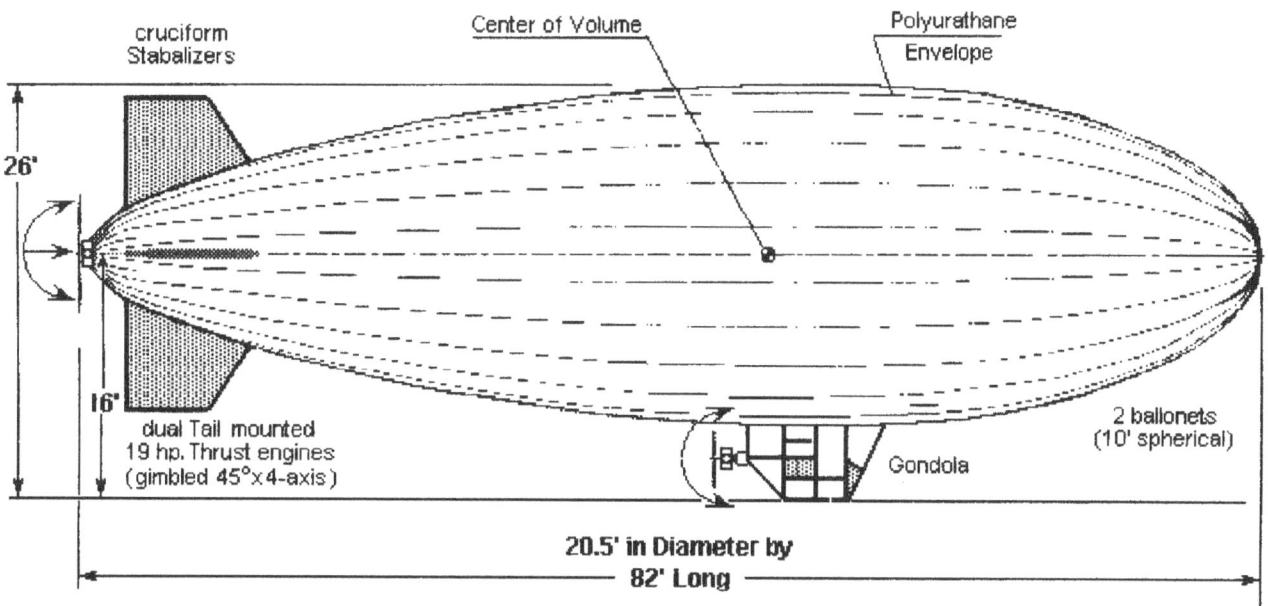

Empty Weight approx. 500 Pounds
.009" matl. fabric has 300% stretch Safety Factor

Available from:
U.S. AIRSHIPS International
153 Little John Road
Statesville, NC 28625
704 / 876-1236

ALL Dimensions approximate;
Final specs. & details
UNKNOWN as of 9/98.
First production units
scheduled for delivery 2/99.

CONTACT
MANUFACTURER
for options & latest prices
— (Ten hours of Flight & Ground training included) —

inside view of frame enclosure

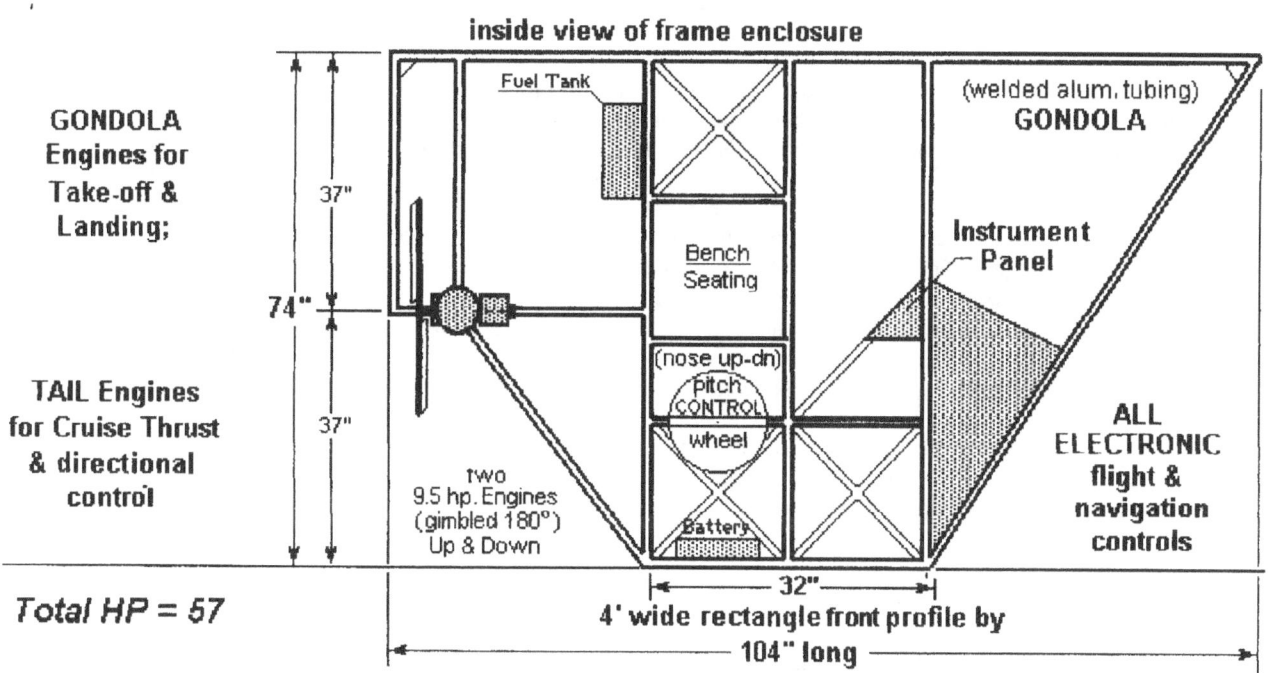

GONDOLA Engines for Take-off & Landing;

TAIL Engines for Cruise Thrust & directional control

Total HP = 57

United States Airships International

153 Little John Rd ~ Statesville NC 28625 ~ Ph/Fax 704-876-1234 ~ E-Mail Airship@Abts.Net

UL-580 Two Man Ultralight Airship

First Flight: March 23, 2001
Location: Ardmore, Oklahoma
Length: 80'
Diameter: 21.17'
Volume: 18,780 cubic feet
Car Engines: Two 10 hp Twin Cyc Engines Electric Start
Tail Engine: One 20 hp Twin Cyc Engine Electric Start with 24v Alternator
Total HP: 40 hp
Electrical System: 24v Aircraft System with Battery, Regulators and DC Converters
Ballonets: One Front and One Rear
Docking System: Nose Docking with Attachable Wheel on bottom of Car
Landing Gear: Twin Floats for Land and Water Operations
Cruise Speed on Tail Engine only: 30 mph
Max. Speed with all 3 Engines at Full Power: 45 mph
Envelope Construction: 12 Gore 9 mil Polyurethane Heat Sealed
Other Envelope Features: Man Entrance Port, 24" Window above Car, Temperature and Pressure Sensors, 78 internal Catenary Lines and Emergency Helium Release
Instruments: **(1)** *Power Control Panel* (Digital), **(2)** *Engine Control Panel* (Start, Run and Choke), **(3)** *Flight Instrument Panel* (Altimeter & Vertical Speed), **(4)** *Envelope Panel* (Temperature and Pressure Meters), **(5)** *Ballonet Control Panel* (24v 3 Stage Blower for Inflating Ballonets), **(6)** *Rear Control Panel* (Mike/Amp/Speaker for Audio from Rear Engine and Rear Fuel Tank Indicators), **(7)** *GPS Panel* (Sectional and Street Color Moving Map), **(8)** *Engine/Flight Control Panel* (Throttles for the 3 Engines) and (Rear Engine Movement Controls and Meters command Actuators to move the Tail Engine L/R U/D for Vector Thrust directional control)
Flight Controls: The Two Car Engines rotate 360 degrees allowing for Vertical Take-Offs and Landings and the ability to hover over a spot even into a headwind. The rotation (in 10 degree steps) also allows for Forward, Reverse, Hold Down (110 lbs) as well as other combinations of control. The Two Car Engines are normally used only for take-off and landing or during tight low speed maneuvers. The Tail Engine runs all the time since it has the alternator and provides Vector Thrust directional control of the airship even with no forward speed.
Other Controls and Features: The ballonets mainly are used to control the envelope pressure over a wide range of ambient temperatures, but can also be used to balance the center of lift if necessary. The Tail Engine Fuel Tank is one gallon and received fuel from a 10 gallon Car Fuel Tank thru a automatic 24v fuel pump in the car, which maintains the same fuel level (and weight) in the Tail Engine Fuel Tank regardless of the fuel usage. An Emergency Battery keep the airship flying with a complete electrical failure. Each engines can be removed in less than <u>one minute</u> and all components of the airship (car, engines fins, etc.) can be removed from the envelope in less than one hour when deflation is desired.
Flight Performance: The airship has a flight duration of 18 hours (in low power cruise on the Tail Engine only) and range of over 500 miles. The 6-1/2 foot tall car allows for the pilot to stand and fly the airship. There is only one flight control in the floor, which is the break release to lock the Car Engines in the desired position. All other flight controls are on the panels in easy reach by the pilot or co-pilot while sitting or standing. All flight and throttle controls lock in the last selected position leaving the pilot free to move around during long duration flights.
Other Models: USAI offers 11 Models of Airships ranging from a basic 38' One Man Airship with 1 Engine up to our deluxe 80' Two Man Airship with 4 Engines.

Building Small Gas Blimps

GONDOLAS

1.3 GON_2
0.7 R

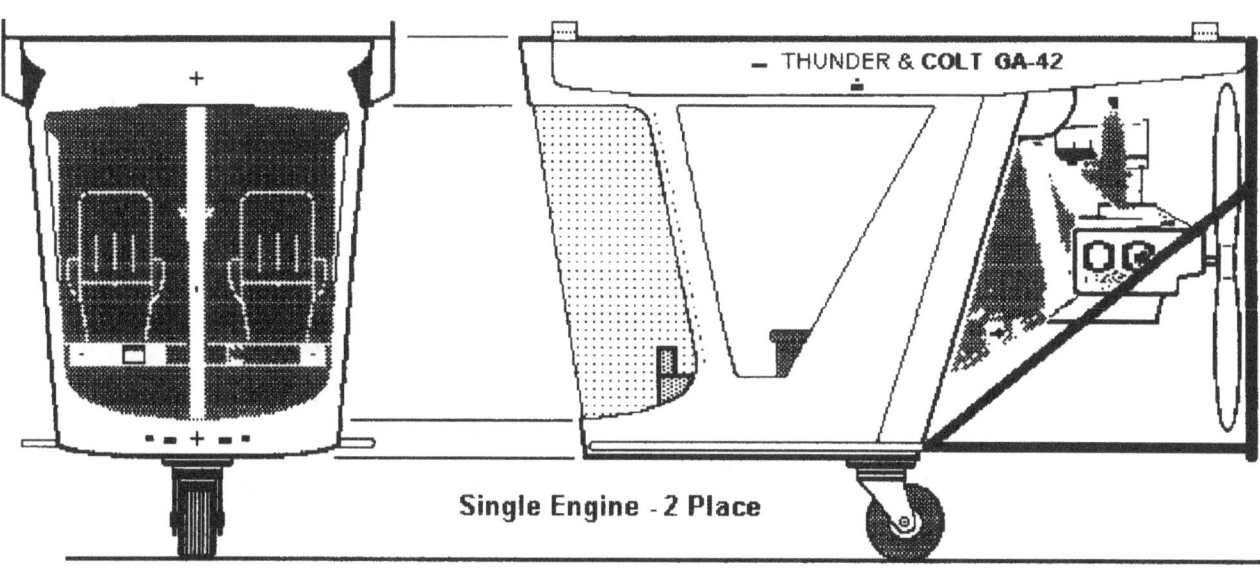

Single Engine - 2 Place

FAA Standard Type Certified
MEDIUM
Size Gondolas

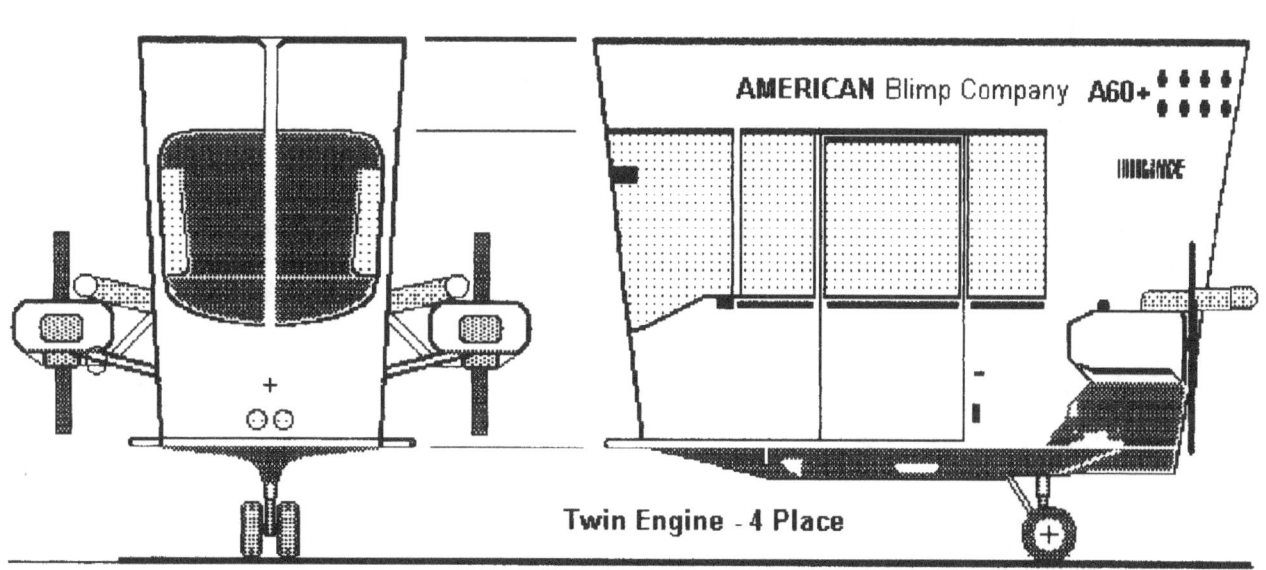

Twin Engine - 4 Place

Building Small Gas Blimps

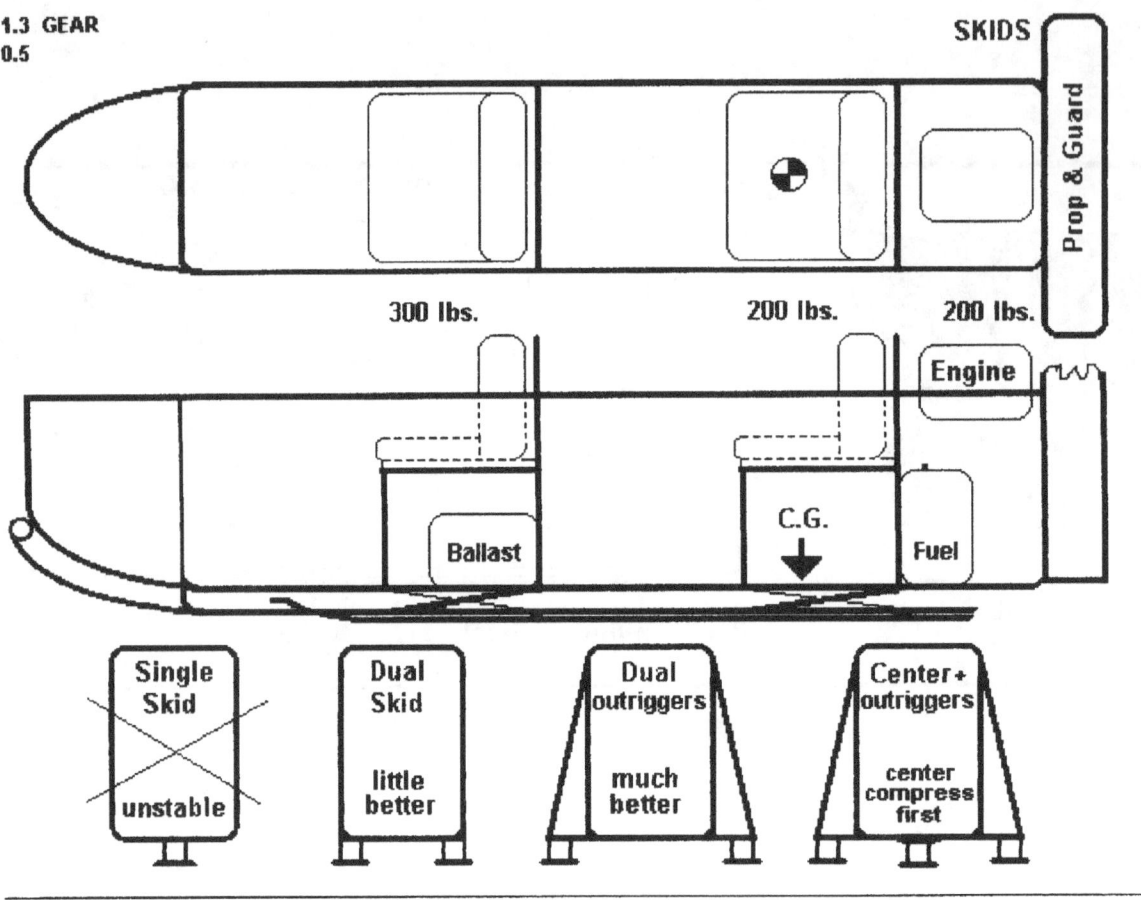

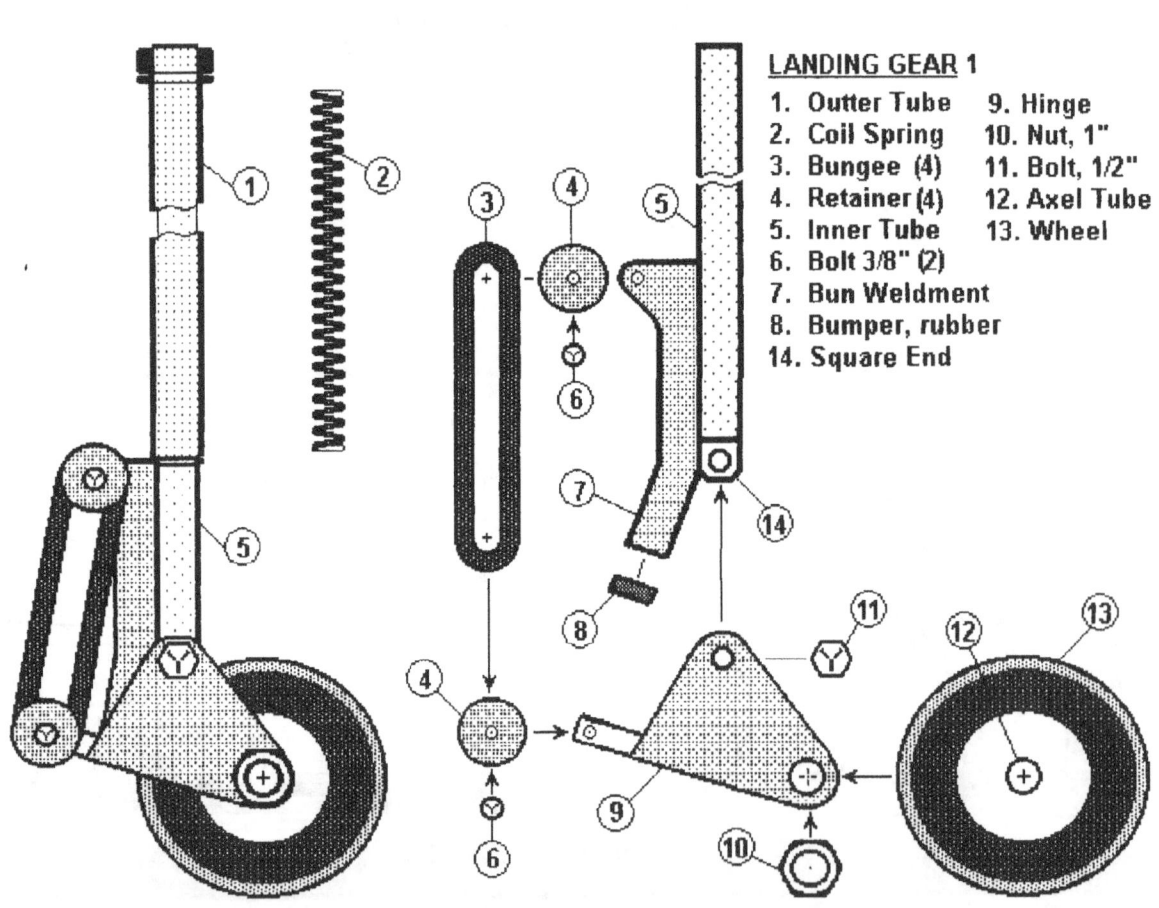

LANDING GEAR 1
1. Outter Tube
2. Coil Spring
3. Bungee (4)
4. Retainer (4)
5. Inner Tube
6. Bolt 3/8" (2)
7. Bun Weldment
8. Bumper, rubber
9. Hinge
10. Nut, 1"
11. Bolt, 1/2"
12. Axel Tube
13. Wheel
14. Square End

GONDOLA cont.

Landing Gear
There are several ways to successfuly approach this. All with great trade-offs, of course. SKIDS offer a very light weight solution and should be seriously considered for blimps of less than 10,000 cuft. Some of the considerations are:
- PRO   Easily designed into the gondola structure;
  - Parts are cheap, and easily found;
  - Easily replaced as they wear out;
  - Very light.
- CON: Requires good landing technique;
  - Limited pilot protection on landings exceeding 1G;
  - Ship will be hard to correct when "off-wind" on the ground;
  - Gondola will tend to tip when heavy on the mast;
  - Ship should be kept secure in a hangar when not in use.

SINGLE WHEEL (full castoring) with it being just aft of the gondola C.G., is the most common. It has been proven the best solution for blimps in the history of airships. Some considerations are:
- PRO: Cross wind landings are easier to manage;
  - Easier to stay "on-wind" on the ground;
  - Castoring wheel allows easy ship handling;
  - 90 degree Wheel-Lock reduces scuffing wear on the tire.
- CON: Gondola must be supported in a customized dolly when off the envelope.

MULTIPLE WHEELS have been used successfully, but the trade-offs are vastly different between "small" and large (usually have prop reversing) blimps. To stay on focas, only small be considered.
- PRO: Easy to move the gondola around the shop or hangar;
- CON: 3-wheels are very unstable on uneven pavement,
  - with, or without the envelope attached;
  - Requires good landing and handling technique;
  - Ship will be hard to correct when "off-wind" on the ground.

SUMMARY: Basically, the best safety considerations save you money in the long haul. By far, the best approach is the single castoring wheel. But for a hangared ultra light, go for a skid, and keep it in a hangar.

DESIGN GOALS: A few features in my "wish book" come to mind. In order of importance, consider:
1. Beefy is better. 2 & 3G landings are not uncommon;
2. Ditches and rabbit burrows tend to rip the gear off;
3. Be liberal with mulitiple shock absorbing features;
4. Wheel must be lockable cross axis while on the mast;
5. Make it easy to fix (esp.a tire) in the field.

Building Small Gas Blimps

INSTRUMENTS

Altimeter

Airspeed
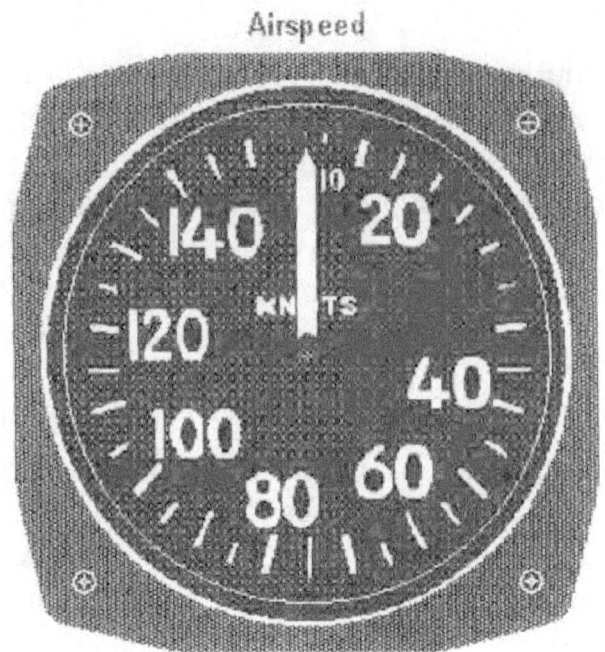

INSTRUMENTS

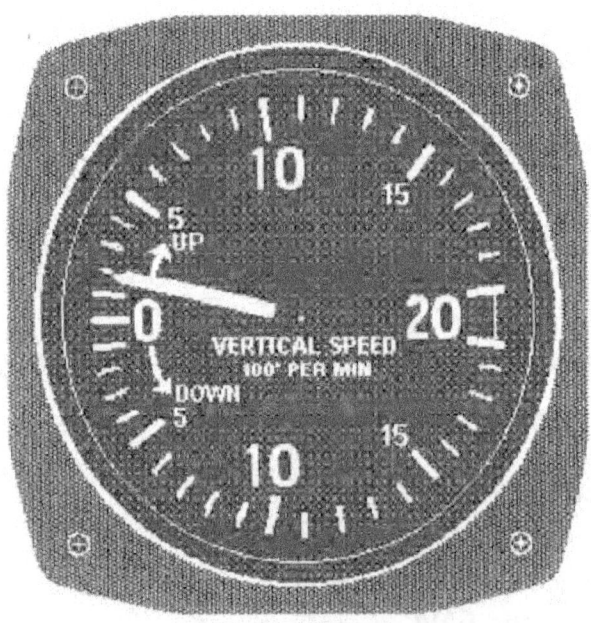
Rate of Climb

Pressure

Building Small Gas Blimps

INSTRUMENTS
R: 12/05/97

NEED; What do you really need? From the FAA's point of view (FAR 91.205) for Visual Flight Rules (VFR) in daytime you must have:

1. Airspeed Indicator
3. Magnetic Direction Indicator
2. Altimeter (0-120 Kts.)
4. Tachometer (ea. engine)
5. Oil Pressure (ea. engine)
6. Temperature (ea. engine)
7. Fuel gauge (ea. tank)
8. Manifold Pressure (for altitude sensitive engines)
9. Landing Gear position (if retractable)

Because an airship is a unique vehicle, you are also going to need:

10. Helium Pressure gauge
11. Ballonet pressure gauge
12. Superheat (temperature inside, vs. outside)
13. Clock
14. Rate of climb
15. Volt / Load meter

WANT; How much can you afford? You can easily put $20,000 worth of instruments in a $5000 gondola. But don't loose sight of your goal to get it off the ground. Besides, people just love midnight souvenirs.

TIPS; You can pay any price for the same instrument, buy from a reputable source. A "new" instrument may give you peace of mind, but most any of them can also be purchased "used", "reconditioned", "surplus", or "junk". Do not think that you are going to get away with just "dummy faces". Your life depends on knowing the truth.

The old standard mechanical faced instruments are cheaper and readily available. If you want the latest electronic instruments, be prepared to double or triple your investment. They do look nice, and make your panel very compact. The BAD point is that when you are flying around in the sunlight, you can't read them in any reflective glare.

If you are just going to put-put around your farm in your "ultralight" airship, you may want to consider some of the "strap-on" gauges available from the manufacturers in ultralight publications.

OTHER; I would like to emphasize the need for communications, mainly with your ground crew, but also airports and other aircraft. For this, you need a reliable transceiver installed on board*, and at least a "hand-held" for your ground Crew Chief.

Since FAR 91.205 also covers required "equipment", you at least owe it to yourself to give serious consideration for:

1. Position & Landing lights
2. Anti-collision strobe
3. Seat belts / shoulder restraints
4. Gyro direction indicator
5. Each electrical device on a separate Circuit Breaker.
6. Emergency equipment.

* An "Aircraft Station" license is required by the Federal Communications Commission for a nominal fee ($10?), but fee can be waived for "Public Safety" reporting. FCC applications are available by mail or from local FCC offices in metropolitan areas.

Building Small Gas Blimps

1.8 HeliPress
1.3

HELIUM PRESSURE
Cockpit Gages

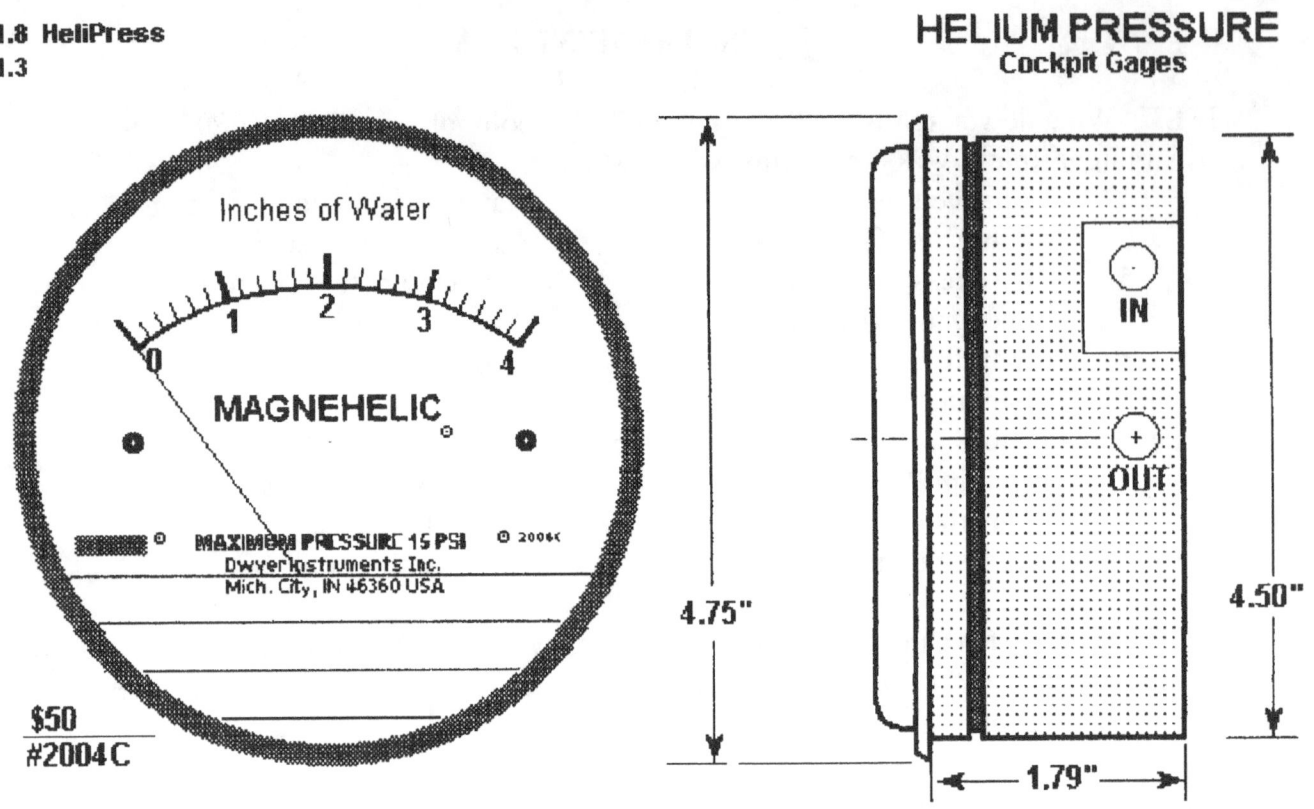

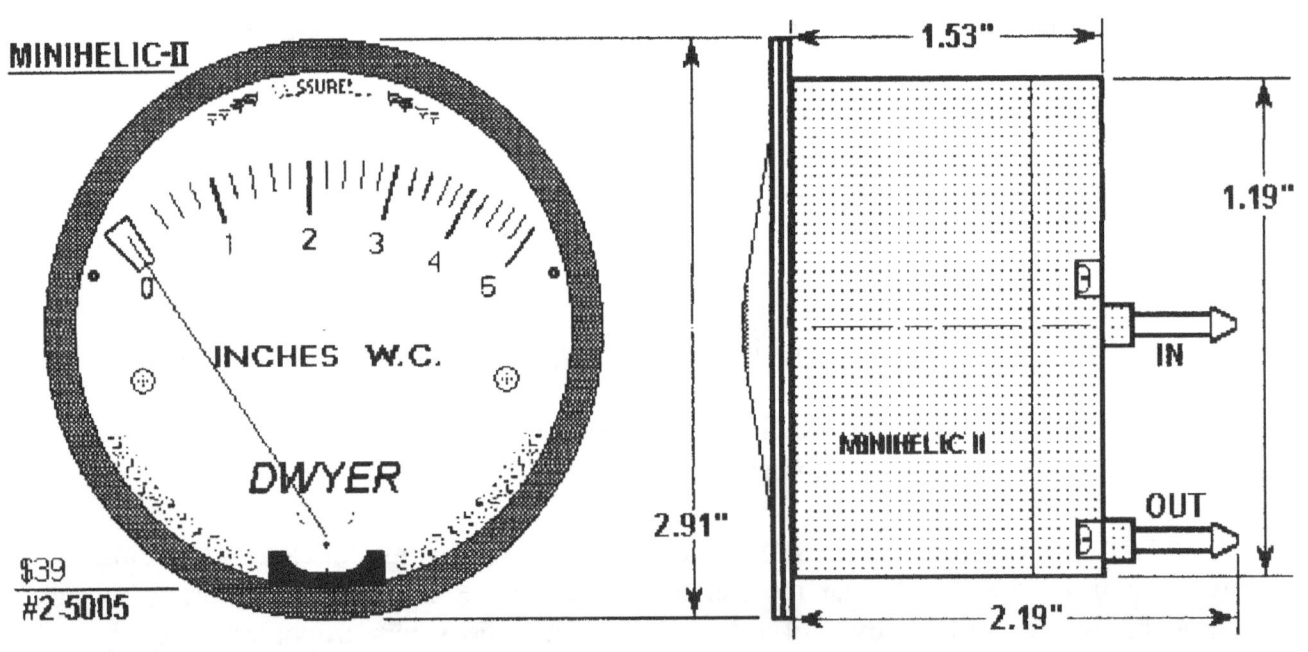

Building Small Gas Blimps

INSTRUMENTS

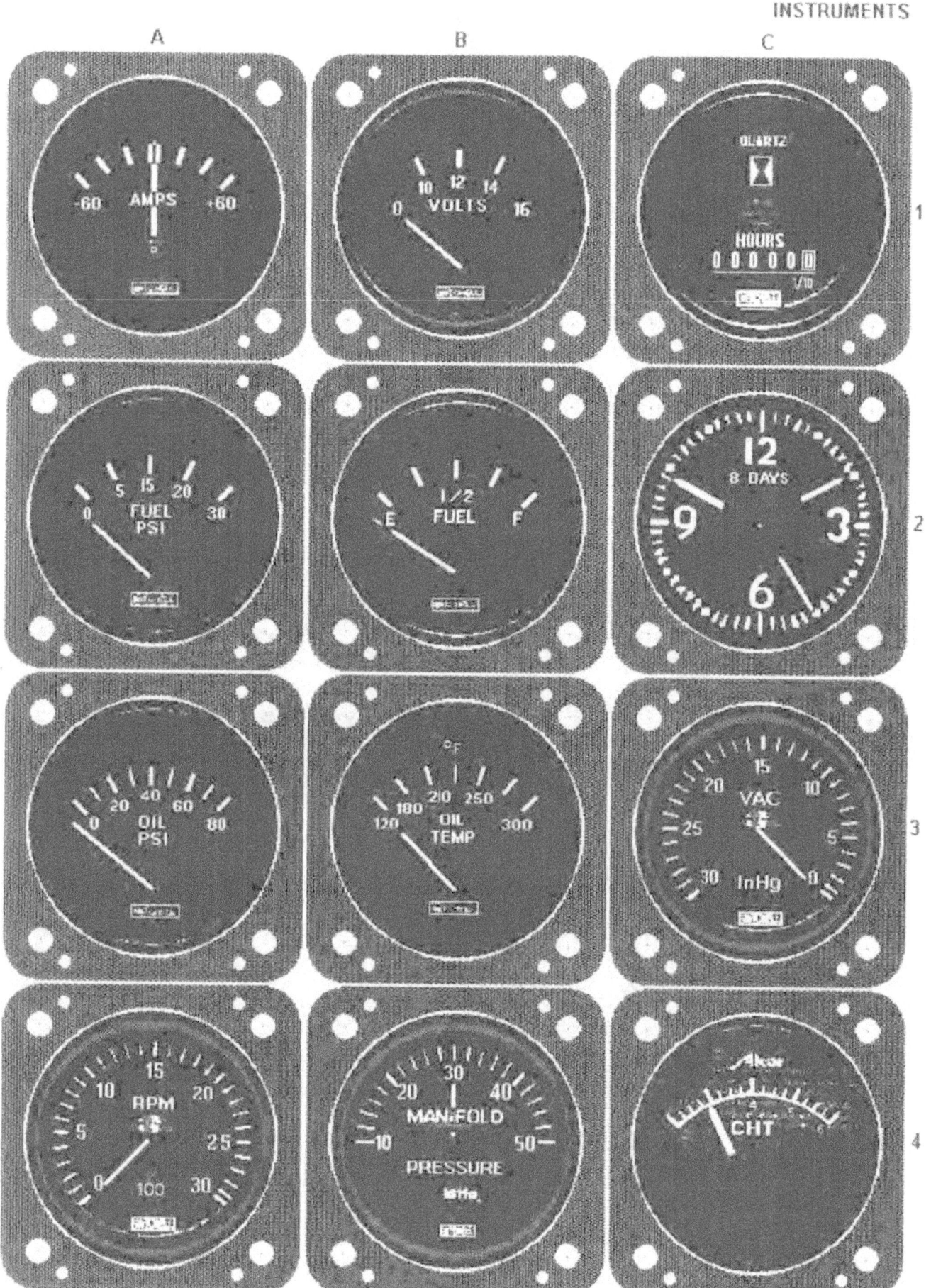

Building Small Gas Blimps

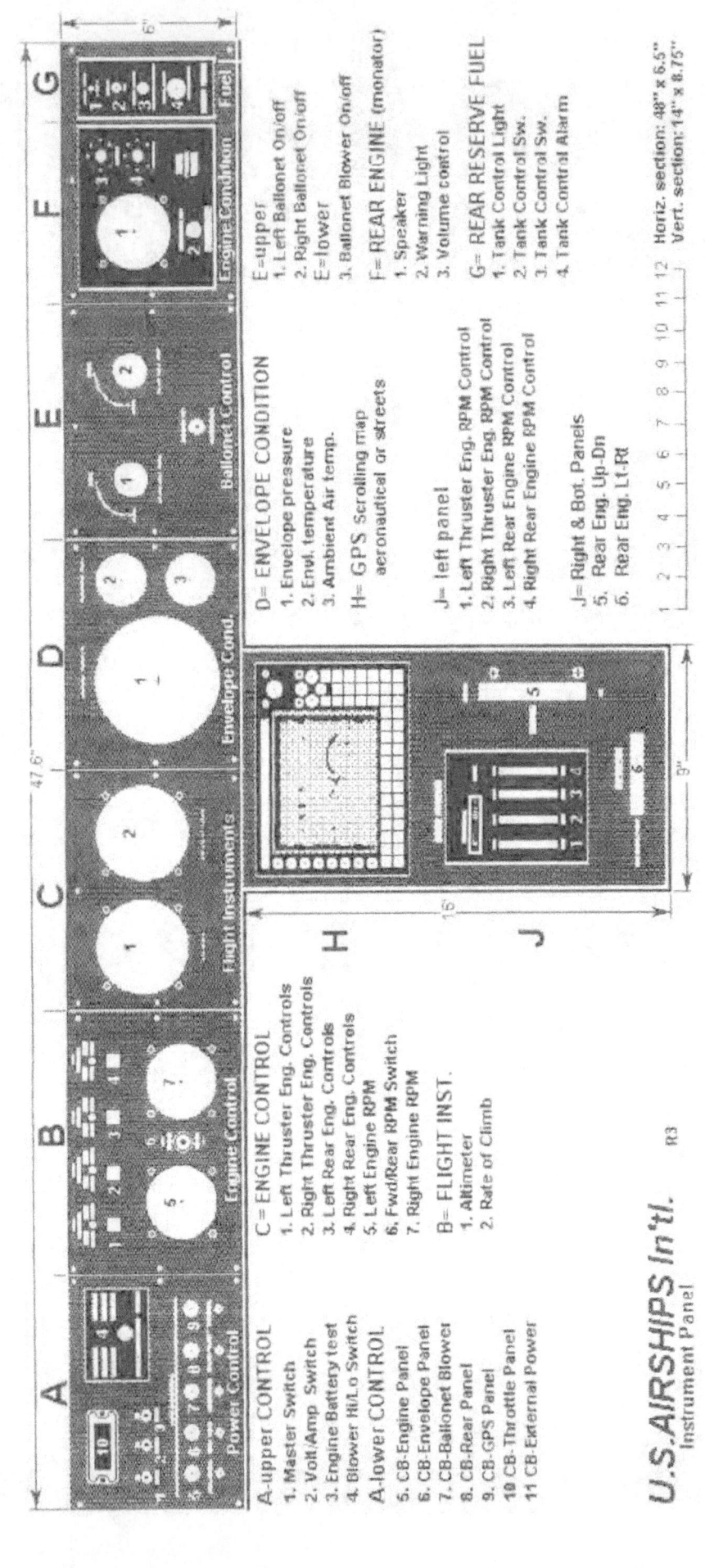

A - upper CONTROL
1. Master Switch
2. Volt/Amp Switch
3. Engine Battery test
4. Blower Hi.Lo Switch

A - lower CONTROL
5. CB - Engine Panel
6. CB - Envelope Panel
7. CB - Ballonet Blower
8. CB - Rear Panel
9. CB - GPS Panel
10. CB - Throttle Panel
11. CB - External Power

C = ENGINE CONTROL
1. Left Thruster Eng. Controls
2. Right Thruster Eng. Controls
3. Left Rear Eng. Controls
4. Right Rear Eng. Controls
5. Left Engine RPM
6. Fwd/Rear RPM Switch
7. Right Engine RPM

B = FLIGHT INST.
1. Altimeter
2. Rate of Climb

D = ENVELOPE CONDITION
1. Envelope pressure
2. Envl. temperature
3. Ambient Air temp.

H = GPS Scrolling map aeronautical or streets

E = upper
1. Left Ballonet On/off
2. Right Ballonet On/off

E = lower
3. Ballonet Blower On/off

F = REAR ENGINE (monator)
1. Speaker
2. Warning Light
3. Volume control

G = REAR RESERVE FUEL
1. Tank Control Light
2. Tank Control Sw.
3. Tank Control Sw.
4. Tank Control Alarm

J = left panel
1. Left Thruster Eng. RPM Control
2. Right Thruster Eng. RPM Control
3. Left Rear Engine RPM Control
4. Right Rear Engine RPM Control

J = Right & Bot. Panels
5. Rear Eng. Up-Dn
6. Rear Eng. Lt-Rt

Horiz. section: 48" x 6.5"
Vert. section: 14" x 8.75"

U.S. AIRSHIPS Int'l. R3
Instrument Panel

Building Small Gas Blimps

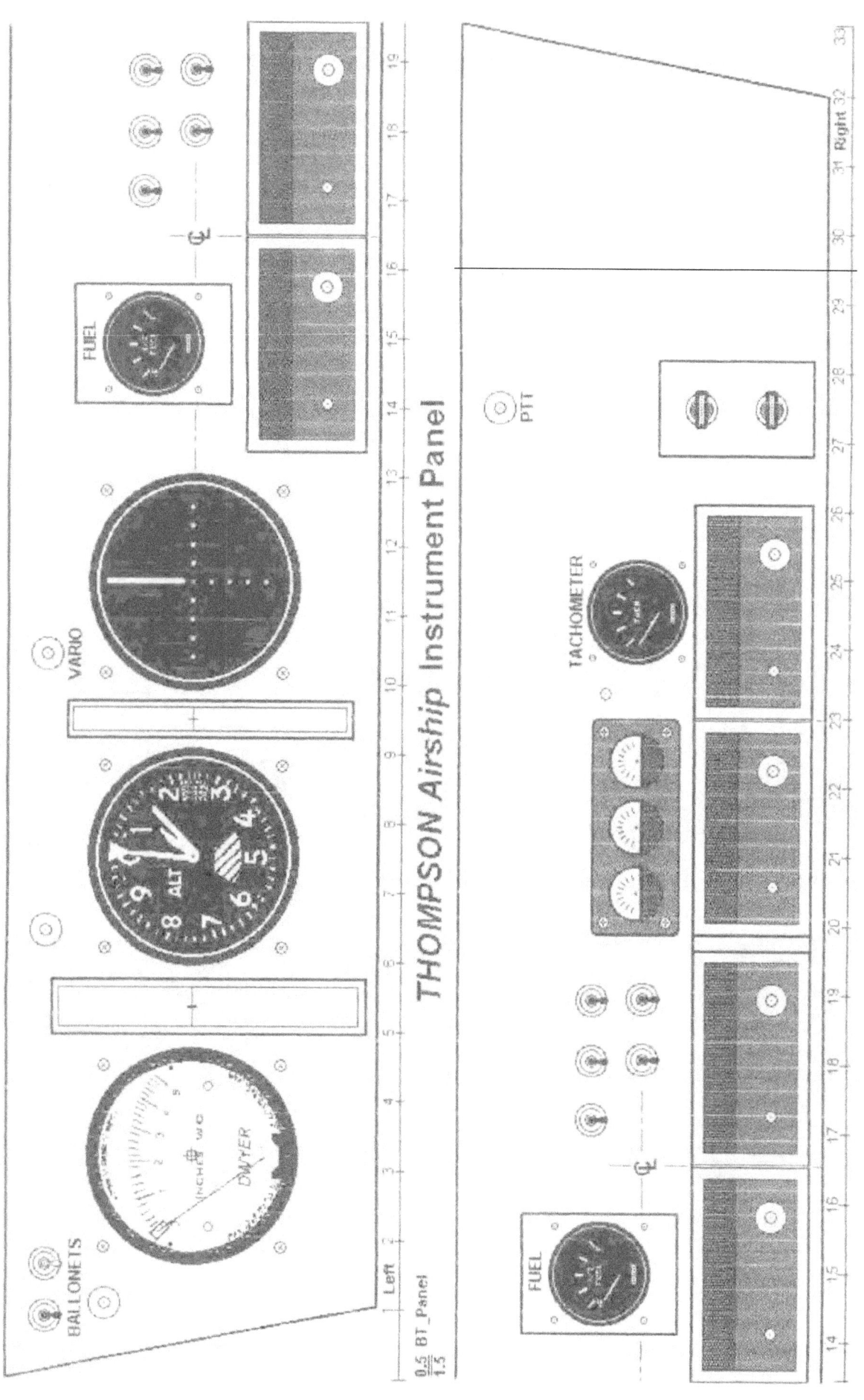

73

Building Small Gas Blimps

Sample INSTRUMENT Panels

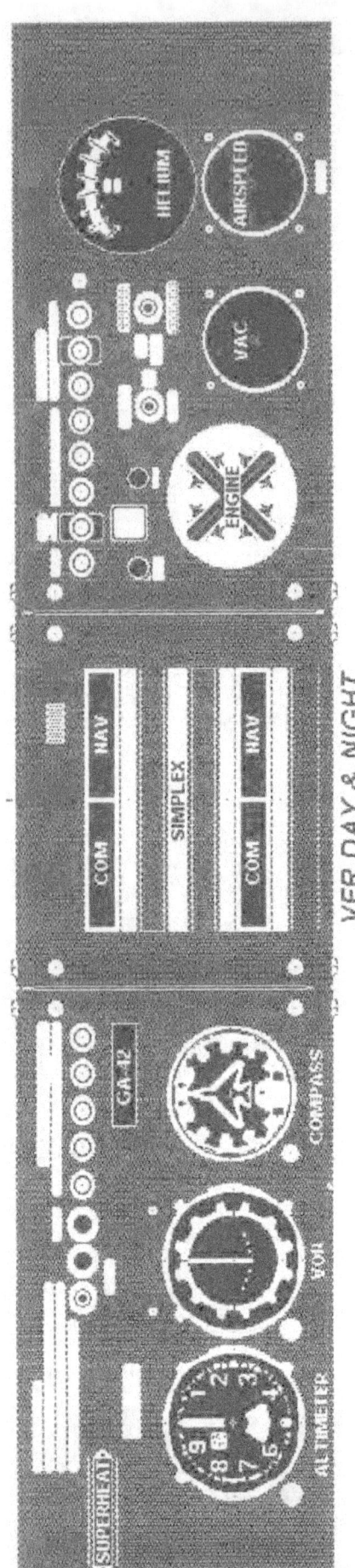

VFR DAY & NIGHT

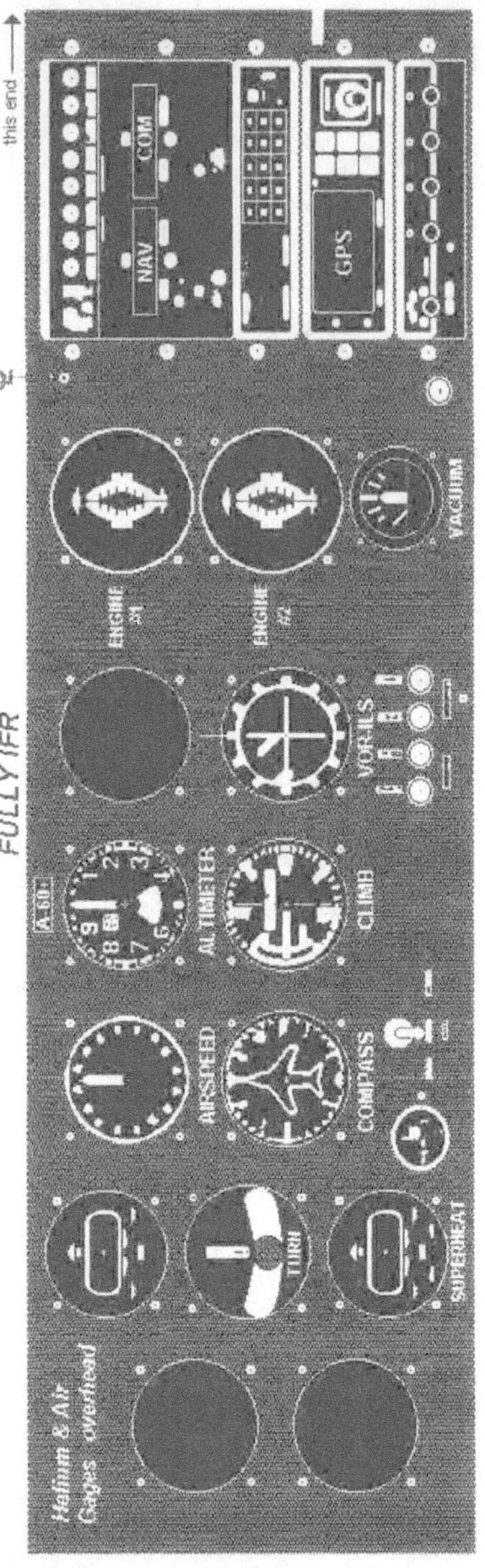

FULLY IFR

74

Building Small Gas Blimps

PITOT - STATIC System

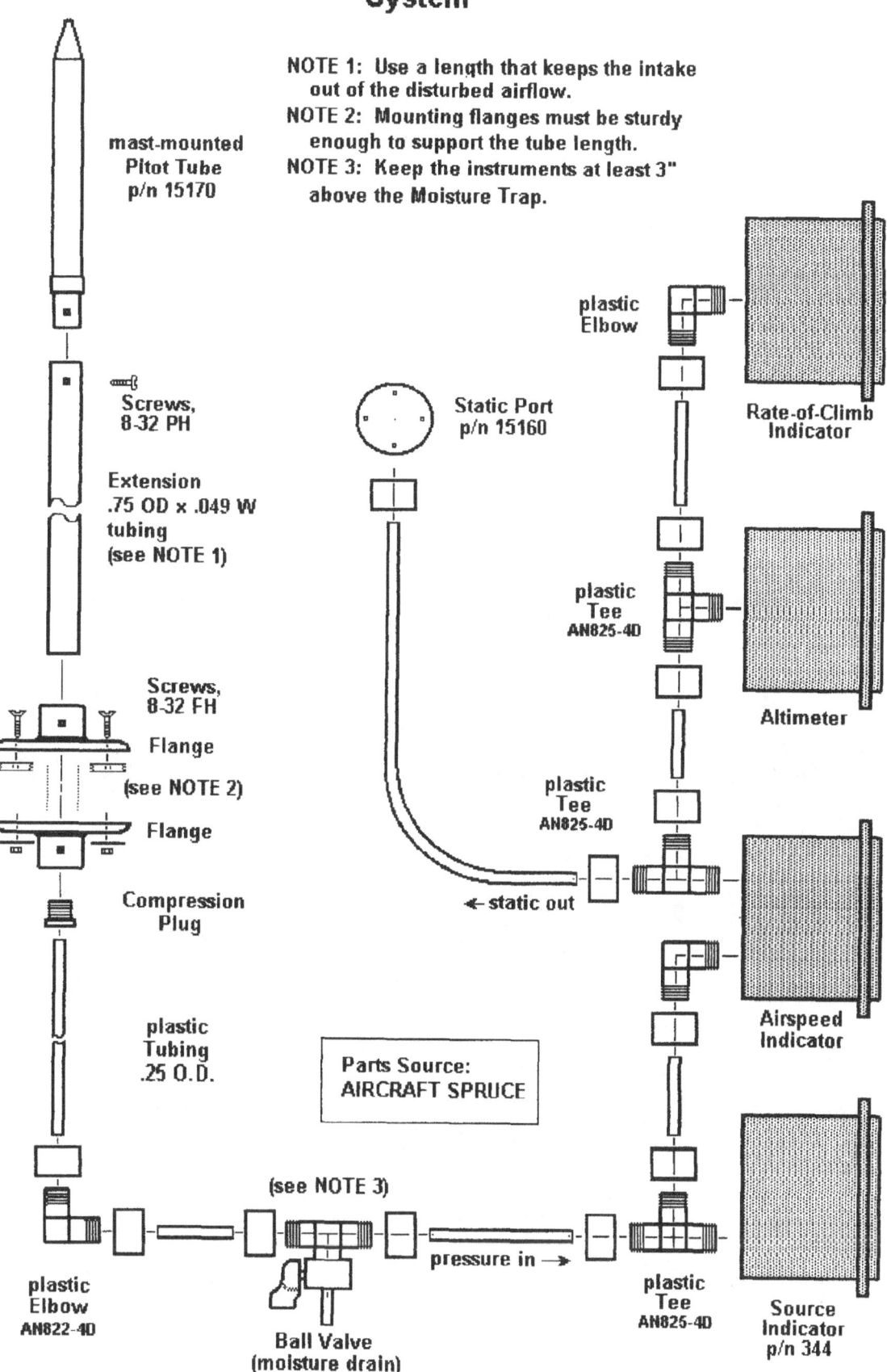

Building Small Gas Blimps

INCLINOMETER Single Axis, Double Ball (4.2 x 6.2")

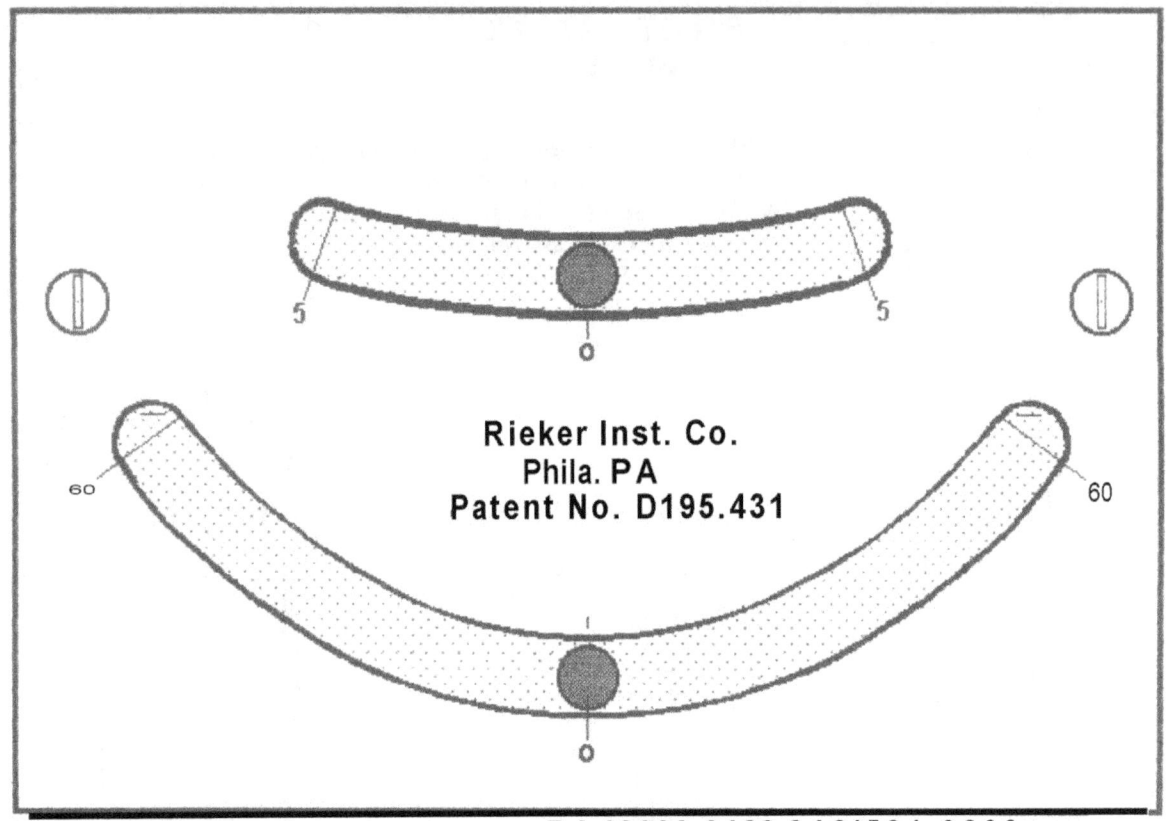

Rieker Inst. Co.
Phila. PA
Patent No. D195.431

Rieker Inst.Co.: **POB 128: Folcroft.** PA 19032-0128:610/534-9000

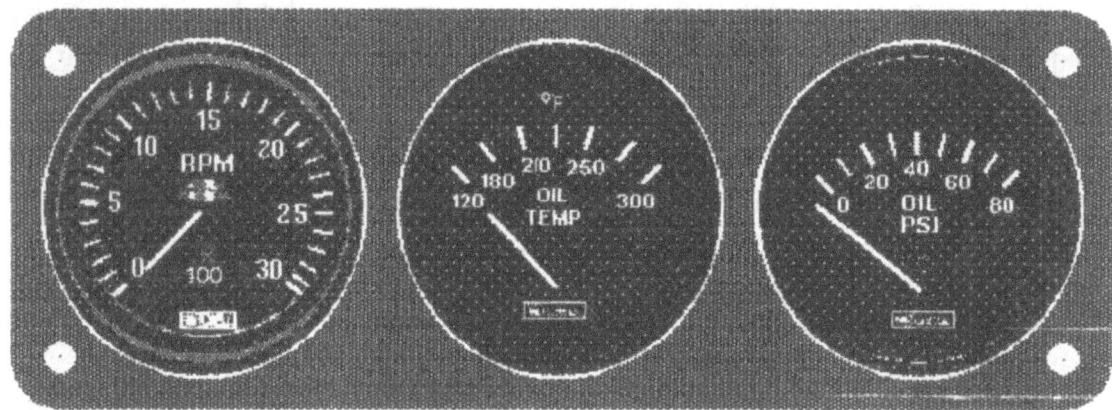

Available in custom combinations of 3-gauge units

INSTRUMENTS
R: 11/15/97
LIGHTS

Well, you don't really need any if you are not going to fly at night. By definition, however, that means from one hour before official sunset (for your Latitude) until one hour after official sunrise. Even an emergency is no excuse for not having them. So plan ahead. I'll list them in order of importance:

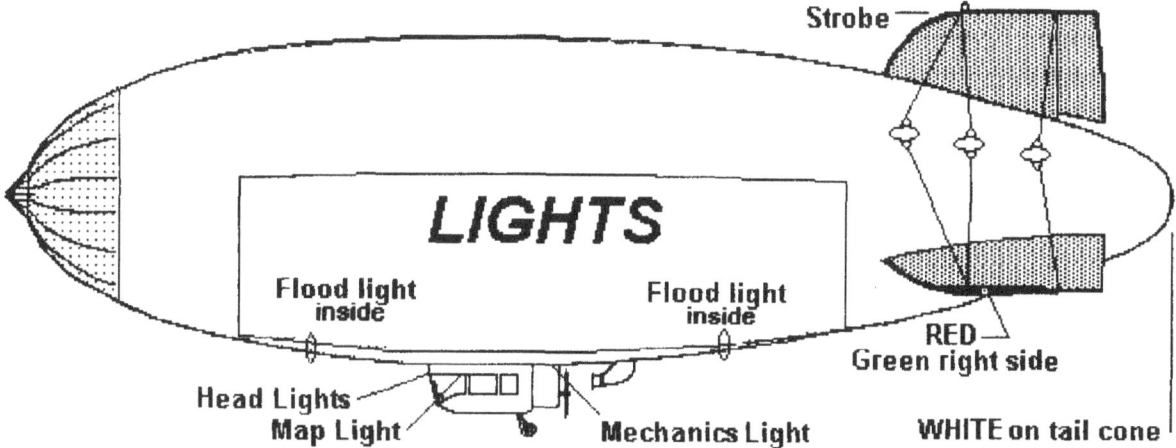

Strobes: allow you to be seen by others, and hopefully keep the other guy from running into you. I only vaguely remember a blimp being hit by another aircraft, and that was another blimp in WW-2. The location of a strobe requires that it be seen from 360 degrees. The bottom of the gondola would be the simplest place to wire it. However, the flash will drive you "buggie", and usually interferes with your radios. The best place is to get it up on the top fin (yea, it'll will take a cherry-picker to replace the lamp). The important thing to remember at this stage is that they come with a bulky power supply (booster) that is **normally designed into the base of the upper vertical stabilizer**. It is that power supply that causes the radio interference.

Position Lights; Red, White, & Green. Allow the other guy to ascertain your direction of flight, and your potential threat to him. Again, they have to be visible from 360 degrees. They are straight forward installations and do not cause any electrical interference. Remember, **RED** is **PORT** (wine) **LEFT**.

Map Light(s) are needed to read your chart for that important radio frequency. Without it, all you can do is call MAY-DAY, if you can read the 121.5 dial. Lights shouldn't be so bright that you loose your night vision, but then it isn't a time to be navigating with a street map either.

Head Lights, well ok; at least you can see that barbed wire fence coming up. But don't expect to see that rabbit hole that will tear your gear off. Remember also that they get hot-hot. Don't put it/them next to anything sensitive, like fabric. Every designer wants to put then down on the chin of the gondola. And they become the first structural failure of the aircraft. EVERY blimp pilot makes a nose heavy landing sooner or later (usually sooner, in his first ten hours). And of course, it just tears them right off.

Mechanics Light over the engine helps greatly with fueling and the preflight inspection. But try finding a loose distributor coil at night.

Flood Lights, forget it. Your generating capacity won't handle the power you need.

Building Small Gas Blimps

R: 10/21/97

INSTRUMENT & CONTROLS
SUPPLIERS

AIRSTAR Discount 2251 Shady Willow Lane Brentwood, CA 94513	800/AIR-STAR f)510/516-1188	INSTRUMENTS, ultralight Basic Engine & flight Also Props & Wheels
ALL AIRCRAFT Parts 16673 Roscoe Blvd. Van Nuys, CA 91406	818/894-9115 retail sales good walk-in stock	INSTRUMENTS, aircraft Hardware, bolts, fittings pilot supplies
BALL Instruments 6595-C Odel Place Boulder, CO 80301	303/530-4940 quality products	INSTRUMENTS, aircraft variometers barographs
CALIFORNIA Systems 790 139th Street San Leandro, CA 94578	800/247-9653 f/510/357-4429 1 stop shop	INSTRUMENTS, ultralight All flight & Engine
CONSOLIDATED A/C Supply 55 Raynor Ave. Ronkonkoma, NY 11779	516/981-7700 Fax:981-7706 major supplier	INSTRUMENTS, aircraft gages, all types electrical
DWYER Instruments P.O. Box 373 Michigan City, IN 46361	219/879-800 quality products ONLY supplier	INSTRUMENTS.helium pressure gages controls
LEADING EDGE 8242 Cessna Drive Peyton, CO 80831	8--/LEAF-INC	INSTRUMENTS, ultralight Basic flight & Engine Radios
LOCKWOOD Supply 280 Hendrics Way Sebring, FL 33879	800/527-6829 f)941/655-6225	INSTRUMENTS, ultralight also Engines & props small strobes
SKYSPORTS Linden Airport, Hangar 1 Linden, MI 48451	800/AIR-STUF f)810/735-1078 great catalog	INSTRUMENTS, ultralight Instrum. specialist
WHELEN Engineering Rt. 145, Winthrop Rd. Chester, CT 06412	800/231-9536 f)860/526-4078 quality supplier	LIGHTING, aircraft Position lights Strobes

Building Small Gas Blimps

1.3 CONTROLS
COCKPIT CONTROLS
GA-42

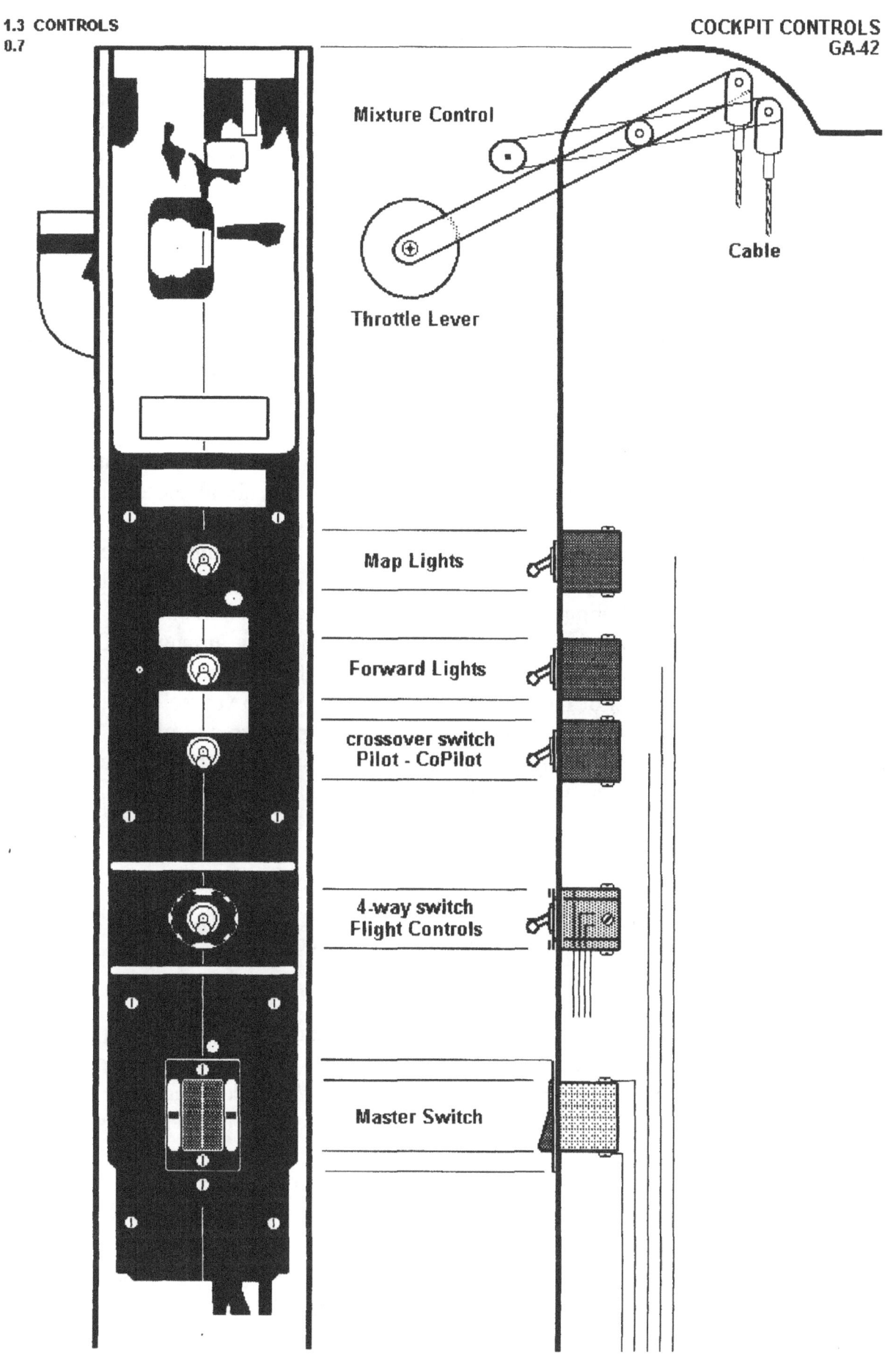

79

ENGINES
R: 12/04/97

I am probably wasting your time talking about this, as by now, you have a used one in your garage that someone gave you. I hope you understand that your life may depend on more than the initial cost. If it quits, you had better have your "free ballooning" technique down pat. However, for the sake of analysis, lets call it a review. So lets consider these in sequence:

Horsepower by definition is the relationship of *torque* to *RPM*. But we can consider it further as the critical part of the speed you expect from your airship. Remember the old rule: Anything can be made to fly with enough thrust and enough money. Now apply the maxim that: To double the speed, you have to quadruple the horsepower (and money). Most (present) large airships cruise at 55-59, medium 45-49, small 20-30. Without knowing what your speed dreams are, or the propeller you are going to match with the engine, lets continue.

Efficiency by definition is the ratio of input to output. But we can consider it further as the ratio of *Weight* to *Horsepower*. Santos-Dumont flew his blimp successfully (1901) with a 47 pound, 8 horsepower engine (5.9:1). Modern "certificated" engines are about 1-pound per horsepower. No doubt your "small" blimp is not going to be restricted by certification and will use a highly touted "ultralight" engine. I suppose that if I had to put one on my aircraft, a small blimp is probably the safest aircraft to put one on.

Reliability is undoubtedly the most important item in choosing an engine. Certificated engines have a substantiated, proven history. Ultralight engines only claims. But engines, like life, are a cast of the dice. You at least, owe it to yourself to talk to other users before putting your money on the table. A subscription to any of the "ultralight" publications listed in the bibliography is the first place to start. The big "bug-a-boo" is in cooling, an additional problem to the slow speed of blimps. I might add the availability of spare parts also.

Types of engines are an easy one to decide. Forget turbines, they suck fuel faster than the ballast can compensate. *Please*, stay away from 2-cycle engines (sorry, I'm partial). Stick to a well known 4-cycle. Most are Air Cooled, but some of the newer ones are now water cooled. At slow speeds it is an important factor to consider.

PROPELLERS

Engines are only half the story, propellers the other half. Unless they are matched correctly, you are just turning a crank. They are available in all sizes and shapes (also in metal, wood, or composites, and in any chord or pitch). *Please* don't go out and just buy a "pretty" one. The prop MUST be matched to the engine. Anyone who says otherwise is dumber than I am. Prop matching (RPM to load and Tip speed) isn't that hard, but loosing a blade is life threatening (3:1 bet it will go through the envelope).

VECTORED THRUST

Yes, vectored thrust is a good idea. It doesn't require much change in flow direction to get noticeable effects. But here is something to think about. It can be costly to find a good actuator with enough controllable force, but it reduces the number of required ground crewmen. It does require some exceptional engineering to eliminate engine vibration feedback.

MOUNTING
Engines & Accessories

<u>Push or Pull</u> (tractor) means little to efficiency, but a lot to cooling. Pusher engine propellers produce vastly different airflow cooling pattern, that can cause great problems at low airspeeds. The ROTAX has particular problems that are not easy to baffle effectively.

<u>Noise</u> is the biggest complaint. It can make you give up flying, or at least make you deaf. OK, so we on the ground don't give a damn about you, but don't fly over when I'm 2 over par and ready to 6-iron out of the sand. One complaint to the FAA and you will be "Operation Limited" to the cow pastures. It is you with your back next to the engine; design with the best muffler system you can buy, and baffle the area around your head. A helmet helps, especially with a tight fitting noise canceling headset.

<u>Vibration</u> is not a big problem in the air, as the envelope absorbs all or most of it. On the ground is another matter. Be liberal with the inclusion of proper load mounts on the attachments. Listen to your engine supplier, but get expert help if in doubt.

<u>Servicing:</u> Don't box yourself in. You at least need to be able to change spark plugs and ignition while it bounces around on the mast.

<u>Fueling</u>; Adding gas is a bothersome chore anytime. Not only bothersome, but downright dangerous on an airship. Don't add to it by making it hard to reach / fill the gas tank Or put the filler mouth so close to a hot engine, that a fire is inevitable. Remember, even on a mast, the ship is always moving, and filling from 2-gallon containers *is* normal. The fueler will always be at risk. Plan ahead. But no matter what, DO NOT **ever** attempt to "hot refuel" (with the engine running).

<u>Wind/Cold.</u> Wind going by you at 30 miles per hour *is* cold anytime. But without a windscreen? Try riding a motorcycle for 2 hours in a T-shirt. Winter flying is out without an enclosed gondola AND an **exhaust heater** installed.

<u>Blower Fan, Engine driven</u> pressure system. A 12 volt pressure fan draws to much power, and has a great potential for radio noise. An engine driven, V-belt fan has been successfully used in medium sized blimps. (see GA-42 pictures)

<u>Air Scoop</u> pressure system is most commonly used on all sizes of airships. It is a little hard to design a scoop into the airstream in front of a small prop. But can be done.

<u>Generator/Alternator</u> of adequate capacity is a must. 35 amp is common, but 50 amp units are available. Even a 50 amp is marginal for internal envelope night lighting.

Building Small Gas Blimps

Ultralight
ENGINE COMPARISONS

Specification	ROTAX-582	PVM V-8*	yours
ENGINE TYPE	2-Stroke, 2-Cyl Water cooled	4-Stroke, 8-Cyl. Oil cooled	
HORSE POWER/RPM	64 / 6500 53 / 6000 44 / 5100	80 / 2400 70 / 2100 60 / 1800	
TORQUE/RPM	55.3/6000 51.0/5000 46.5/4700	175/ 2400 175/ 2100 175/ 1800	
HORSEPOWER/WT.RATIO (actual)	.69 H.P/lb.	1.0 H.P/lb.	
FUEL CONSUMPTION Max Load Cruise @ 75% power 65% 55%	7.95 GPH 5.96 GPH 5.16 GPH 4.37 GPH	4.89 GPH 3.67 GPH 3.17 GPH 2.68 GPH	
SPECIFIC FUEL CONSUM.	.82 Lbs/HP/Hr.	.41 Lbs/HP/Hr.	
FRONTAL AREA	242 Sq.In.	184 Sq.In.	
ENGINE WT. (bare) no prop or accessories	62 lbs.	62 lbs.	
EXHAUST TEMP.	1300 F	900 F	
TIME BETWEEN O/H Recommended overhaul	250 hrs	3000 hrs.	
3000 HOUR OPR. COST	$22,487	$15,660	
RETAIL COST	$2795	$3600	

NOTE: I've never seen a **ROTAX** go 50 hours between O/H, nor have I ever even seen an operating PVM V-8 (made by *EVESTAR* Tech.). Beware of wild claims. 1991 data.

Building Small Gas Blimps

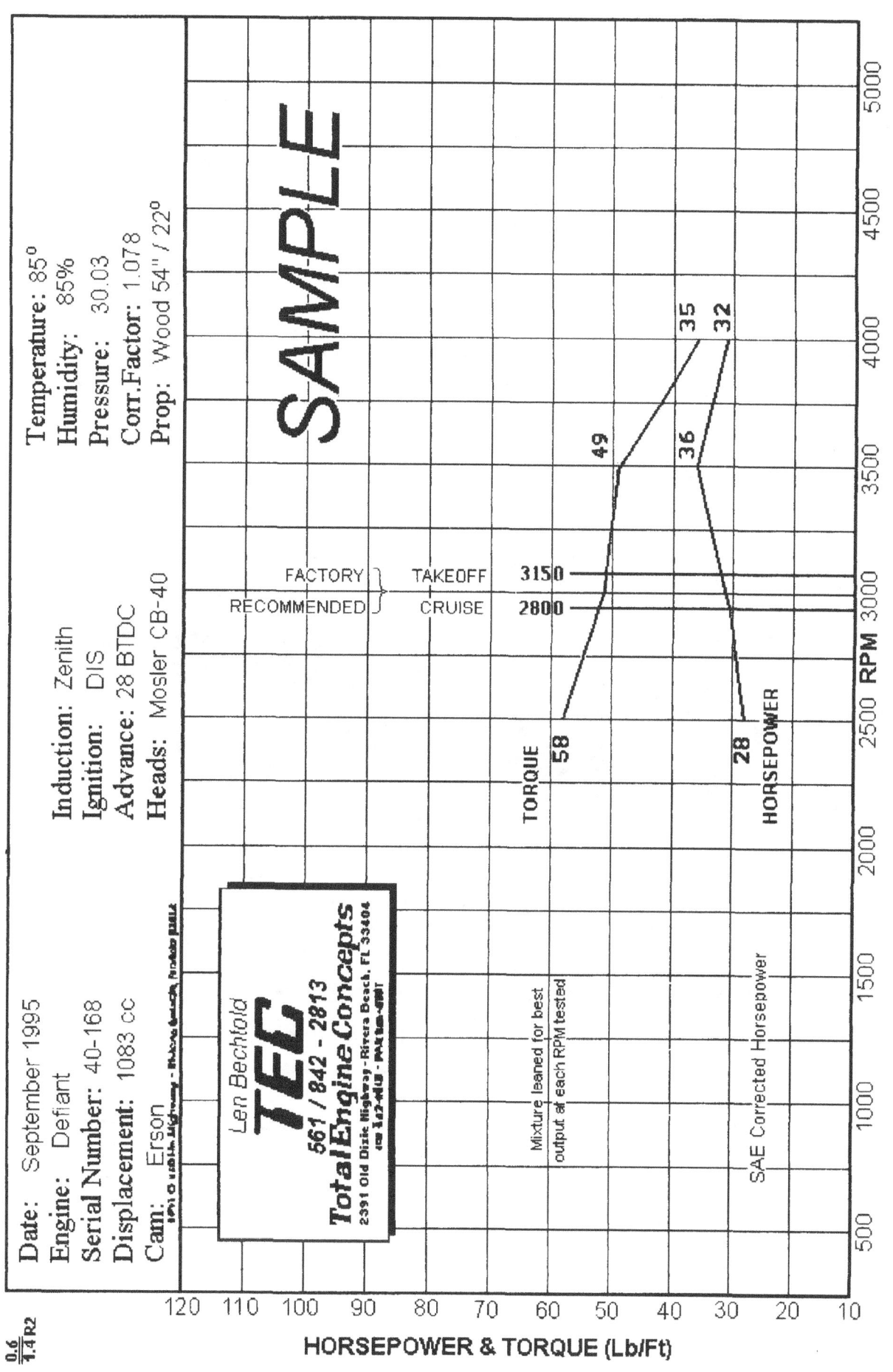

83

DUCTED FANS

My basic statement is: Forget it. They are more problems than they are worth. The rigidity and tolerances are not reasonable, even with weight and money "no object". A light weight "Prop Guard' to protect the crew, OK. But a ducted fan, NO. Even the big ships are doing away with them. Now that I said that, you probably still want to consider it. So I at least owe you a little more detail.

Current powered airplanes have no use for use ducted fans because of the higher speed ranges sought. But actually, low powered, low speed (less than 80 kts.) blimps are the best possible (theoretical) aircraft to use one on. (see chart)

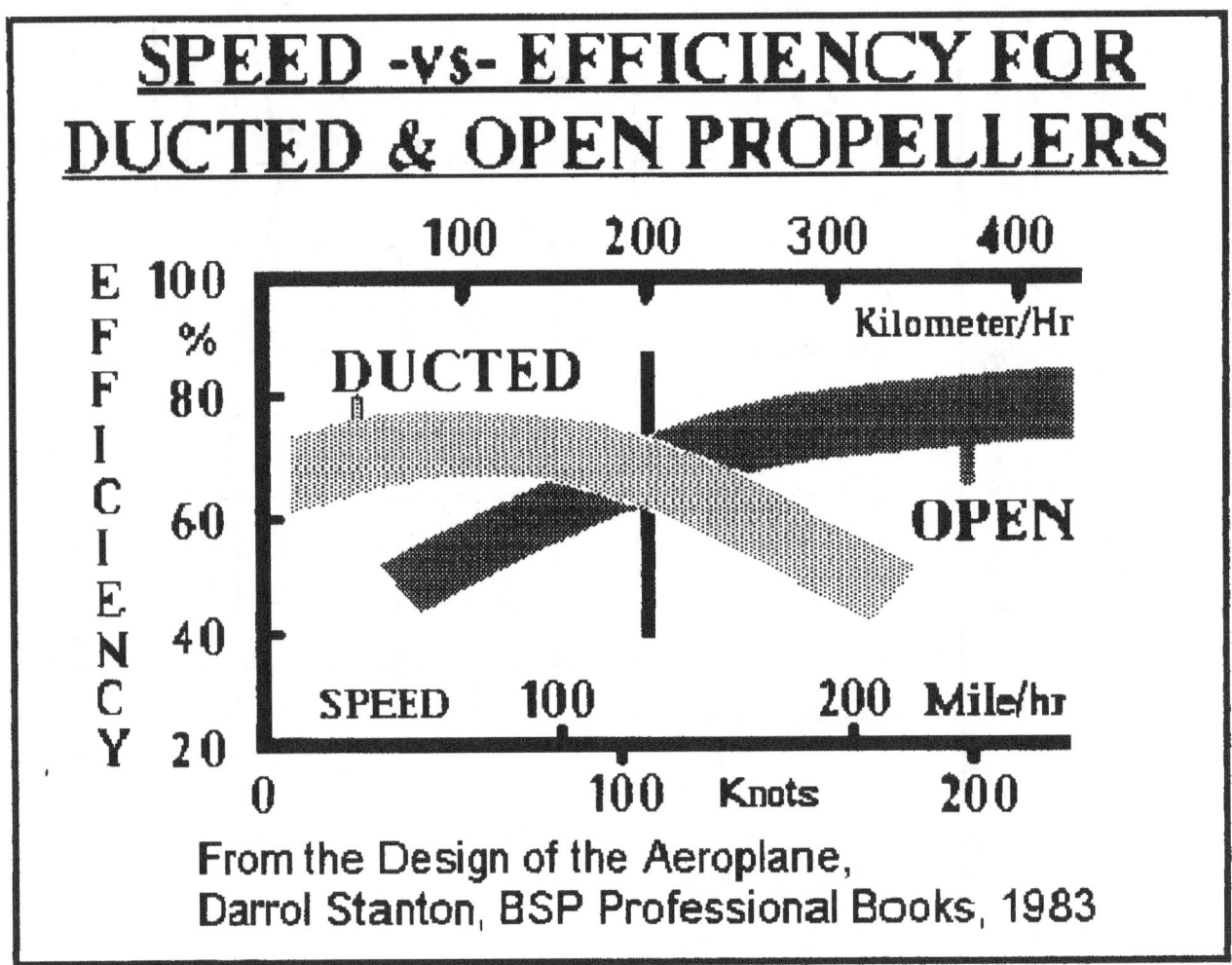

Where ducted fans loose their value is the weight associated with an efficient design. But the fat width of the line on the chart only represents 5 to 15 thousanths of blade tip clearance inside the duct. Obviously, the prop blade(s) must be square tipped because of the way the airflow rolls toward the tip. .005 tip clearance is hard enough to hold, but if the duct wobbles, the prop & duct will be damaged The exact relation is:

$$To = (Ad*Rho*Pi)^{1/3} (550*Maf)^{2/3} (Hp_p*PD)^{2/3}$$

Where: To is in lb/ft; Rho is in slugs/ft^3; M is in slugs/sec; Hp_p*PD is in ft.;
Hp_p is in horsepower.

Since both Ad and Maf increases due to the duct around the propeller (disregarding the supports). It would reduce relationship between To and Hpp and D changes. This allows either the To to increase, the inputted Horsepower to be reduced, the propeller diameter to be reduced, or any combination of the three.

As the speed increases, the drag from the duct increases and counteracts the gains from increased exit area. The exact relationship is:

Dd= 1/2 Rho $V_{1/2}$ S Cdd

Cdd depends on the shape of the duct along the cross sectional parallel to the air flow. Typical values of Cdd range from 0.008 to 0.0100 [6]. At speeds above zero the net thrust becomes

T(v) = To - Dd(v) - Dp(v).

Where: T(v) = thrust at speed v; To = Static thrust; Dd (v) = Drag from the duct at speed v; and Dp (v) is drag from the propeller at speed v.

Propeller drag Dp is the same as a non-ducted propeller. In open propellers the diffuser ratio (Od) is a half. In a ducted propeller it could be as high as 2. Assuming the design of the prop hub supports didn't increase the air flow, of course.

Lest I forget, wood props are out with ducted fans. Moisture easily affects the blade balance. Composit erodes quicker than metal, but aluminum maintains the best stablity.

Building Small Gas Blimps

R: 11/06/97

2 CYCLE ENGINES
UILTRALIGHT SUPPLIERS

Supplier	Location	Phone		Type
ADD-AIR 500 SW 21st Terr. B105 Ft.Lauderdale, FL 33312		954/791-8277 Fax/792-2514	53	Distributor KONIG Overhaul
AMW Engine Co. 8 Schein Loop Beaufort, SC 29902		803/846-2167 Fax/846-2169	53	Manufacturer CUYUNA 6 models
BOMBARDIER Engines A4623 Gunskirchen Austria		07246271		Manufacturer ROTAX 3/4/500 series
CALIFORNIA Power 790 139th Ave. #4 San Leandro, CA 94578		510/357-2403 Fax/356-4429 Major supplier	6-8	Dealer/ROTAX Overhaul All parts
DAWSON Sport	Zepher, FL	813/788-6885	74	Dealer/ROTAX
GREEN Sky	Orwell, OH	216/293-6624	49	Dealer/ROTAX
J-BIRD P.O. Box 438 Kewaskum, WI 53040		414/626-2611 status UNK	55	Re-Manufacturer KAWASAKI
KODIAK Research 909 Kal Lake Road Vernon, BC V1T 6V4 Canada		604/542-4151 R. Shettler		Distributor ROTAX-Canada
KONIG Motorenbau KONIG Berlin Germany				Manufacturer 3 & 4 Cyl. radials
LEADING Edge 8242 Cessna Drive Peyton, CO 80831		719/683-5323 Fax/683-5333	58	Distributor/ROTAX All parts supplier good A/C catalog
LOCKWOOD Supply;	Sebring, FL	941/655-5100	26-7	Dealer/ROTAX
MIDSTATE Aircraft	Pahokee, FL	407/924-3488	33	Dealer/ROTAX
MOUNTAIN Aircraft	Eleanor, WV	304/586-3619	11	ROTAX Drives
MISSISSIPPI-Light	Lucedale, MS	800/247-7652	50	Dealer/ROTAX

Building Small Gas Blimps

R: 11/06/97

4 CYCLE ENGINES
UILTRALIGHT SUPPLIERS

CARR Precision 6040 N.Cutler Circle Portland, OR 97217	503/735-9980 Fax/285-0553 ONAN-26	57	Manufacturer CARR-Twin KOHLER-Twin
CLASSIC Sport 7111-152 "B" Ave. Edmonton, ALB. T5C 2B9	403/475-0703 Fax/473-1266	57	Manufacturer MOTO-GUZZI 65hp/149#
FLITE-LIGHT 200 Cventry Drive Lexington, SC 29072	803/356-8587 Fax/951-3396		Manufacturer EMDAIR-CF TWIN-60
HUMMEL Aviation 509 East Butler Bryan, OH 43506	419/638-3390 Fax/636-1642	58	Manufacturer HALF-VW
LIMBACH-Flugmotoren Kotthausener Str. 5 D-5330 Konigswinter 21 Germany	02244-2322 Fax/-6976 used on A-60+		Manufacturer LIMBACH L-2000 series
MORAVIA Aircraft 9102 Herlong Road Jackson, FL 33210	904/783-0490 Fax/783-0047	58	Manufacturer WALT-MIKRON 65 hp/132#
PRO-VW Machine 6614 Southeast Blvd. Derby, KS 67037	316/788-4625 Fax/	58	Manufacturer PREDITOR 35 hp/85#
SADLER Engines P. O. Box 953 Couperville, WA 98239	360/678-6994	58	Manufacturer SADLER-RADIAL 65 hp/108#
TENNESSEE Aircraft 125 Dearfield Drive Normandy, TN 37360	615/455-4516 Fax/455-1309 Major supplier	24	Dealer Wood good reputation
TOTAL Engine Concepts (TEC) 2391 Old Dixie Highway Rivera Beach, FL 33404	561/842-2813 good reputation Len Bechtold		MOSLER wood props ultralights

Source: Ultralight Flying, July 1996 Issue 243.

Building Small Gas Blimps

R: 11/02/97

PROPELLER
UlLTRALIGHT SUPPLIERS

AIRSTAR Sales	Brentwod, CA	510/516-1186	37	Dealer
AIR-TEC	Reserve, LA	504/536-3994	4	Dealer

COMPETITION Aircraft 916/268-3048 62-3 **Manufacturer**
11110 Gopher Mine Trail   Ultra brand
Grass Valley, CA 95949

GREEN Sky   Orwell, OH   216/293-6624   49   Dealer

HOFFMAN Propellers   **Manufacturer**
Kuepferlingstr, 9   Quality products   FAA Type Certified
Rosenheim D-8200 Germany status UNK   All standards

INVOPROP Corp.   310/602-1451   62   **Manufacturer**
15903 Lakewood #103   Fax/602-1374   Carbon fiber
Bellflower, CA 90706   Variable pitch

LOCKWOOD Supply; Sebring, FL 941/655-5100 26-7 Dealer

MT-Propellers   011491   **Manufacturer**
Wallmuhles.Atting   202-8433   All Kinds
G-09429 Straubing GERMANY Gerd Muhlbauer   great products

TENNESSEE Aircraft   615/455-4516   24   **Manufacturer**
125 Dearfield Drive   Fax/455-1309   Wood
Normandy, TN 37360   good reputation

TOTAL Engine Concepts (TEC) 561/842-2813   Dealer
2391 Old Dixie Highway   good reputation   Wood, ultralights
Rivera Beach, FL 33404   Len Bechtold   Recommended

WARP Drive   800/833-9357   62   **Manufacturer**
909 North 40th Street   Carbon fiber
Clear Lake, IA 50428   Ground Adjustable

WINDESIGN   714/943-1695   Dealer
Valley Airport   major supplier   many brands
Perris, CA 9259?   retail   Engines also

MIDSTATE Aircraft Pahokee, FL 407/924-3488 33 Dealer

Source: Ultralight Flying, July 1996 Issue 243.

Building Small Gas Blimps

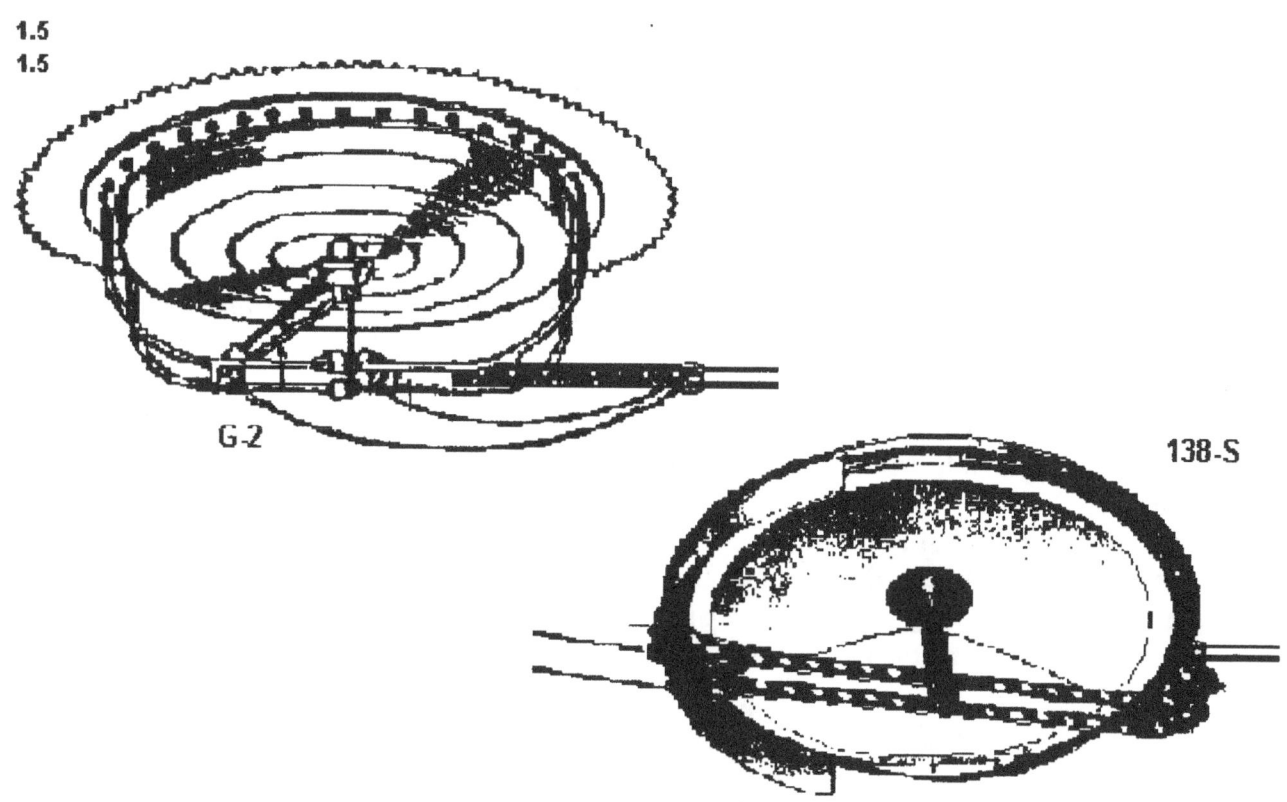

Helium / Air
VALVES

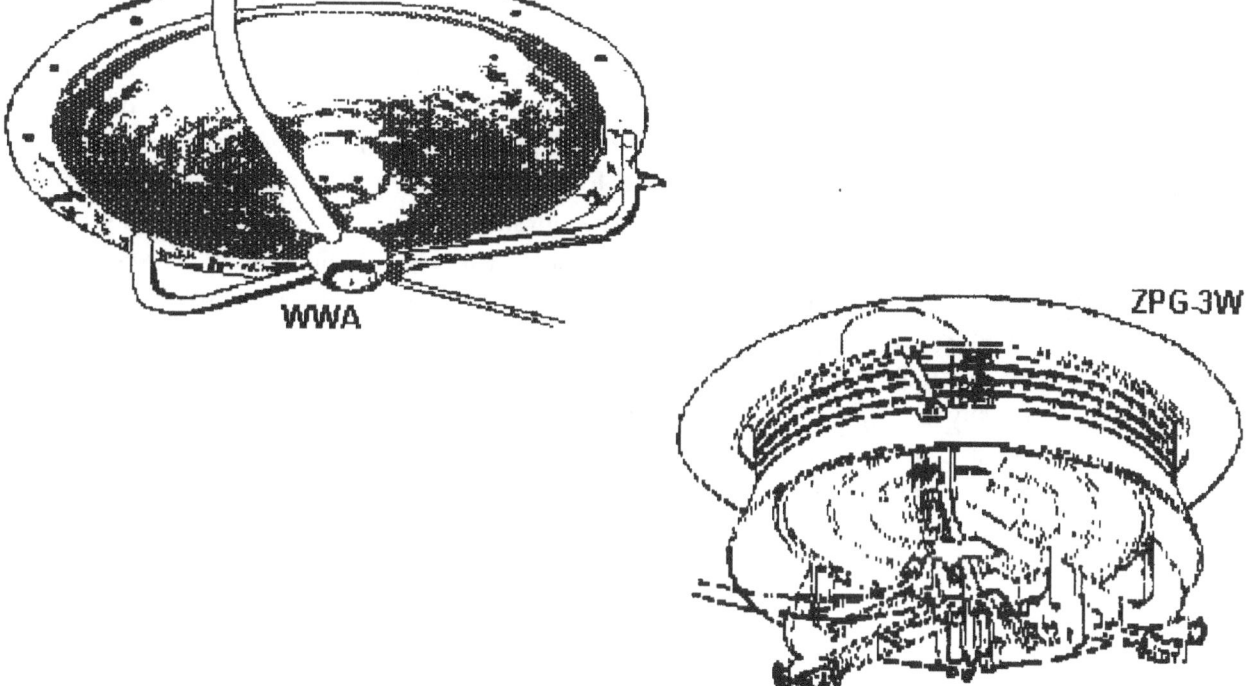

Building Small Gas Blimps

PRESSURE FORCES
And Functions of the Gas

The gas contained within the envelope or an airship ordinarily is considered to be the source of buoyancy, and while this conception is sufficiently accurate for general purposes, a more precise understanding of the function of the gas is essential to the airship designer.

The true source of buoyancy is the surrounding air and not the confined gas. If the gas could be removed, leaving a vacuum, without collapse of the container, the lift or buoyancy of the airship would be increased by an amount equal to the weight of gas abstracted. Although the specific weight, and hence the buoyancy of air is small, being only .07635 lbs./cu.ft. in the standard atmosphere at sea level, its pressure is very great, being 2,120 lbs./sq.ft. in the same conditions. No structure enclosing a vacuum and displacing a volume of air of less weight than itself could resist this pressure. The cell or container for displacing the air must therefore be filled with gas, having an absolute pressure approximately equal to that of the surrounding air, and as light as possible, because the weight of the gas diminishes the useful lift just as much as any other item of weight in the ship. Strictly speaking,

The true function of the gas is to maintain the internal pressure.

Definition of gas pressure. In practice, the term "gas pressure" is used to express the difference between the absolute internal gas pressure *and* the external air pressure. A superpressure of 2 inches of water is considered large in a nonrigid or semirigid airship and automatic gas relief valves are commonly set to blow open at less than 0.5 in. of water superpressure.

Example. Find the pressure at 70 ft. above the lowest point in a container filled with gas lifting .060 lb/cu.ft.(Helium), with a bottom ind. pressure of 1.5 inches of water.

From equation (87), Outside Pressure (po) = 1.5 x 5.2 = 7.8 lbs./sq.ft; and from equation (86), p = 7.8 + (70 x .060) = 12 lbs./sq.ft., or <u>2.31 inches of water*</u>.

Effect of longitudinal inclination upon gas pressure. In nonrigid and semirigid airships, the gas space is either wholly undivided by transverse bulkheads, or the bulkheads are of little effect in preventing equalization of the gas pressure on each side of them. The vertical distance between the lowest and highest points of the envelope increases with the longitudinal inclination *of the* airship, and since the lowest point must always have sufficient pressure to prevent collapse (usually about 1 inch of water in nonrigids), the largest gas pressures occur in the inclined position. The envelopes are usually designed for the gas pressures occurring at 30° inclination.

Gas pressure an asset or a liability. In pressure airships! Some gas pressure is an absolute necessity to prevent collapse, and resist compressive forces. In all types of airships, excessive gas pressure must be avoided by ample provision of gas valves to liberate the gas when climbing rapidly, or when thrown violently upward by strong ascending air currents, such as in a thunderstorm.

Note*: One inch of water = 5.2 pounds per sq.ft. of surface area.

Airship Design, p179

Building Small Gas Blimps

PRESSURE SYSTEM
R:12/04/97

There seems to be a lot of misinformation on what its function is, and how complex it should be. Basically, the helium will expand (or contract) with a change of temperature, or a change on (atmospheric) pressure. The two major things that effect airship operation the most are the Suns' effect on Temperature, and Altitudes' effect on the pressure. The penalty for not providing for this natural occurrence is that at:

>Low Pressure: The envelope sags and looses its shape;
>>Control surfaces droop and loose effectiveness;
>>Control surfaces want to flutter, causing drag to increase;
>>Control cables become slack and ineffective;
>>Nose softens and starts to invert causing the ship to mush;
>>Gondola weight puts uneven strain on the envelope suspension.
>
>High Pressure: Will cause a catastrophic failure of a (glued) seam.
>>Once helium is valved off, you can't recover it;
>>Lift (helium) can't be recovered when you need it most.
>>Helium is expensive.

So, finally we get to the bottom line. How you conserve the helium directly affects the cost of operation. One (369 cu.ft.) bottle can cost you $70, but only lifts about 6.5 pounds. Think about that; then think about what technology, time and money is worth the investment.

Pressures, minimum, maximum, and ideal must be discussed at this time. I have said elsewhere that the new envelope is tested to a pressure of 6 inches of water pressure. 5" is only a fraction of one pound per square inch (PSI), but it is enough to determine overall strength. It will never be expected to reach that pressure again in its life. Normally, it will remain between 2.0 and 3.0 inches weather flying or masted. But 2.5" is the ideal. At 1.5" it takes a hard push to deform the envelope; at 2.5" a thump with your finger resonates through the entire envelope.

Ballonets: effectively provide for part of this pressure change (valves and blowers, the other). How big they should be, is no mystery if you can remember two rules of thumb.

>Helium expands 1% for each 3 degrees of superheat (and conversely);
>Helium expands 1% for each 375' of altitude (and conversely)

I can tell you that most airships have a ballonet capacity of about 25% of total envelope volume. They have functioned nicely for 80 years but still can only cross the Continental Divide (5280' at Deming) under ideal conditions. BUT if YOU only want to put-put around the seashore at 1000' you can do with a lot less.

One. Two. or More ballonets? Again there are trade-offs to consider. Each ballonet gives that much more potential for an air leak to contaminate your gas (purity). Two ballonets (fore & aft) allow you to control the pitch of the nose, and allow the pitch to give you extra (dynamic) lift. But then if you are only building an "ultralight", how much of your total volume do you want to sacrifice? So, since we are only thinking small (less than 60K), lets design for only one. A couple more rules of thumb are in order:

Building Small Gas Blimps

PRESSURE SYSTEM cont.

10% is the minimum ballonet for a cool climates (at less than 1000'MSL);
2 % more for each 1000' and or each 3 degrees over 70 ambient

Valves have to meet at least four requirements

They have to be reliable and easily pressure adjustable;
They have to seal well and open easily;
They have to be the optimum size and light in weight,
They have to be set to open & close automatically at specific pressures.

The first three are design (and redesign) problems. The "automatically" part is a safety item that can have senous consequences. Several airships have had seams split, and at least two were lost from high stresses in a (bent) "sausage" mode. Understand that there are weather forces that can cause rapid (uncontrolled) climbs. 500 feet per minute (fpm.) is very common, I have experienced 1000 fpm. 1500 fpm has been recorded, 2000 is possible. So when we use that word "automatic" we mean that the **forward** ballonet **valve** should be spnng loaded to **open at 1.6", close at 1.2"W.P.** The aft (or second, if there is another) at 2.8 and 2.4. **The helium at 3.5 and 3.1.** The opening orifice of the Ballonet valve(s) and the Helium valve should be able to handle a 1500 fpm climb. Just to throw some numbers around and better expose you to the hazards. Check this.

Assume a 30,000 cuft. airship; and an average expansion rate of 1% per 375' altitude. Climbing at 1500 fpm, the flow rate out of the ballonet must be 4% of the total volume, or 900 cuft./minute. The other side of the coin is that you need a valve with an aperture of 67 sq inches (9.2" diameter, or 2 at 6.5" dia.).

Aperture(Ap)=Sq.Root of (Flow Rate/Water Pressure * Air Density) in sq.inches.
(Air Density = .07651 Lb./cuflt.)
Aperture Diameter(Ad) = 2 times the Sq.Root of (Ap / Pi) in inches

Blowers for maintaining pressure are needed both on the ground and in the air. Obviously, we are talking about two different blowers, one engine driven, and one electric motor driven. Since much higher pressures are needed in the air, the engine driven (fan or scoop) can be designed into the system. On the mast, only 1.5" W.P. is needed to keep the ship tight, and an AC powered centrifugal does the job well. Weed blowers don't last worth a damn. But "on the road" a little gas driven *HONDA* generator powers the AC centrifugal blower nicely.

T-Box or Flow Chest is used on 2-ballonet airships to direct the air pressure forward or aft as desired. It requires a built-in tunnel inside the envelope. Direct flow of a blower to a single bayonet does need a floppy door to act as a check valve installed. Otherwise the blower will be in continuous operation.

Pressure Controller is a small portable box, used on the ground, that is manually set to actuate the blower automatically when pressure is needed.

Building Small Gas Blimps

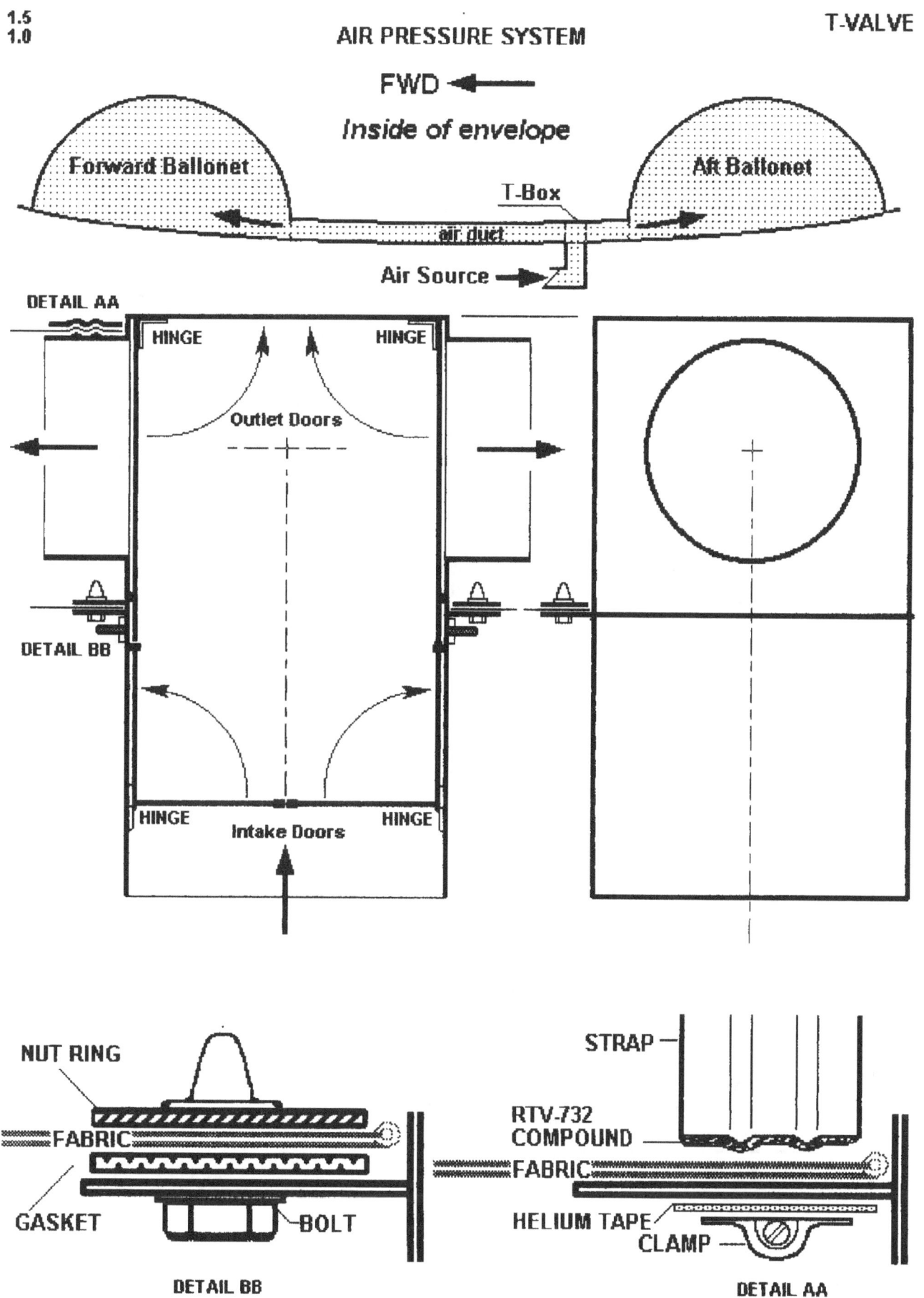

Building Small Gas Blimps

Surplus VALVE ASSEMBLY
Helium / Air

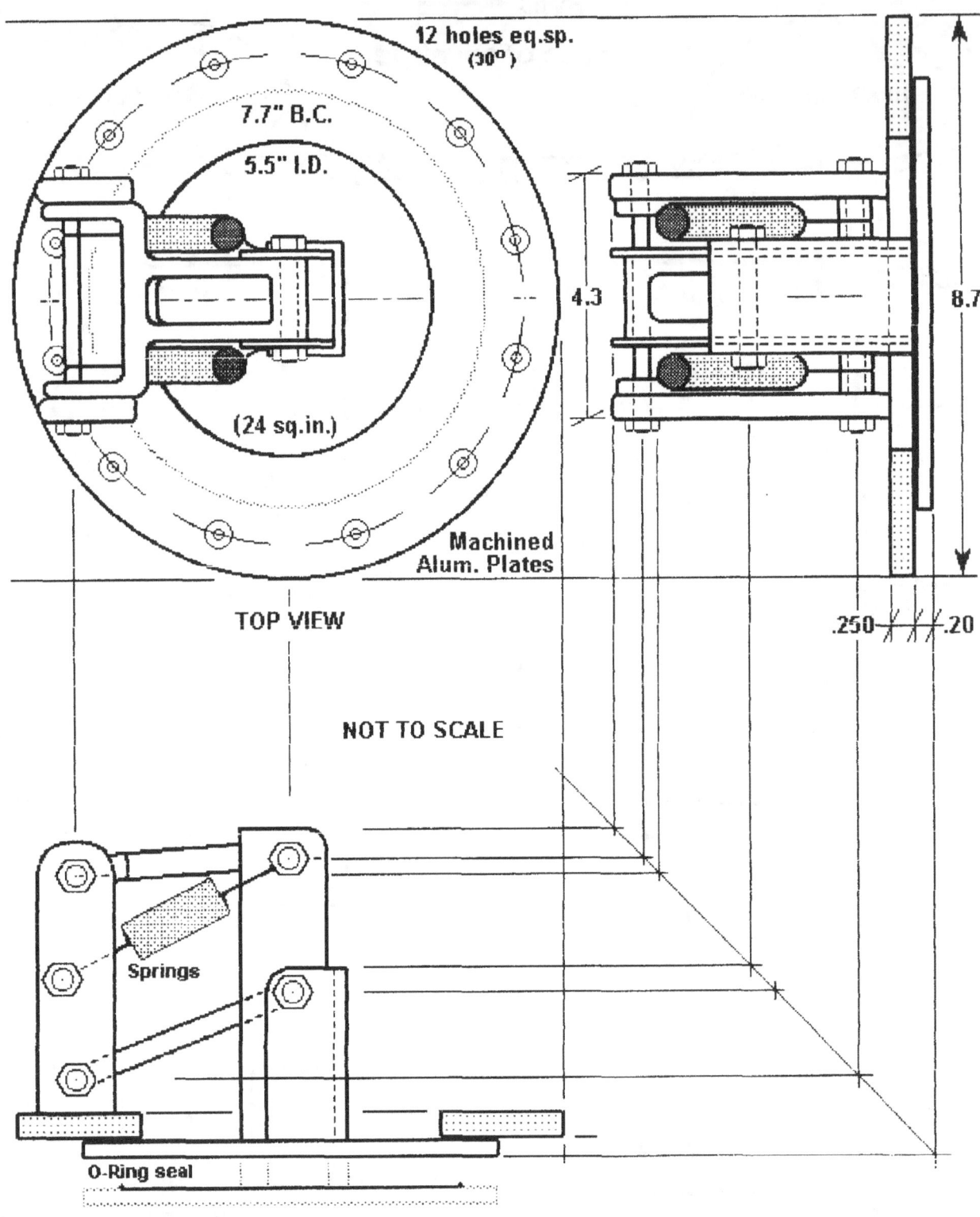

WARNING: These surplus units seal poorly. Replace springs and seals.

Source:
Govt. Surplus Sales
Tucson, AZ
(B-52 pressure valve)

Building Small Gas Blimps

Helium / Air Valve
COVER

1.5 COVER 11
1.0

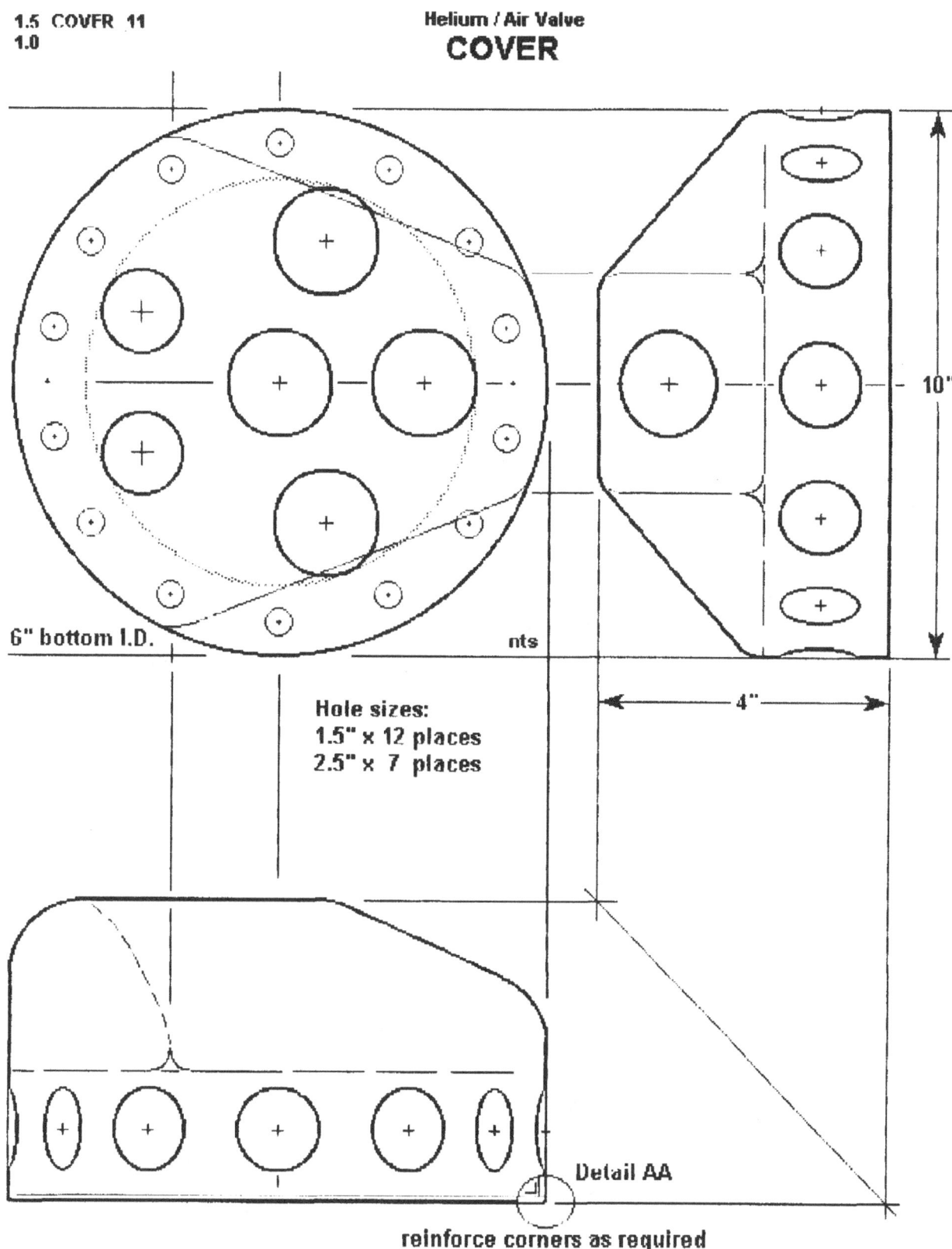

6" bottom I.D.

Hole sizes:
1.5" x 12 places
2.5" x 7 places

10"

4"

Detail AA

reinforce corners as required

Manufacturing Sequence:
Fabricate Base & Dome separately;
Fabricate Base 1" diameter oversize;
Drill 3/8" bolt holes on req. bolt circle;
Bond Dome to Base in position.

Parts List:
2 oz. Fiberglass Cloth
Mold Release
Resin

Building Small Gas Blimps

1.5 Port_11
1.0

envelope
11" PORTS

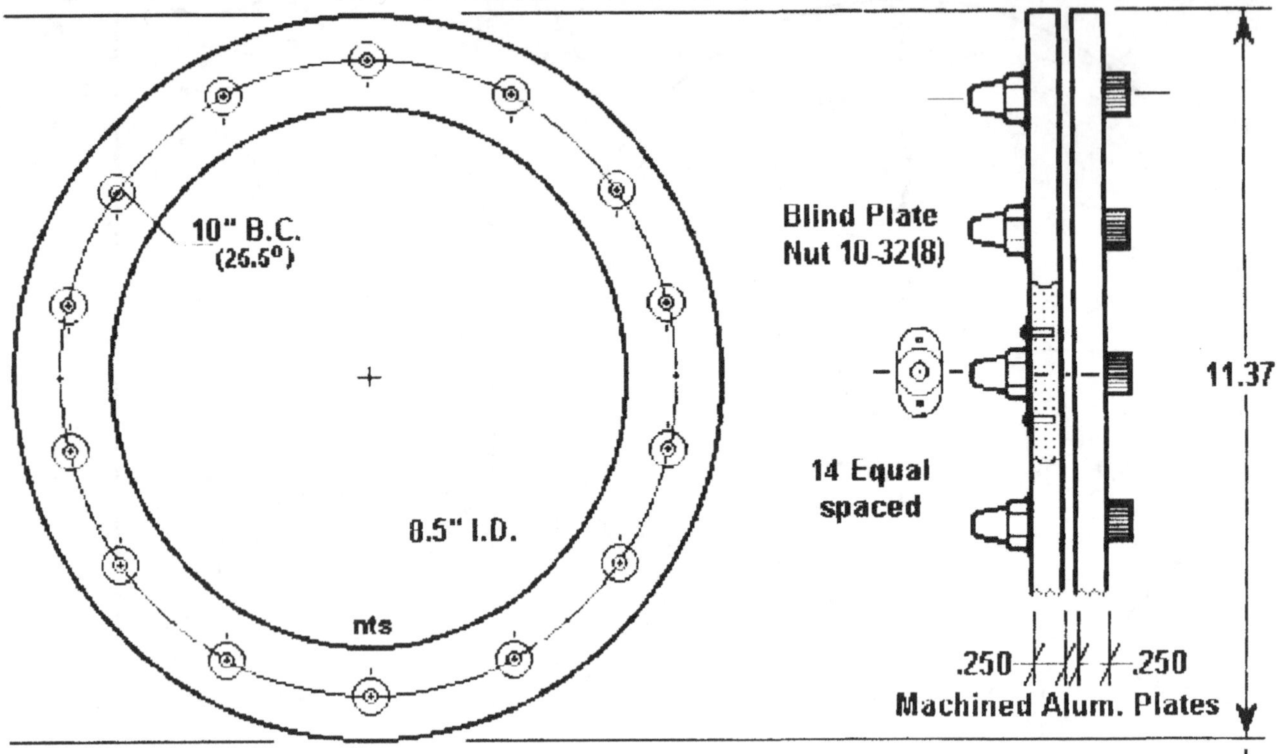

- 10" B.C. (25.5°)
- 8.5" I.D.
- nts

Blind Plate
Nut 10-32(8)

14 Equal spaced

11.37

.250 | .250
Machined Alum. Plates

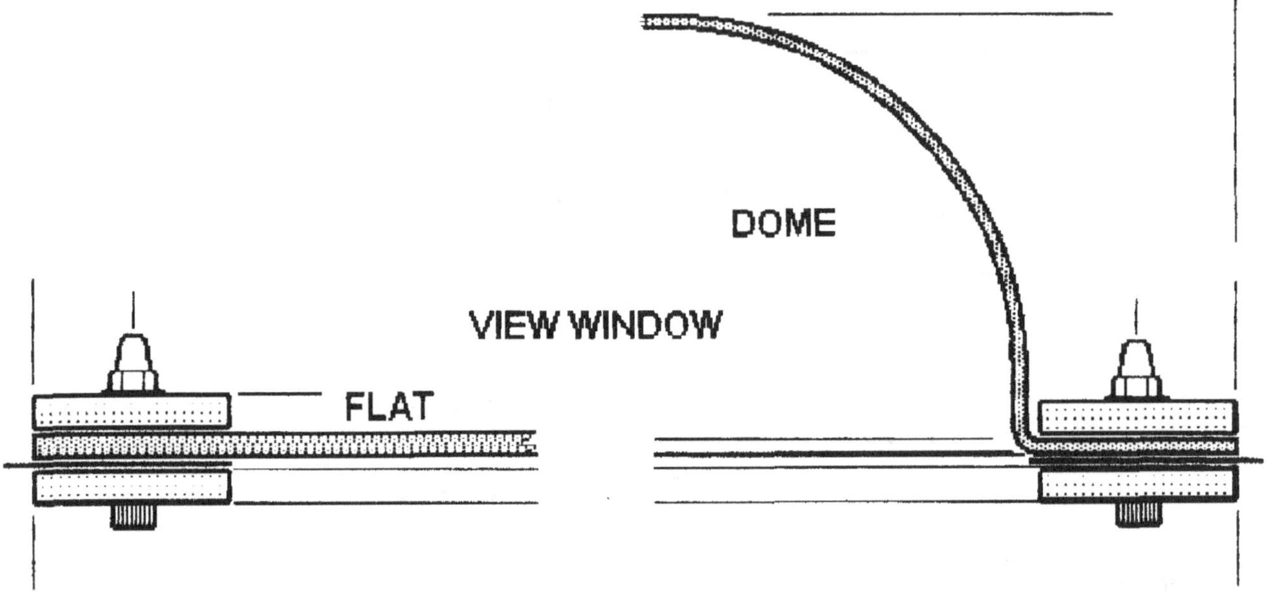

DOME

VIEW WINDOW

FLAT

Parts List:
3/16 or 1/4" Lexan x
11.37" dia. Flat plate

Parts List:
1/8 or 3/16" Lexan
heat blown to 8" hemisphere

96

Building Small Gas Blimps

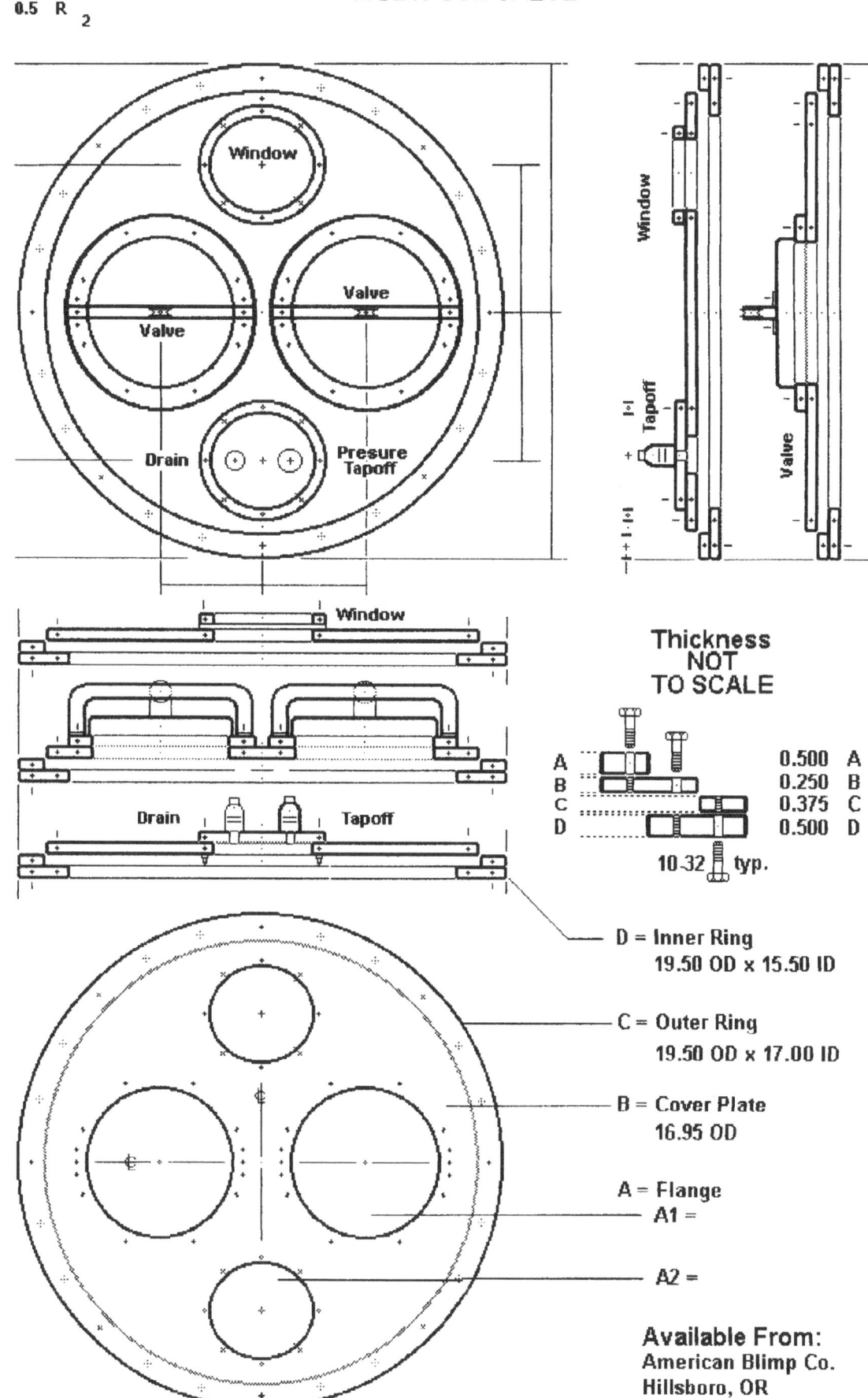

Building Small Gas Blimps

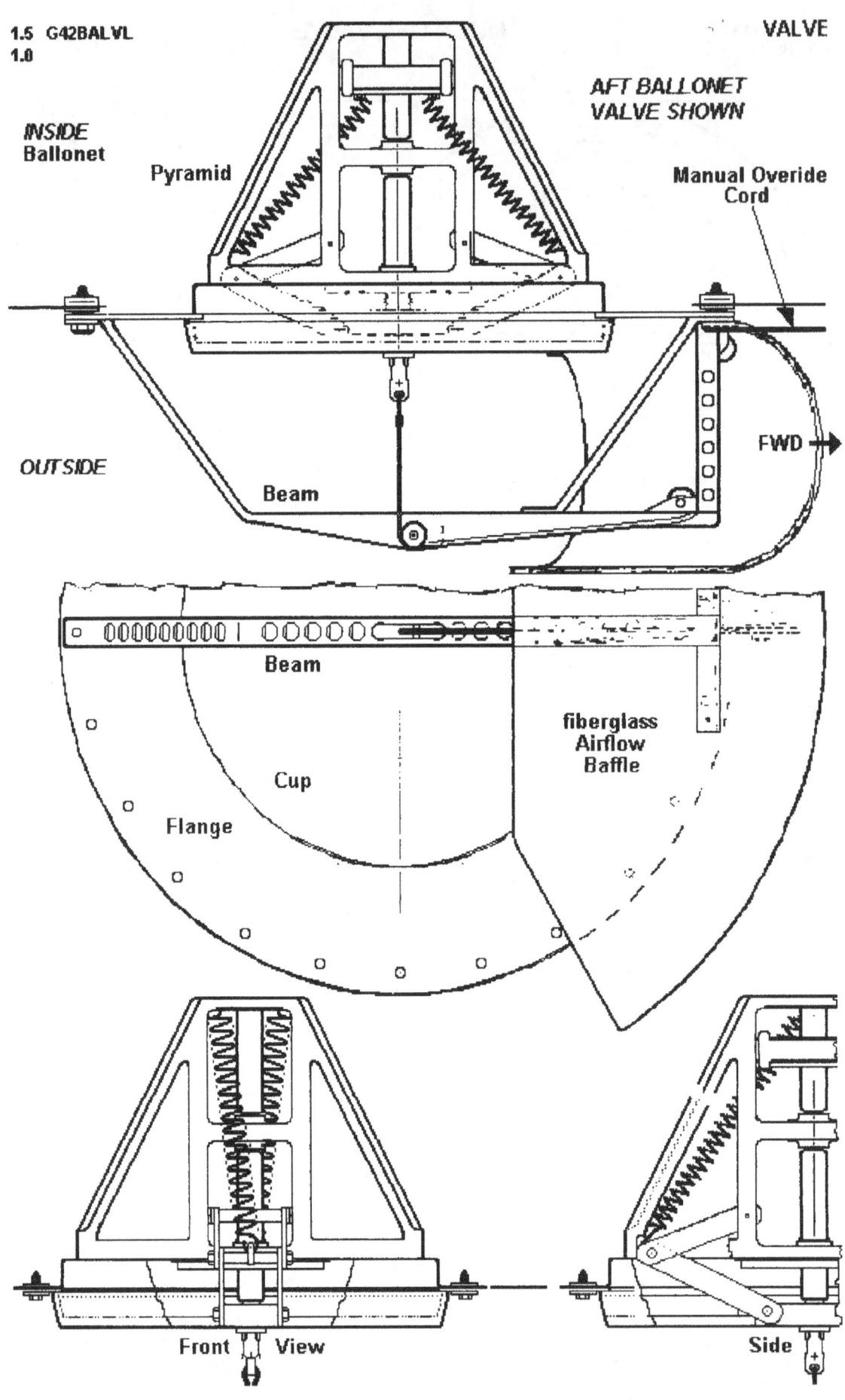

HELIUM VALVE

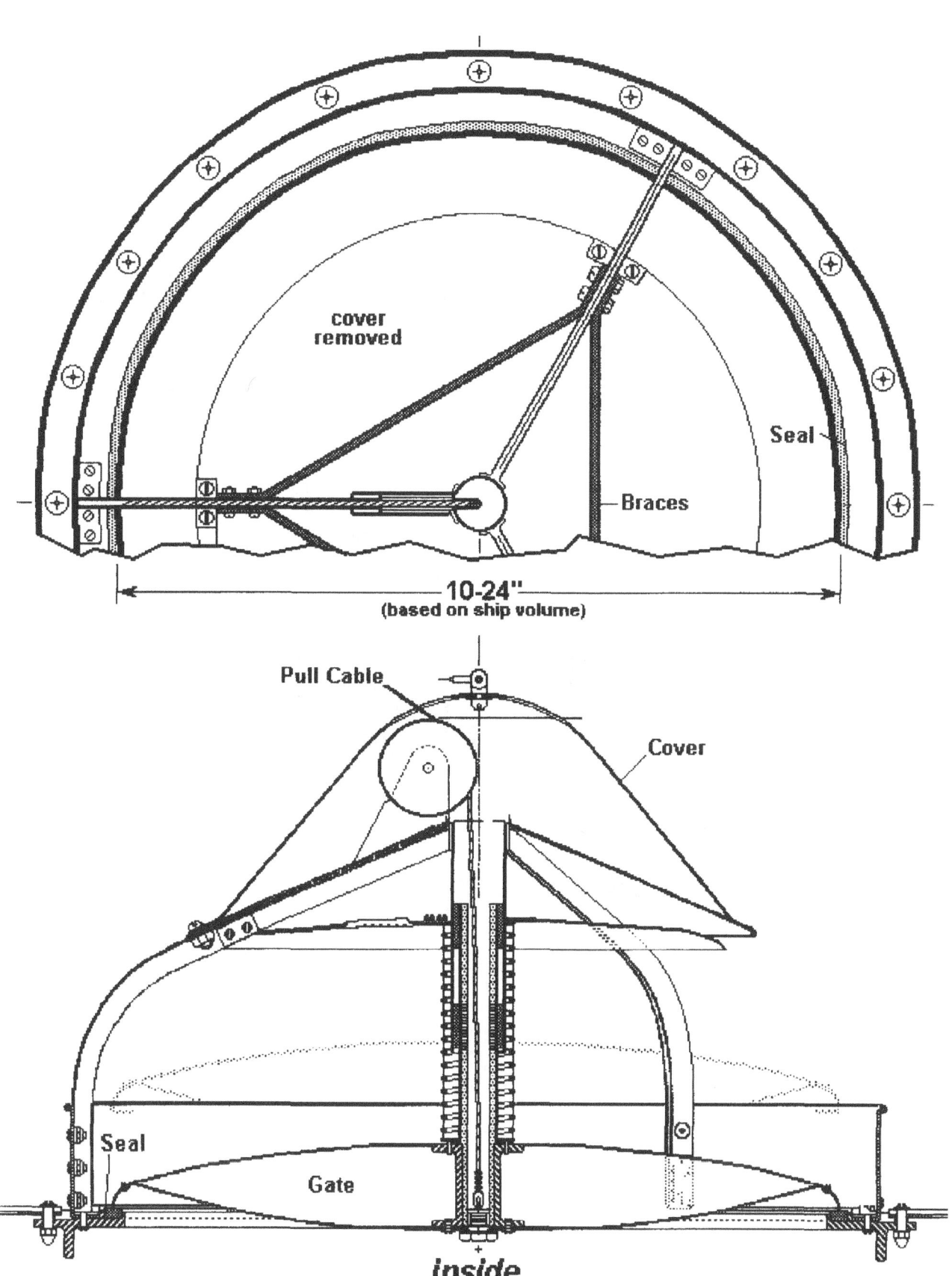

Building Small Gas Blimps

1.5 Fn_Cover
1.5   R2

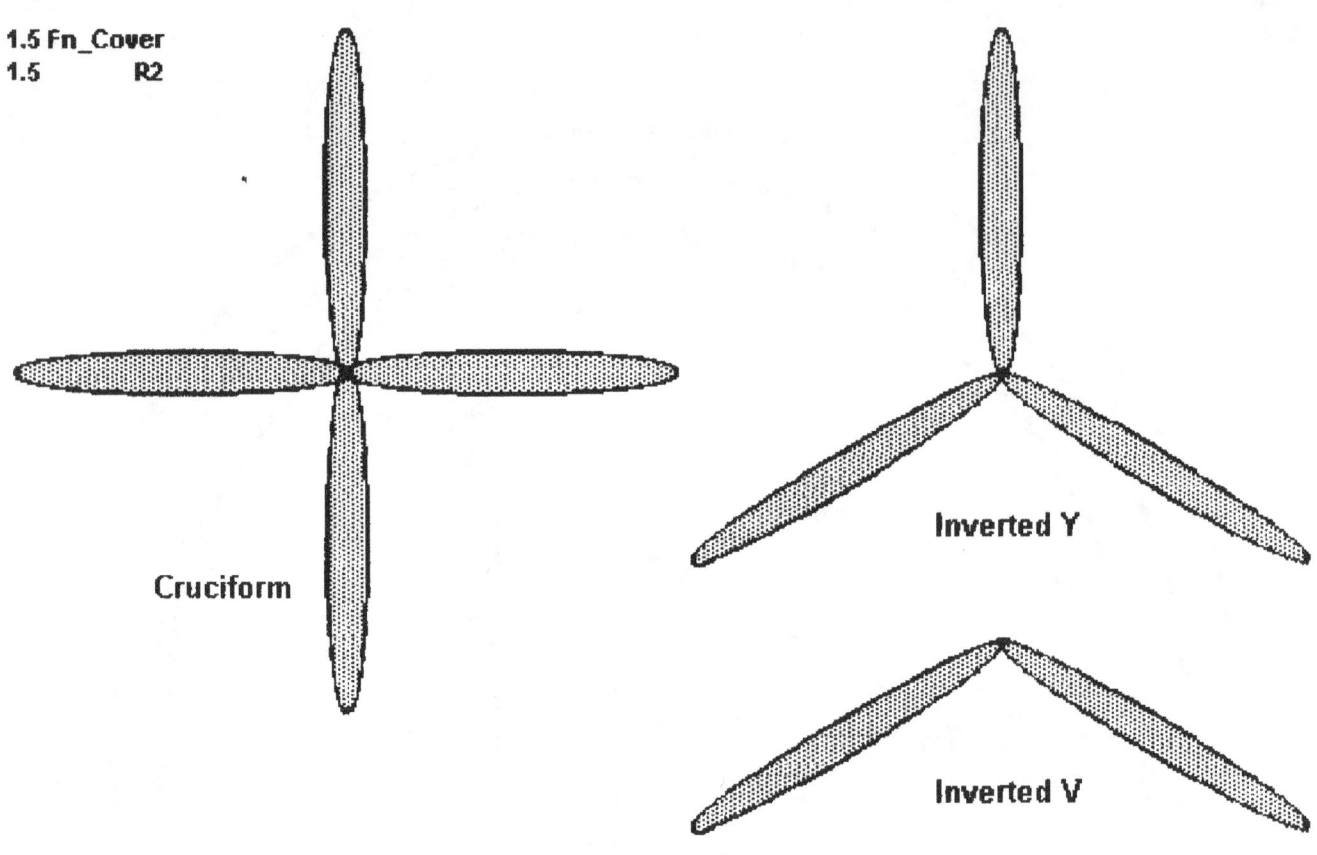

Cruciform

Inverted Y

Inverted V

FINS

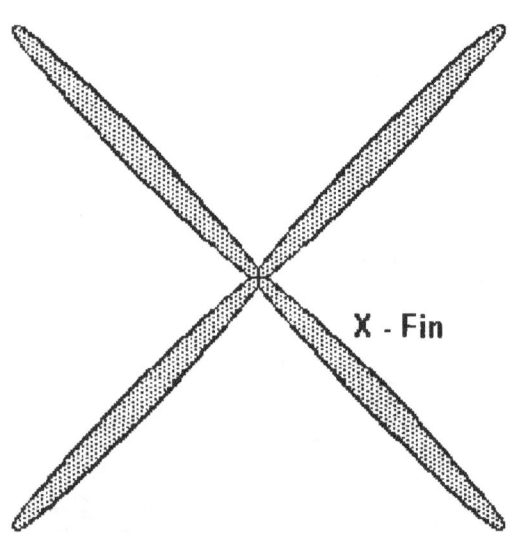

X - Fin

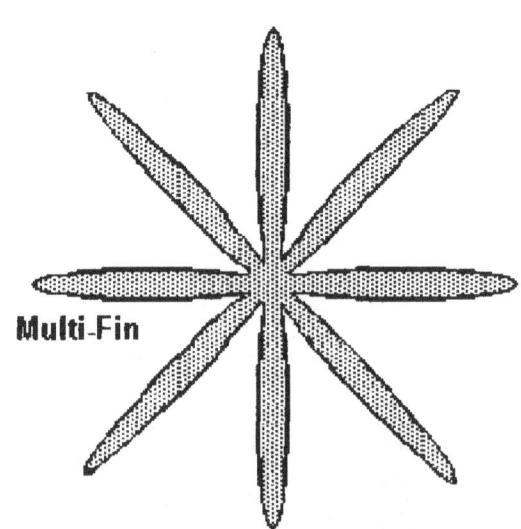

Multi-Fin

TAIL FINS
R: 11/04/97

Definitions are first on the list, so we understand apples from oranges. Your *FINS* consist of two independent structures needed to accomplish two different tasks. The STABILIZER to keep your massive marshmallow (blimp) on a straight line, and CONTROL SURFACES to change direction. The control surfaces have two functions; ELEVATORS (horizontal) for climb & descent, RUDDERS (vertical) for left & right turns. Now that you understand that, lets move on to the voodoo science part.

Yes, I am poking fun at it. Because on a small blimp we are going to compromise it to death. We are trying to compromise Low Speed Aerodynamics with our ability to get off the ground. If we build it to its ultimate efficiency, the technology and cost don't coincide with our realistic goals. Lets list some "wish-book" targets:

> Effective at slow speeds
> Light in weight, but strong;   Easily attached, and removable;
> Keep us stable on a straight line (with minimal "mushing")
> Efficient in changing direction, without applying a lot of input force;
> Able to withstand side loads (gusts), with no flutter in flight.

HOWEVER, to get ALL of these, we will end up with a *HELICOPTER*, so I'm going to let you make the compromises.

The compromises don't seem to be too bad if you stay in the range of 9 to 29 knots. Less, you rapidly loose effectively of your control surfaces; more, the more force it takes for control displacement. On the low end you can gain some with a more aerodynamic fin (Reynolds number); and on the high end by offsetting the leading edge of the control surface (putting the hinge point into the surface). The higher the speed, the more sophisticated the design must be. Even the Goodyear blimp has trouble at low airspeeds: and are not interested in higher airspeeds.

The area of the stabilizers are the key to flight stability, but length should exceed width by 3:2. How the airflow is directed over the surfaces not only effects control, but greatly affects the drag (and horsepower required). 500 horsepower won't necessarily move the ship at 60 mph.; but 60 hp. can move you at 100 mph.

No one cares if you use an inverted "Y", X, +, multiples, or derivatives thereof, for your tail configuration. But all of the configurations require the same total area for a specific size. I have seen a beautifully done 8-fin blimp, but the work to build and rig them was out of reason.

Another tip, use a well made fixture to make all of your fins the same, and interchangeable with the control surfaces (and actuators, if used). Composites work great, but still can't beat the cost of welded aluminum and polyester fabric. Welded aluminum makes a better looking structure, but remember that the weld weakens the T-aluminum. Please don't try to make a complete bolt-together structure.

You might want to give thought to a design that allows a 3-piece disassembly, for shipping, storage, or ease of repair. Or just to get them out of your basement, or through the workshop door.

Building Small Gas Blimps

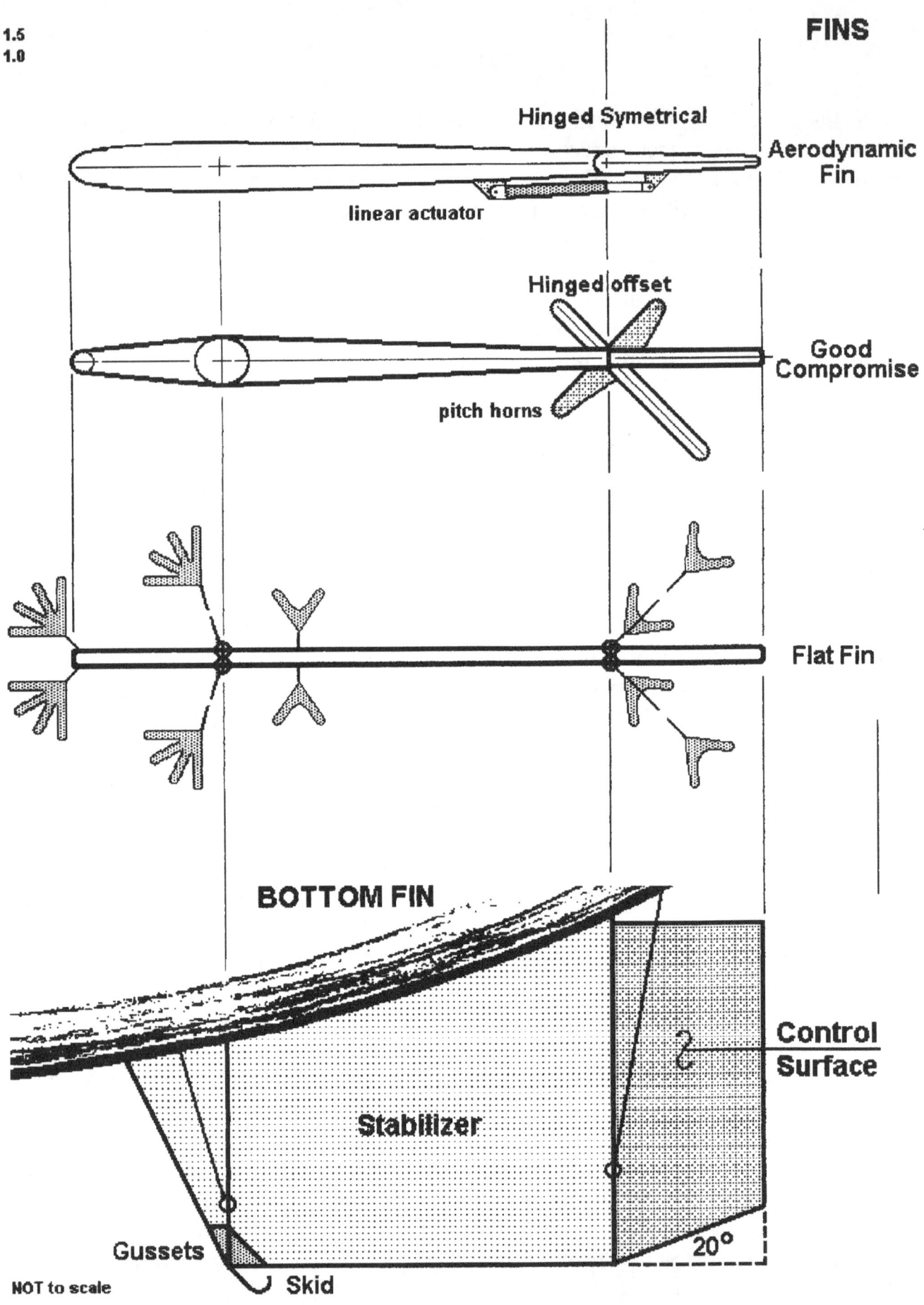

TAIL FINS cont.
R:12/04/97

<u>Concepts,</u> basic & simplified
1. Consider a fin as two sub-assemblies, the stabilizer as the rigid part, and the control surface as the movable part.
2. The rigid part dampens roll, pitch, and yaw oscillations (mushing).
3. The moveable parts (rudder & elevator) point the nose (steering).
4. Effectively determined by a smooth air flow over an adequate fin area, and control response dependent on airspeed over adequate control surfaces.

There are many exotic formulas for calculating the optimum area of fins, and enough variables to dazzle even an Einstein. But in real life, nothing is perfect. Especially a small blimp putting along at 30 mph. So lets use a rule of thumb for the <u>total</u> fin area:

$$AF = V * AR * .003$$

where AF=Fin Area; V=Volume; AR=Aspect ratio (length / diameter).
sample: AF = 30,000 * 3.8 * 3% = 342; Control surface is 33% of that = 93.2 sqft.
   AF = 30,000 * 2.8 * 3% = 252;   "   "   "   = 84.0

Note: Fin Area is not set in stone; plus or minus 1% will be only slightly noticeable.

Kinds: Aerodynamic; airfoil shaped like a wing section.
   Best response, common on larger airships; **Very heavy**.
   Non aerodynamic; Simple/flat, one thickness; good on small airships.
   Prone to flutter at speed.

Styles: Cruciform; like a cross +, horizontal and vertical stabilizers.
   Most common, easiest to rig. **Lower fin easily damaged**.
   X-tail; Good design, rarely damaged; **Hard to rig**.
   Inverted-Y & V: Good basic design, rarely damaged; Not too hard to rig.
   Total surface area same as X & +, but individual size and
   weight get very heavy.

Authors choice

Go with a flat cruciform on a 30,000 cuft. (30K) blimp. Apply some creative design to strengthen the bottom fin and reduce overall flutter. This is not an assembly to sub-contract. Nobody wants to do it anyway because it is a dirty/messy job, time consuming, and requires a certain amount of skill and attention to detail.

Critical Decision time

How are you going to make your control surfaces move? Mechanically or electrically? Whatever method (electrical or cable) you use, it had better be in your design before you start cutting metal. The old reliable is still cable and pulleys. But the new technology is "fly by wire" using linear or rotating actuators. Both methods have drawbacks. Cable gets slack at low envelope pressures, and electric things still go "pouf".

The actuators seem to be winning out though. It is easier to run a couple of electrical wires to an actuator than rig a complex pulley horn. However, this is only true on cruciform fins, as all of the others require complex mixing systems.

Cruciform vs. X-Tails

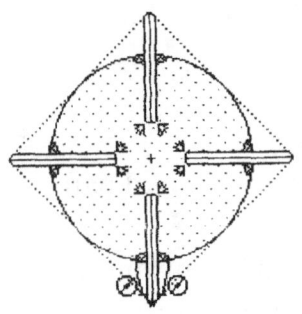
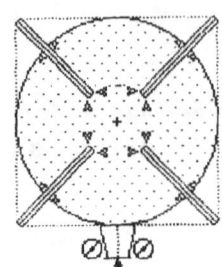

FIRST, Why do most airships use Cruciform-Tails (+)?
Because they are easiest to understand, rig, and assemble. The most effective aerodynamic control, for the least money. They have been around a long time, and will continue to be preferred as long as man continues to fly.

SECOND, Why would you consider an X-Tail (X)?
The greatest benefit is that it greatly reduces the chance of damage to the lower fin(s) when taking off or landing, especially when heavy. A hard landing can virtually destroy a lower "+" fin, and/or tear the restraints (finger patches) from the envelope.

THIRD, Lets consider some factors in selecting an X-Tail:
Efficiency & Effectively: The same mathematics apply to both "X" & "+" types, in straight & level flight AND in a turn & climb. If the fins are otherwise the same, the total amount of aerodynamic drag (and reaction) is the same in all cases.
Rigging of stabilizing fins: The positioning of the fins at 45 degrees to the major axis is not a big deal. The fine "tweaking" of alignment however, can be expected to take twice as long, due to the absence of the normal visual queues. Even a slight twist can be quite annoying in flight.
Rigging of control surfaces: One can understand that with offset stabilizing fins, that both a rudder and elevator are compromised, and now have to share equally in the changes to directional control. Using an all-cable system of input to output, a "mixing box" is required to make the changes proportional. Since the mixing box uses multiple pulleys and tension's, it is fairly complex. The introduction of hydraulic mixers (138-S) and pneumatic mixers (A40b) have produced some improvements, but has compromised redundancy, and thus safety. Proportional electric mixers have not had much success yet on airships, as of this writing. But I will be glad to rewrite this if/when it does*.

Summary:
Pilots view; I like "X" tails because the controls are much smoother, and don't require as much force. With a single "joy-stick", boosted mixers eliminate the need for pitch wheels and rudder pedals. Flys noticeably better, allowing for heavier running takeoffs.
Mechanics view; I don't like "X" tails because it is so hard to get to, and access the upper fins. The cable mixers allow for some redundant safety, but any hydraulic or pneumatic system line leaks cause an immediate compromise of safety.
Engineers view; Birds fly effectively with coordinated control surfaces (and no rudder). If your volume is sufficient for the added weight, "X" tails are a good marketing point.
FAA's view; We take a skeptical (nervous) view of non standard flight controls. But we are here to help you, no matter how long it takes, or how much it costs you.

---

* GA-42 had "+" tail and used all electric linear ball-actuators.

TAIL FINS cont.
R:12/04/97

Fin Criteria:
1. Keep them light but strong
2. Make them all the same (minor change req. on lower cruciform)
3. Make them easy to handle (but not fragile).
4. Make them in dismantible sections, no dimension exceeding 8'
   (for storage, handling, crating, and shipping).
5. Don't allow any part to rattle (fit loose)
6. Make sure the fins are attached to the envelope under tension
7. Install the guy wire cables with the fins under tension fore & aft.

Getting started:
Plan on using 1.5" x .060 round tubing, and 1.5" x 3/4" rectangular tubing of 6061-T6 aluminum. Make fixture(s) sufficient to tack weld four duplicate stabilizers, and 4 duplicate control surfaces. Fixtures are needed if you expect any semblance of quality. The lower vertical stabilizer needs a gusset in the leading edge, and the lower rudder needs to be clipped 30 degrees to the trailing edge.

Control surfaces: As a passing thought in cruciform fins, it may be of interest to the builder (and the pilot) that an UPPER rudder is not really needed. Stabilizer yes, it help in preventing mushing (nose constantly moving off center). But a single LOWER rudder is about as effective as two. Your turn radius may be greater, but you'll never notice it.

Mixers: for multiple and angled control surfaces are not practical for small blimps at this time. Present technology seems limited to the addition of a hydraulic system. It is only a matter of time now that someone will market an electrical (black box) one that will integrate the actuators.

Building Small Gas Blimps

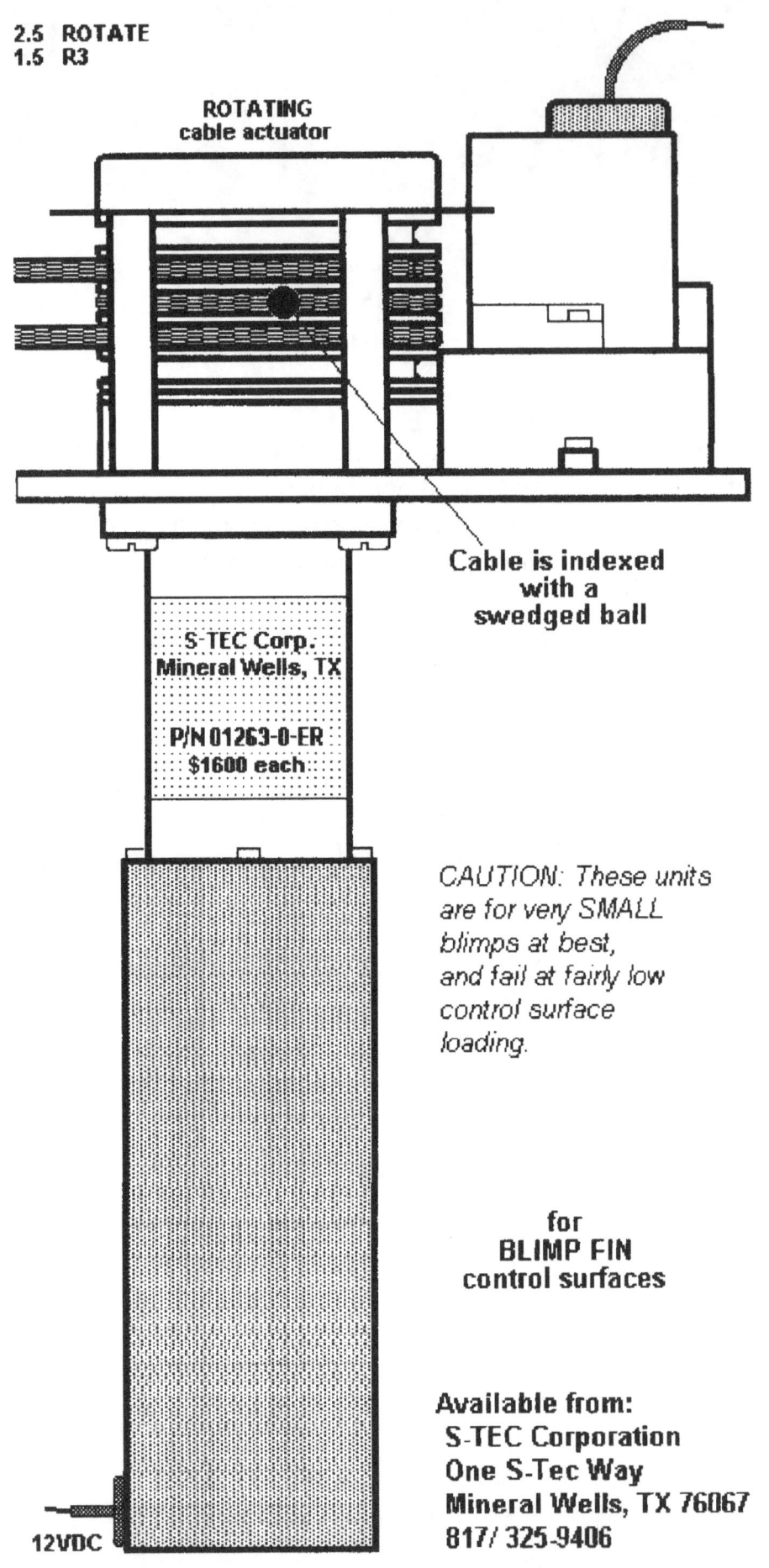

Building Small Gas Blimps

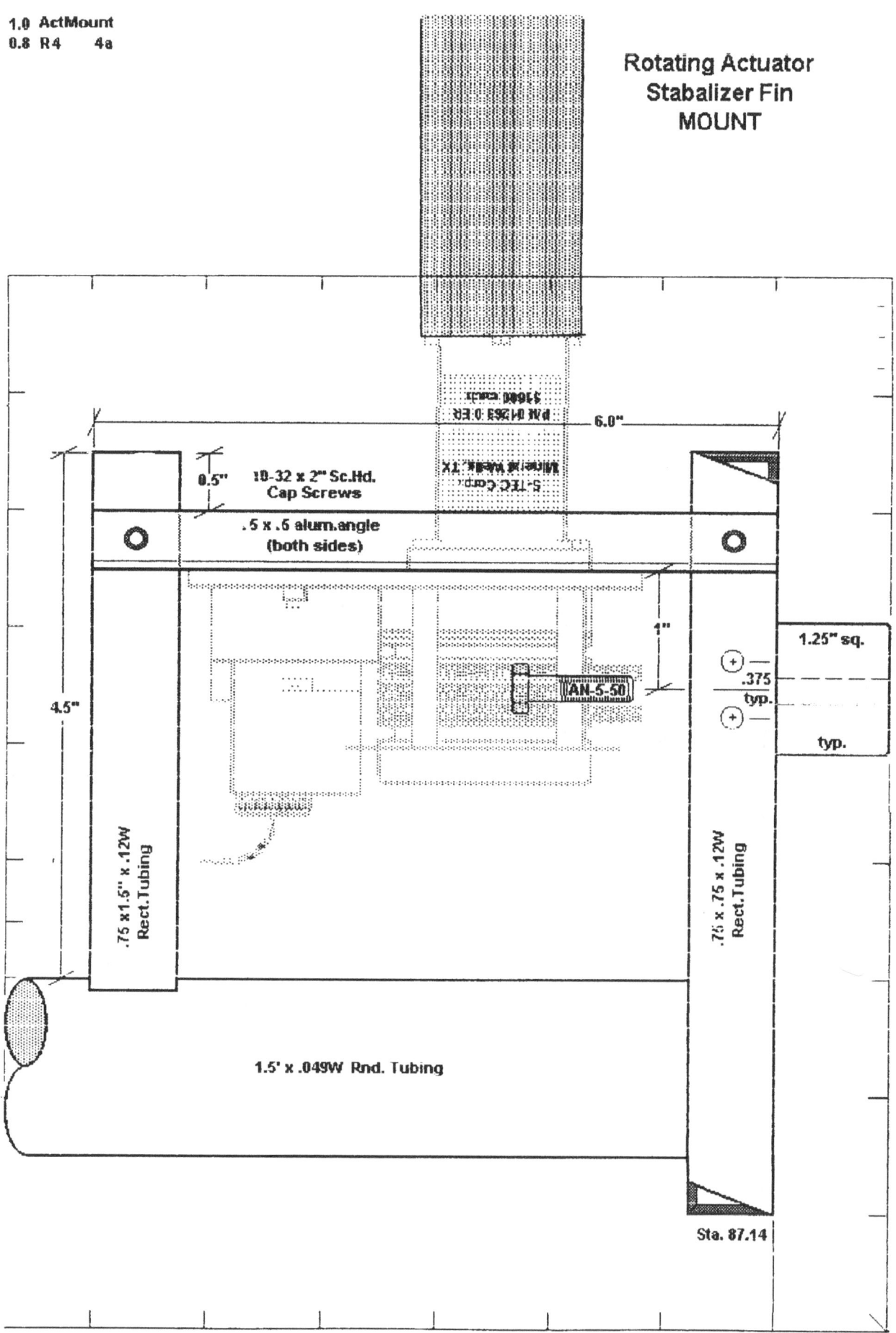

Rotating Actuator Stabalizer Fin MOUNT

107

Building Small Gas Blimps

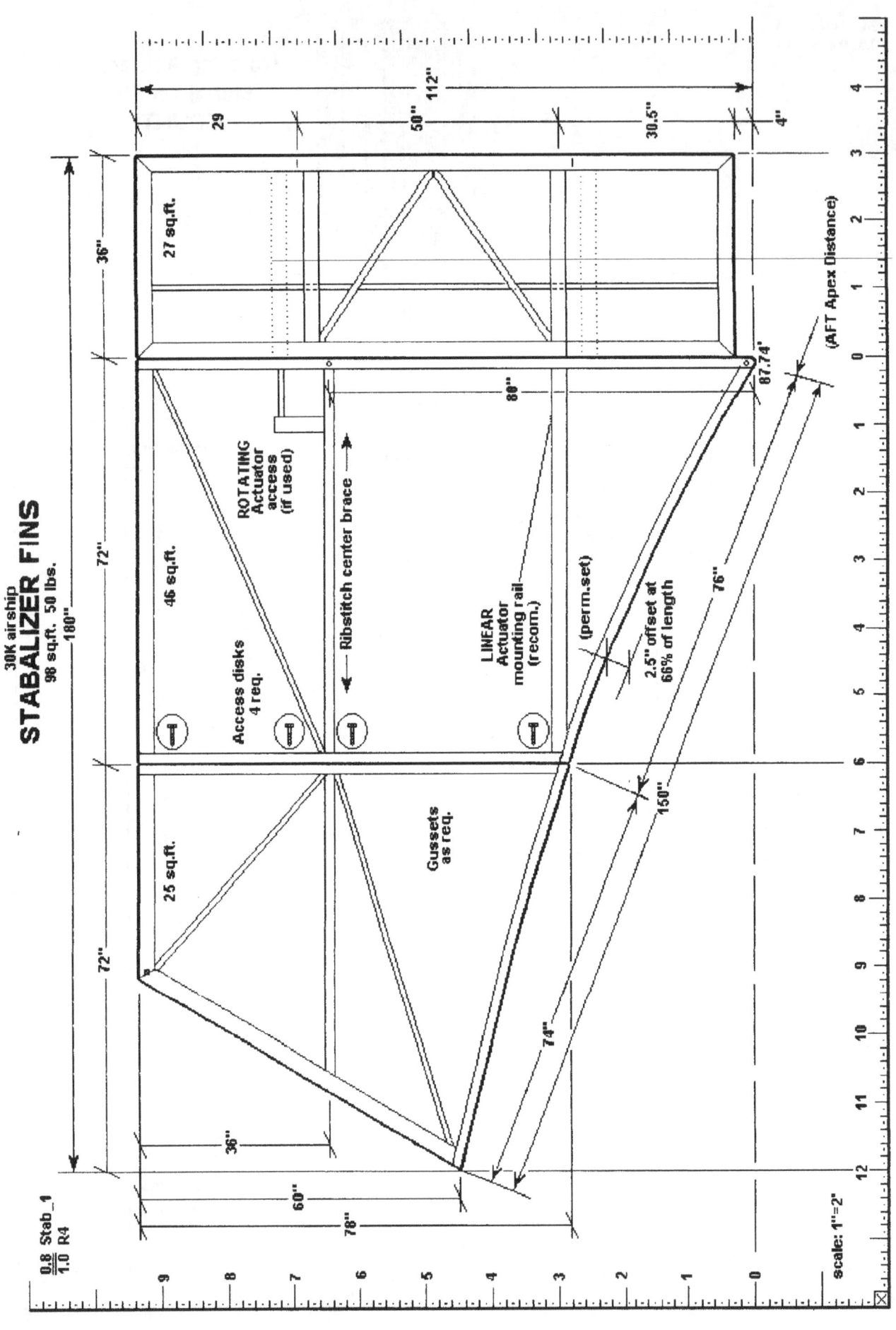

108

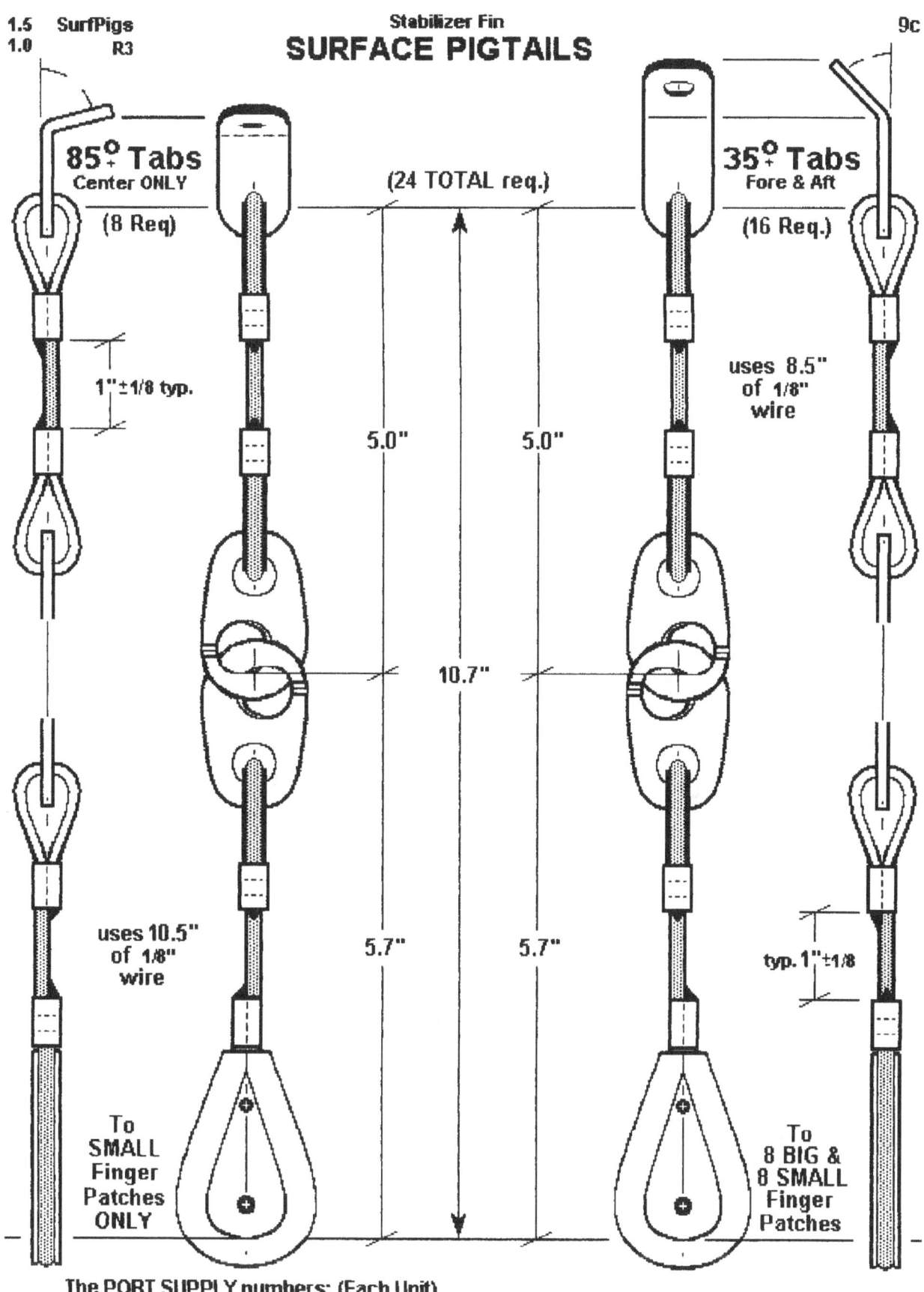

Building Small Gas Blimps

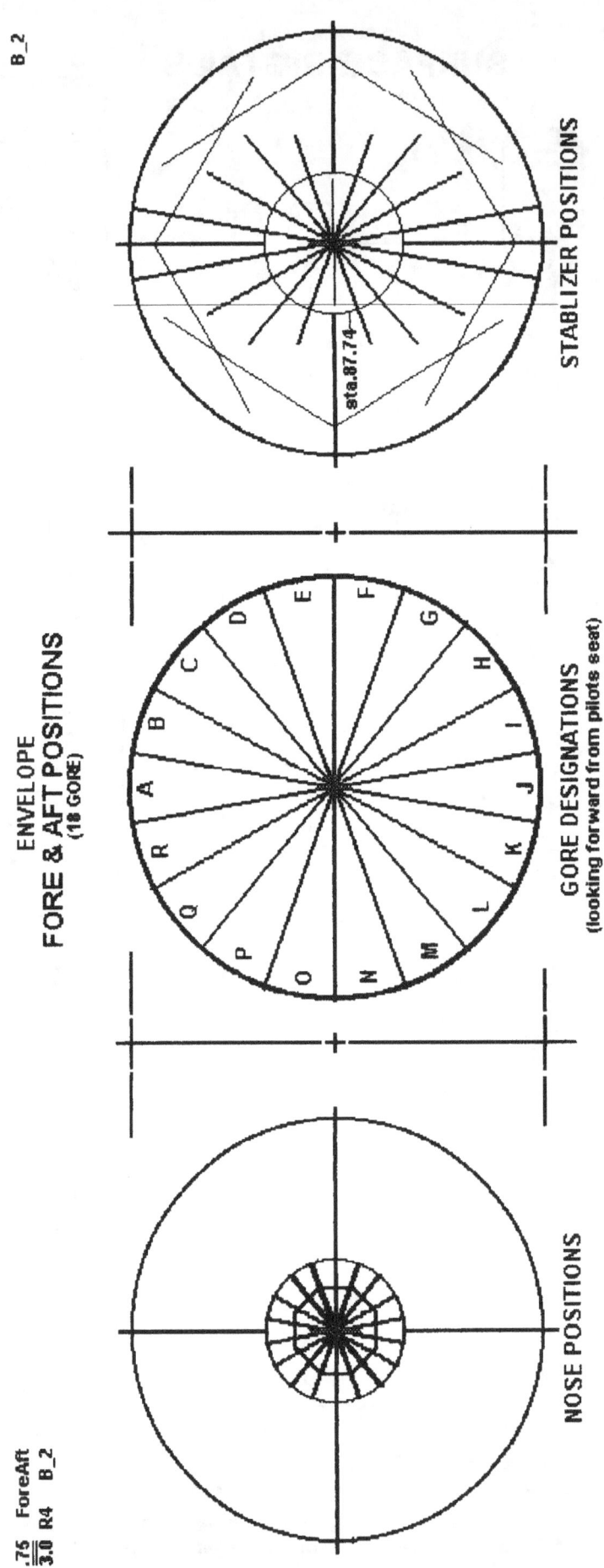

Building Small Gas Blimps

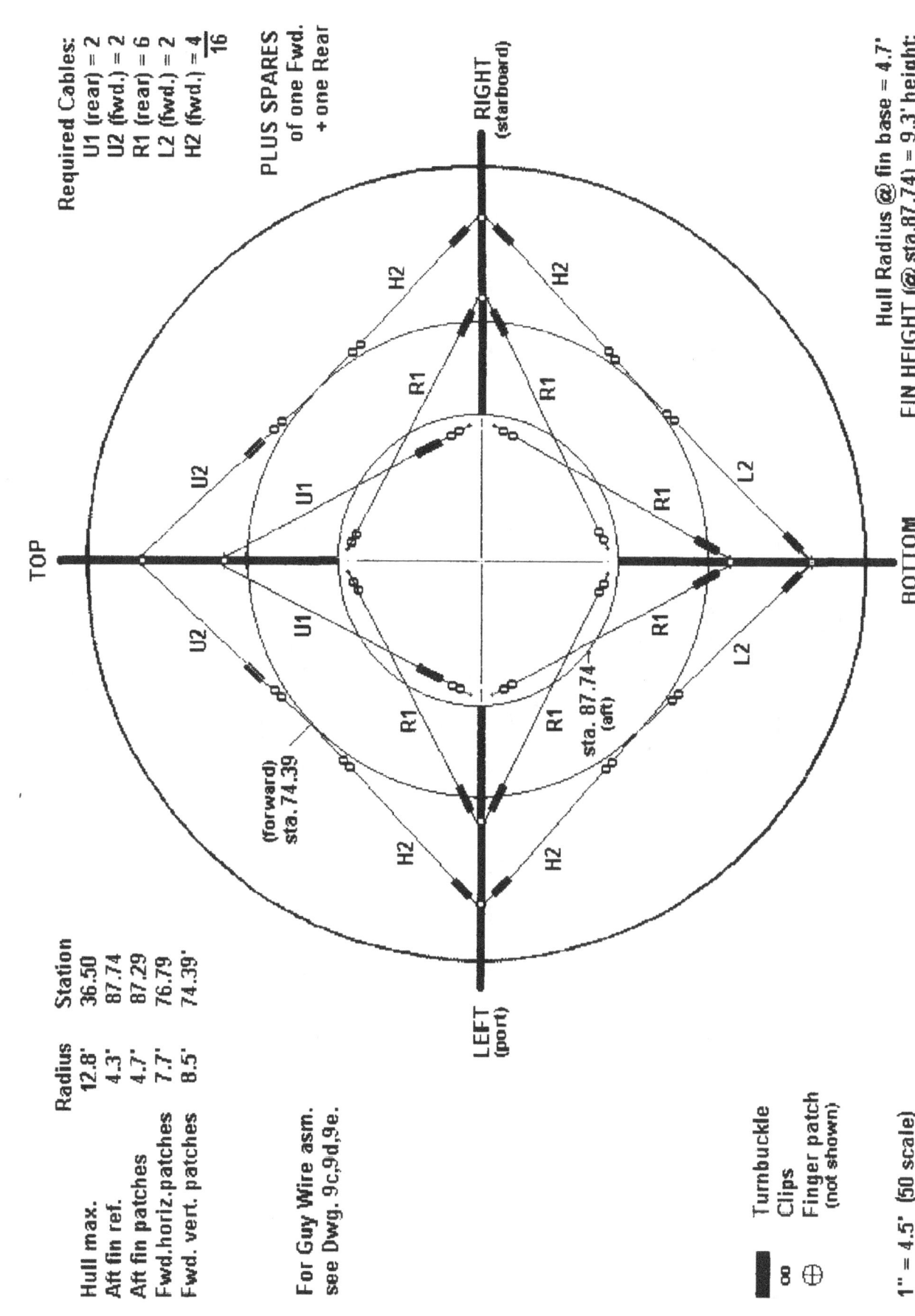

111

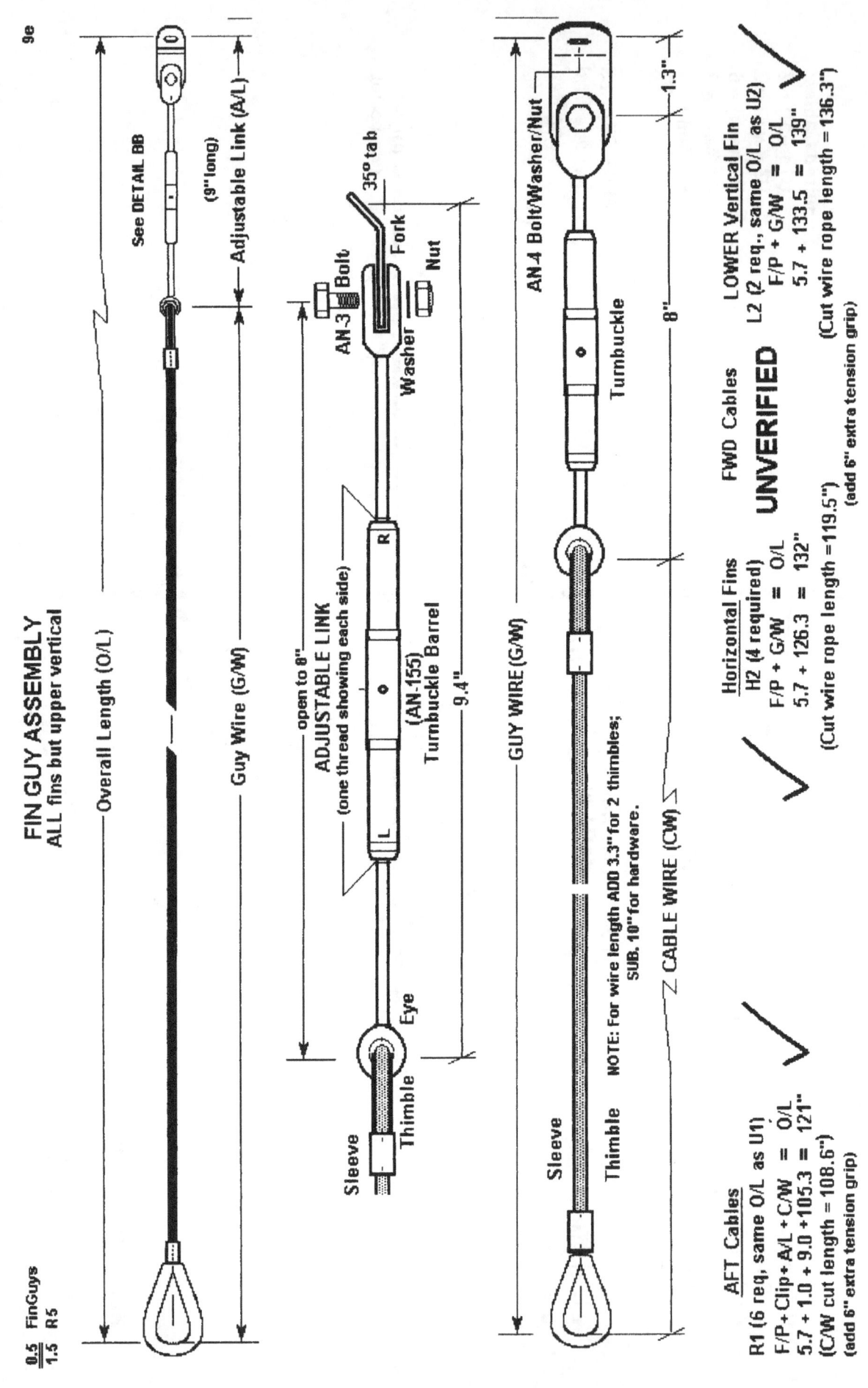

Building Small Gas Blimps

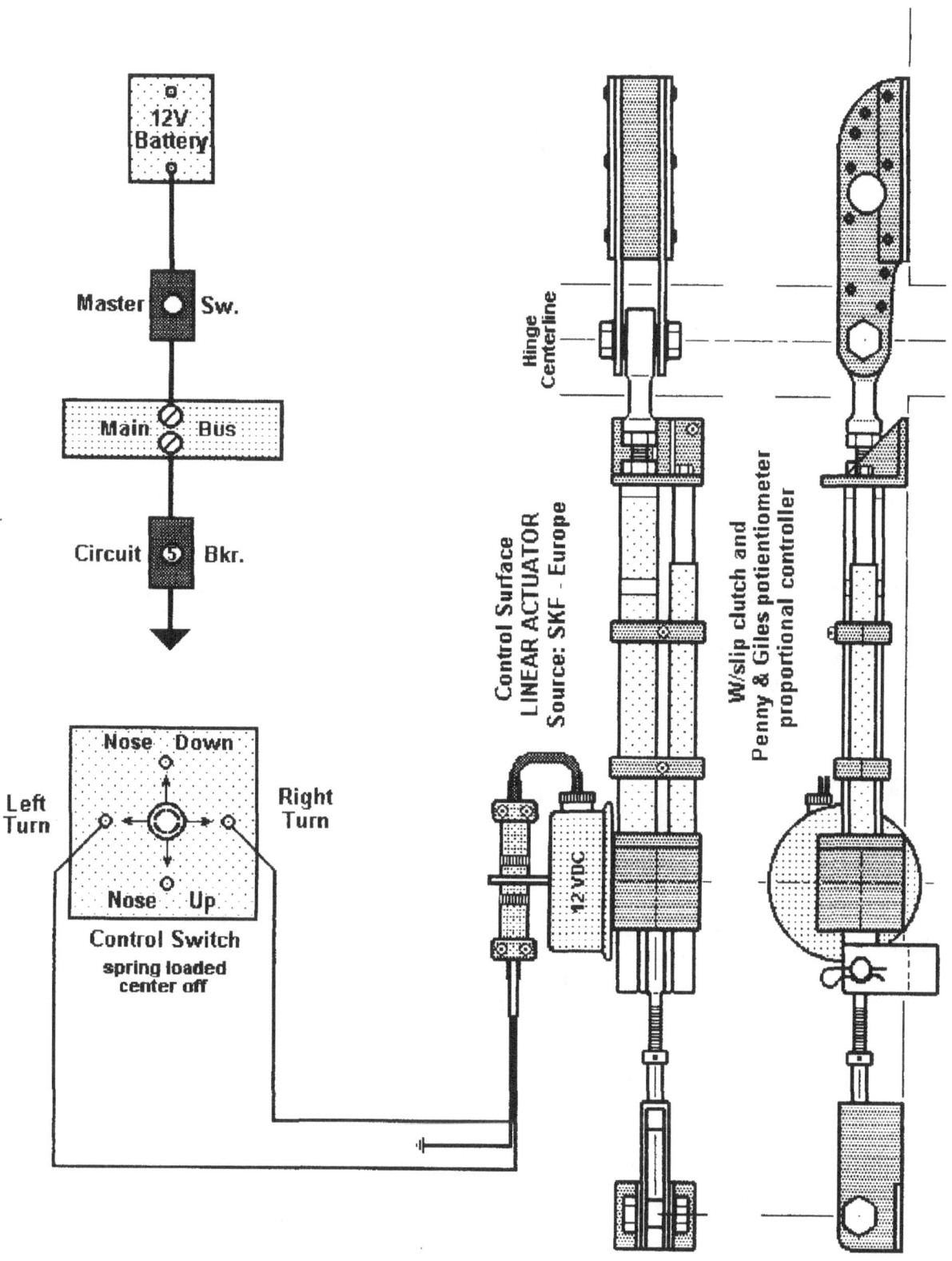

Building Small Gas Blimps

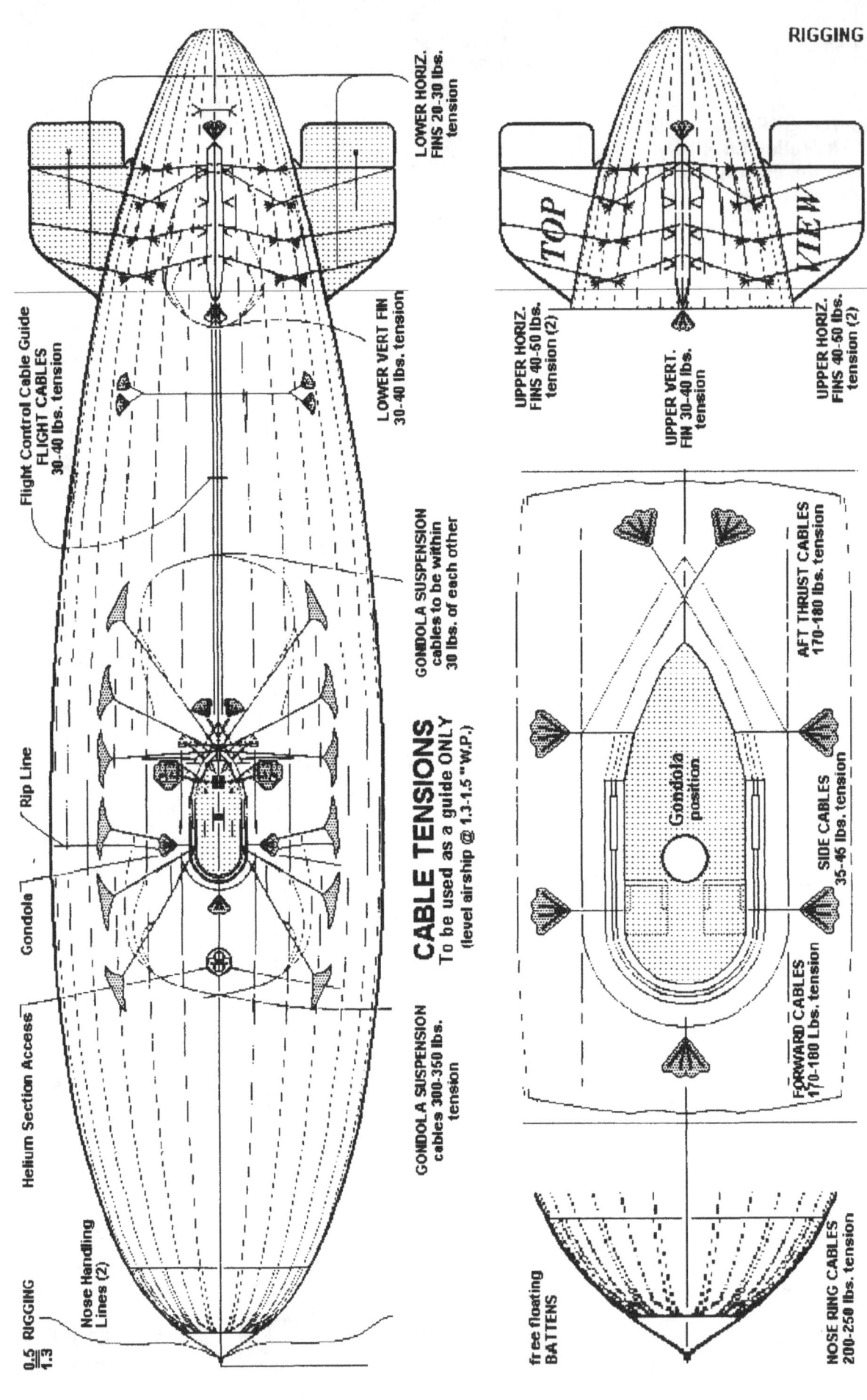

114

Building Small Gas Blimps

Technical Data Ball Drive Actuator 85615/85616

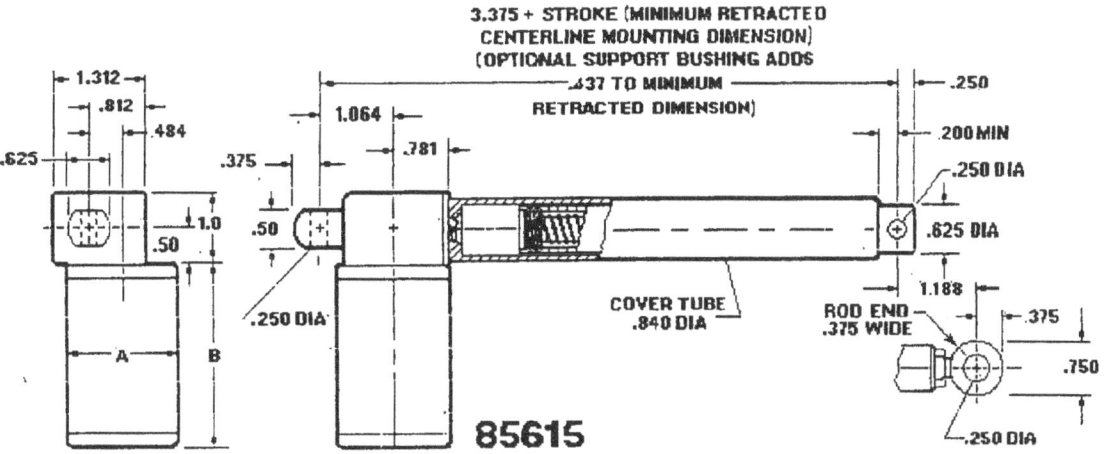

Load Capacity	100 lbs. (dynamic) 600 lbs. (static) Note: 100 lb. load rating can be extended in certain applications.			
Stroke	1, 2, 3, 4, 5, 6, 7, 8, 9, 10, 11 and 12 inches are stock units; however any stroke may be specified from 1.000 to 12.000 inches with no limitation on interim dimensions.			
Stroke Speed	DC Motors: Gear Ratio 12,24 VDC (PM) 7500 RPM Note: Stroke speeds are in./sec. with no load speed shown first and 100 lb. load speed shown second.	7.5:1 2.00/1.60	15:1 1.00/.80	30:1 .50/.40
Gear Ratios	7.5:1, 15:1, 30:1 are standard ratios.			
Motors	DC Motors: 1) 12 VDC Permanent Magnet Brush, 1.6 diameter, 7500 RPM (no load), Intermittent duty cycle (60%), 2 wire, enclosed construction, .4 amps no load, 3.4 amps at 100 lbs. with 15:1 ratio.	2) 24 VDC Permanent Magnet Brush, 1.6 diameter, 7500 RPM (no load), Intermittent duty cycle (60%), 2 wire, enclosed construction, .2 amps no load, 1.7 amps at 100 lbs. with 15:1 ratio.		
Ball Drive	Epicyclic Ball Screw 85206 with integral freewheeling at ends of stroke. Screw OD: .331 Screw root dia.: .280 Advancement/rev.: .125 Centerline pin dimension: .927 + stroke Nut OD: .625	Nut mounting: 9/16-18 thread. Materials: heat treated alloy steel with bearing races Rockwell 55C. Efficiency: 90%		
Gear Reducer	Single stage worm gear type with 626 ball bearing to accommodate ball screw thrust. Housing: die cast aluminum 380 alloy. Bearings: Ball Drive shaft supported on sleeve bearing and ball bearing.	Worm and gear materials: heat treated steel. Lubrication: permanently lubricated with SP500 synthetic grease.		
Mounting Dimensions	85615: minimum retracted centerline mounting dimension is 3.375 plus stroke. For example: a 3.500 stroke unit will measure 6.875 (3.375 + 3.500) in the retracted position and 10.375 (6.875 + 3.500) in the extended position. A retracted dimension can be specified greater than the minimum. For example: a 3.500 stroke unit could be 12.000 in the retracted position and 15.500 (12.000 + 3.500) in the extended position. Design options add to the minimum retracted dimension: Support bushing: .437 Rod end: 1.188	85616: minimum retracted centerline gearbox mounting hole to nut shoulder is 2.875. For example: a 3.500 stroke unit will measure 2.875 in the retracted position and 6.375 (2.875 + 3.500) in the extended position. A retracted dimension can be specified greater than the minimum. When the adapter for trunnion mounting is added to the nut, the minimum retracted centerline gearbox mounting hole to the centerline of the .190 holes is 3.062 (2.875 + .187). When the alternate nut is reversed, this dimension will be 2.201.		
Weight	85615: 1.3 lbs. (6 in. stroke)	85616: 1.0 lbs. (6 in. stroke)		

Building Small Gas Blimps

RIGGING

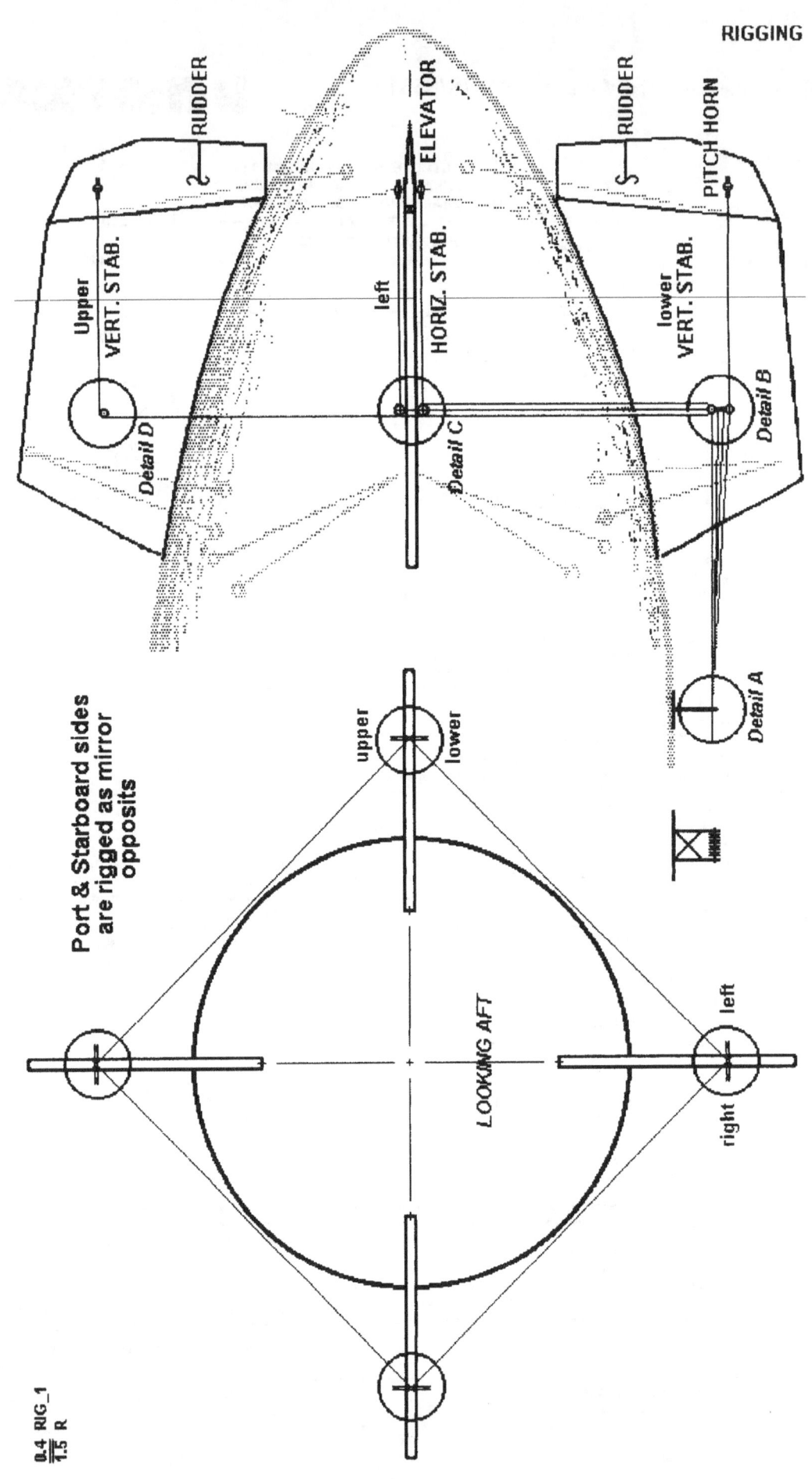

116

Building Small Gas Blimps

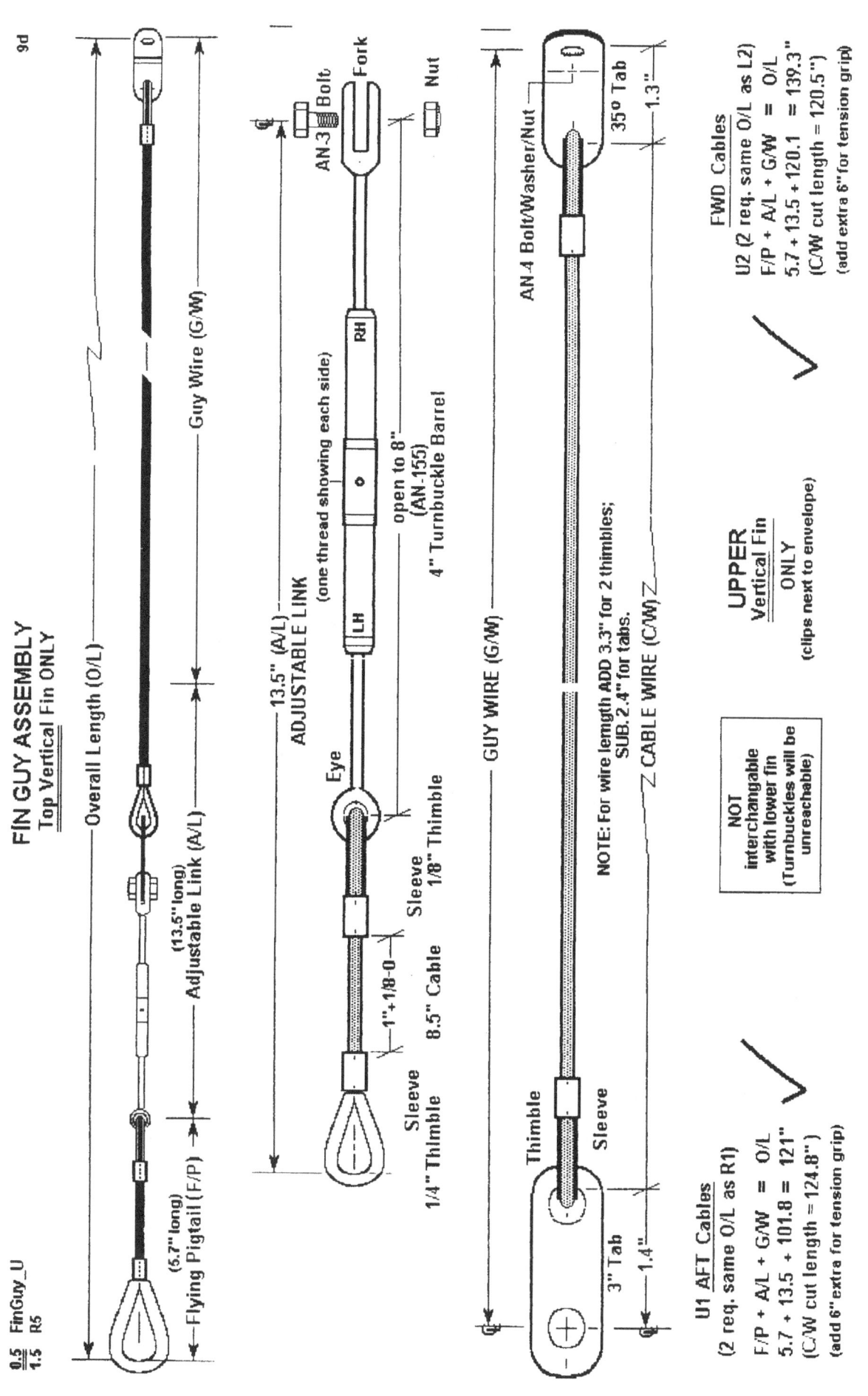

117

Building Small Gas Blimps

CABLE GUIDE
Details

RIGGING

1.5 RIG_2
0.7 R

NOTE: In sun and rain exposed fin areas, use Sealed Bearing, Aluminum Pulleys MS20220A-1 shown.

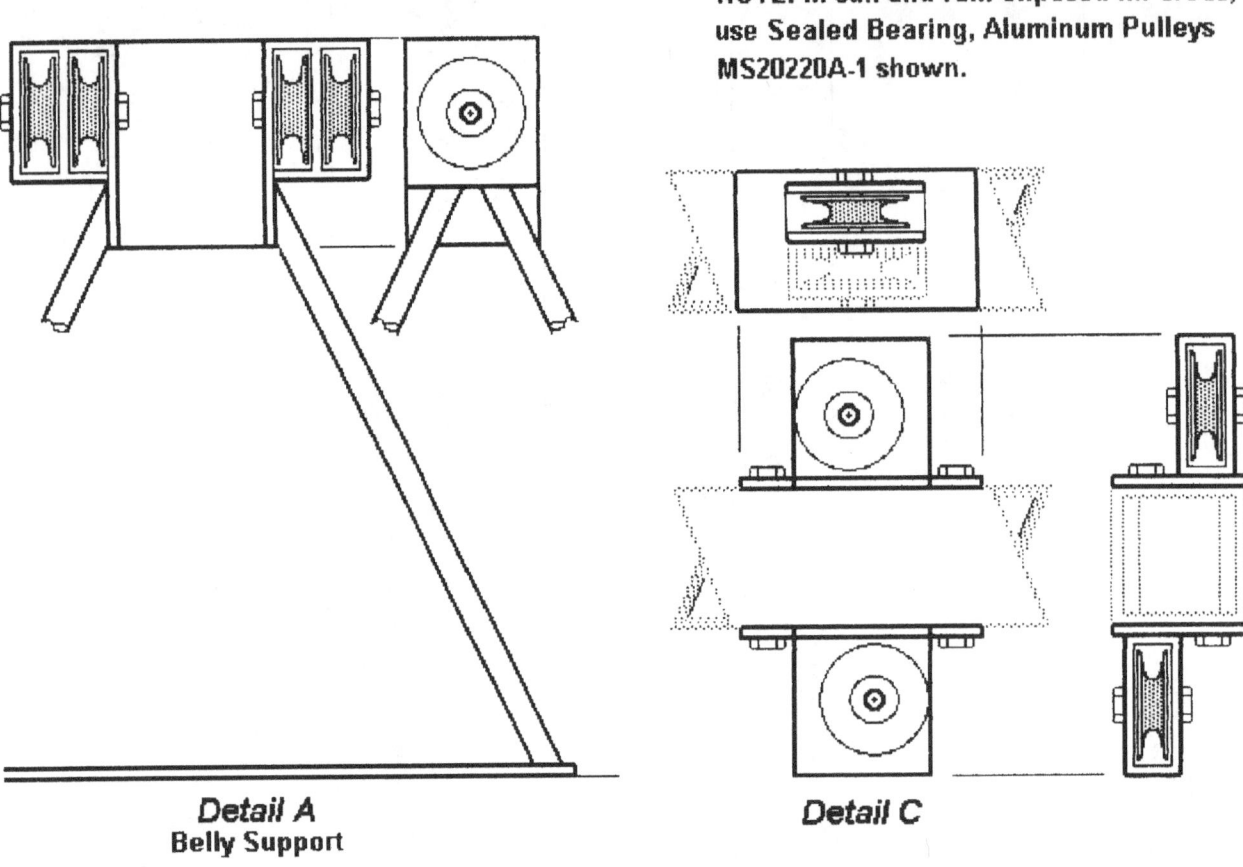

Detail A
Belly Support

Detail C

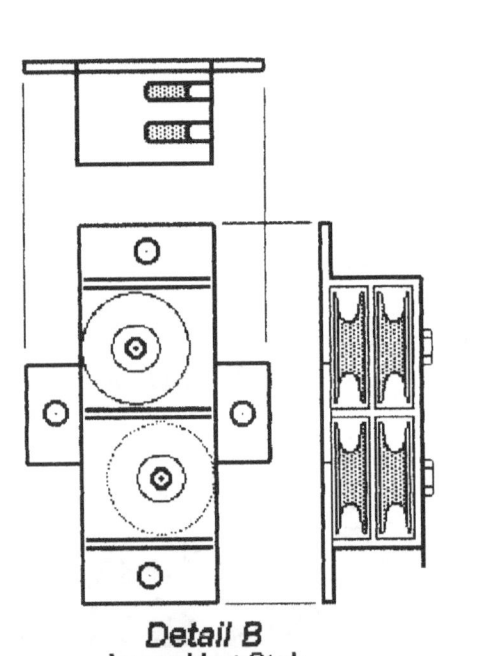

Detail B
lower Vert. Stab.

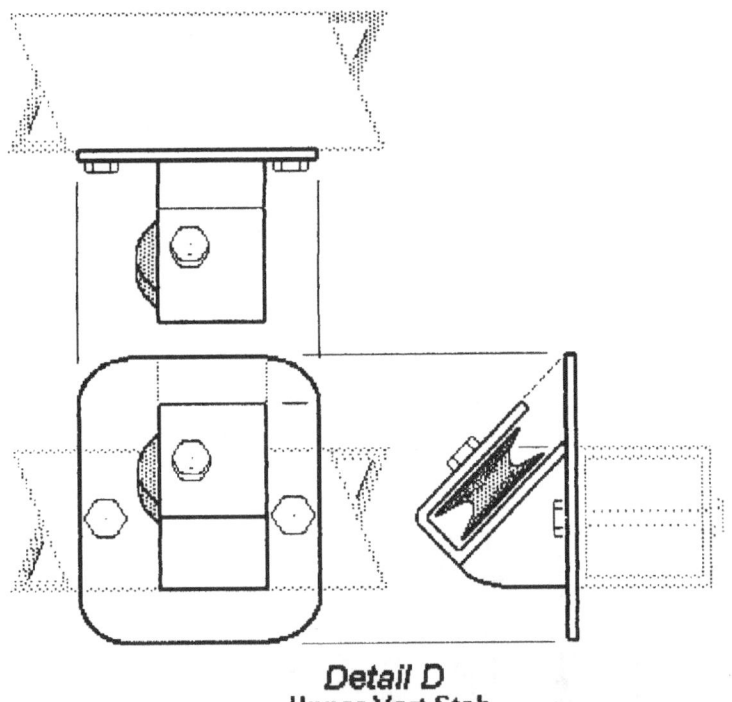

Detail D
Upper Vert. Stab.

INVERTED-Y RIGGING
R: 12/10/87

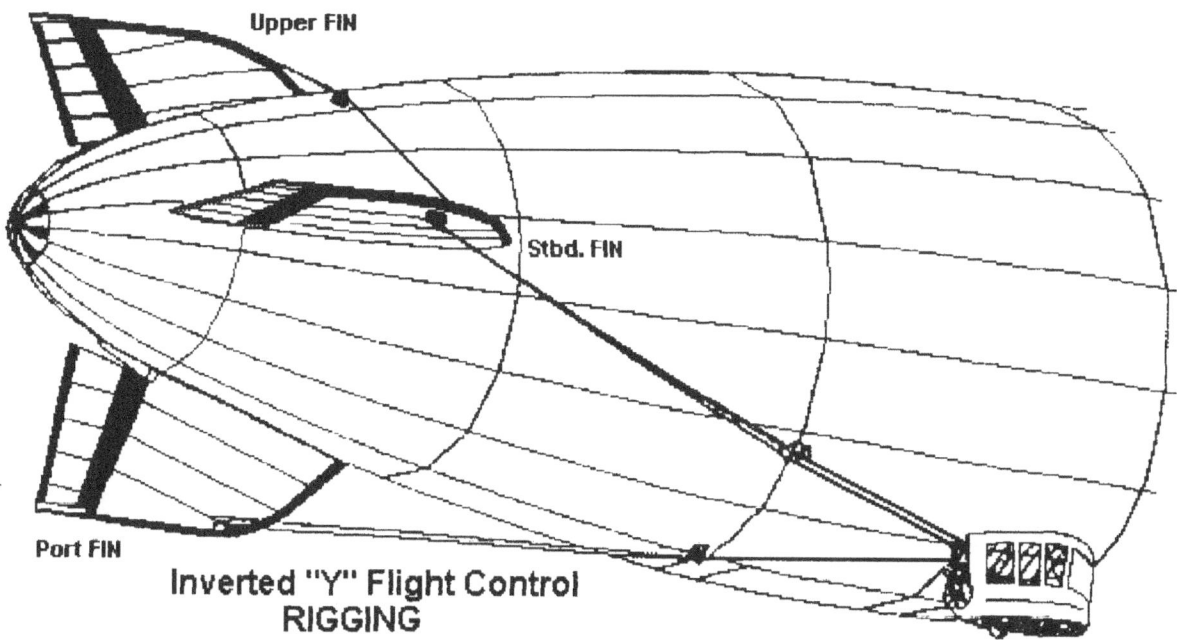

Inverted "Y" Flight Control RIGGING

On the 138-S, flight control is from a single stick mounted in the floor in front of the Captains seat. This conventional stick is connected under the floor by push-rods to a hydraulic boost system, a mechanical mixer, and out to the 3 control surfaces by stranded steel cable. The the event of a hydraulic failure, the system can be overridden by the pilot (with considerable force, Ed.). * Grace 138-S shown.

The external portion of the system from the gondola to the respective control surface is routed asymmetric. The two starboard (stbd.) ruddervator and rudder control cables run aft along the side of the envelope, through pulley standoffs, through the fin, and to the rudder pitch horn. Lateral movement of the stick moves only the rudder surface. Fore & aft movement moves only the ruddervators. Diagonal movement moves all three control surfaces.

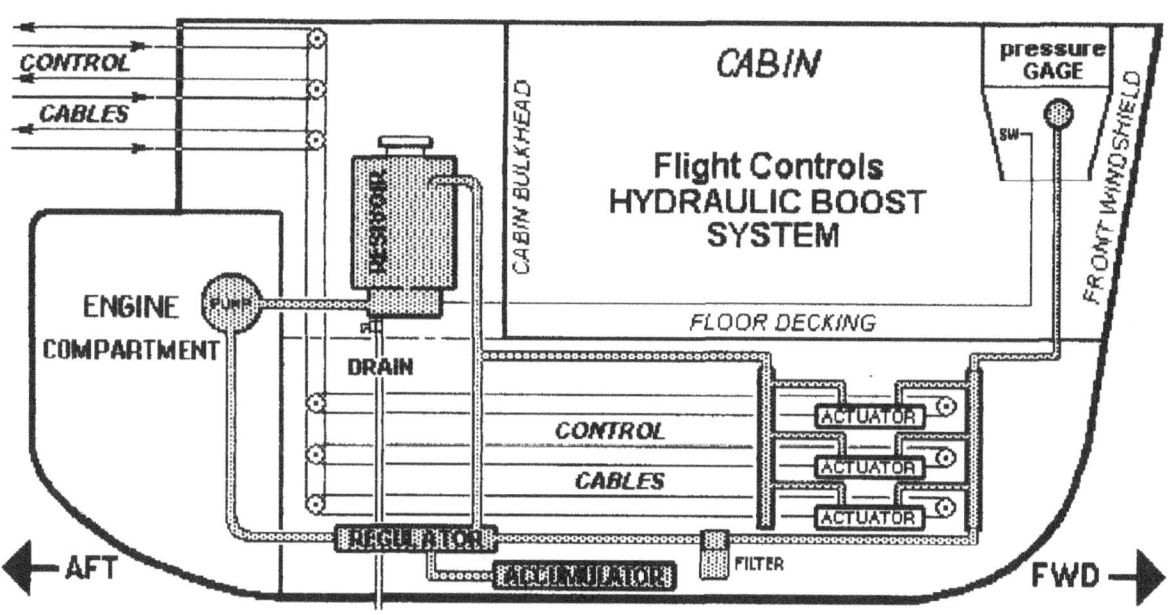

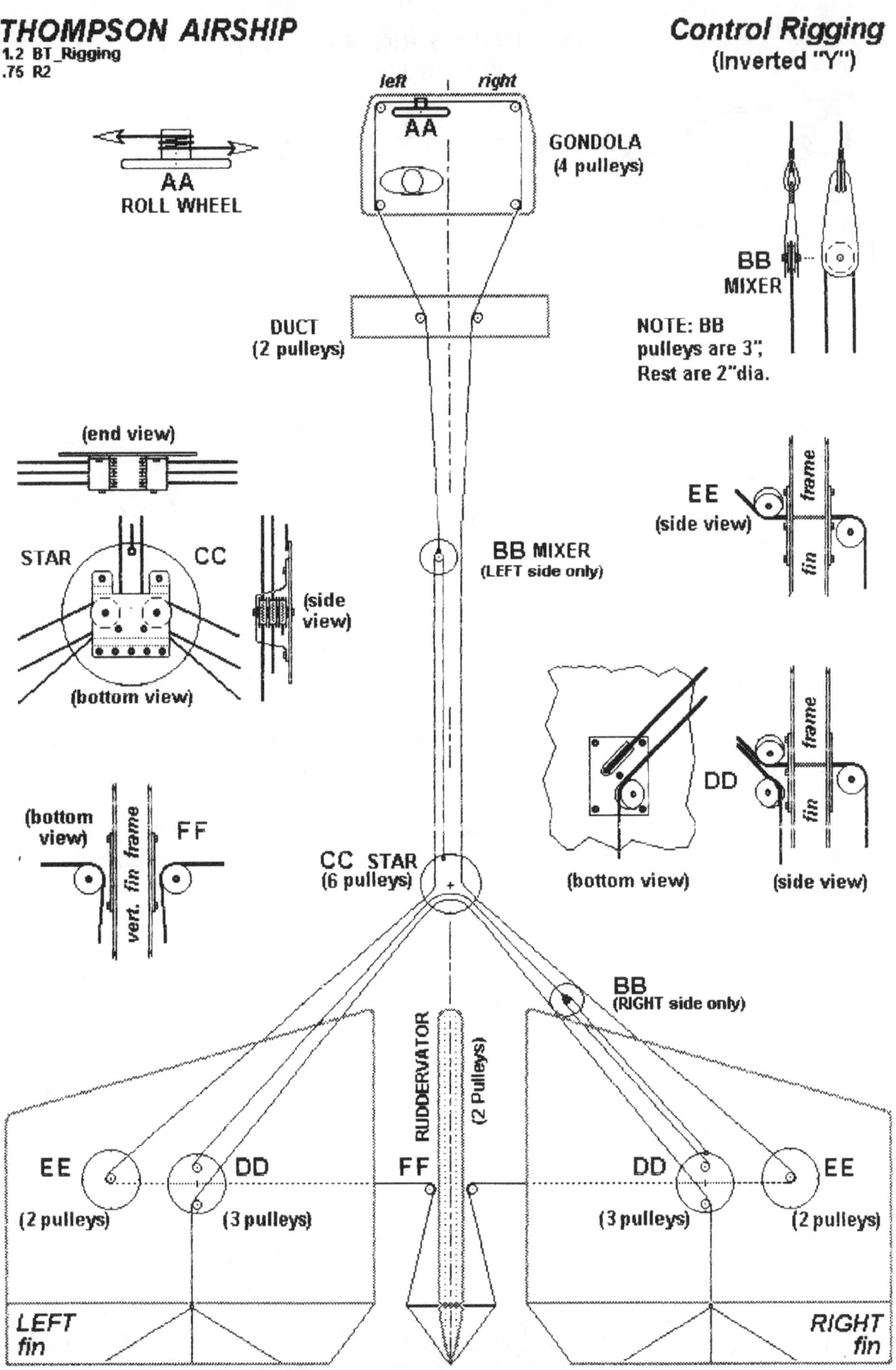

Building Small Gas Blimps

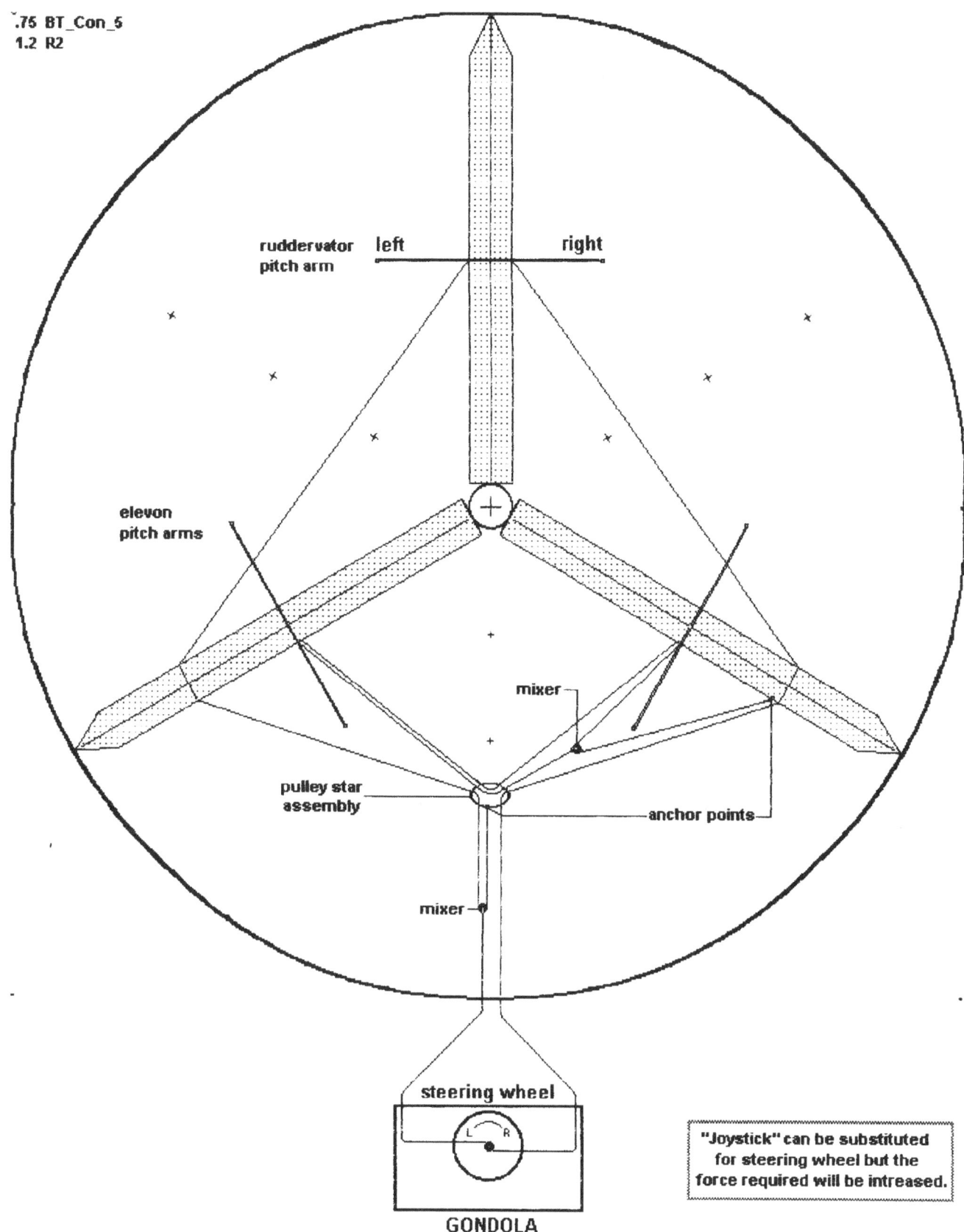

Thompson "inverted Y" Control Rigging
(looking forward)

Building Small Gas Blimps

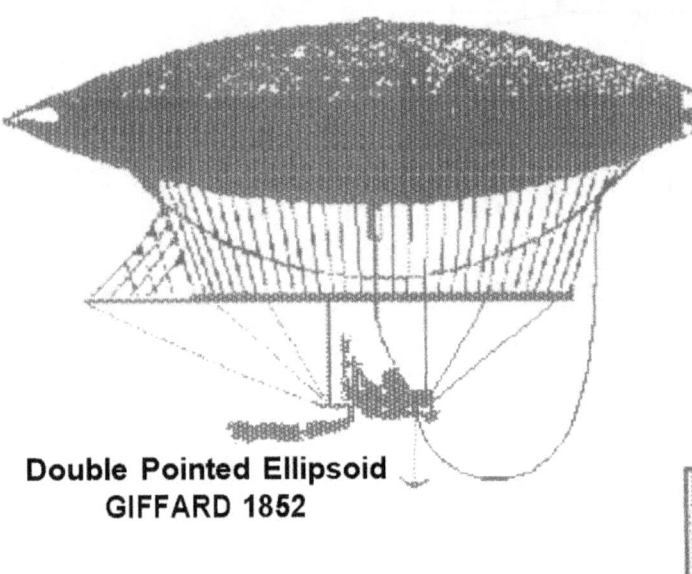

Double Pointed Ellipsoid
GIFFARD 1852

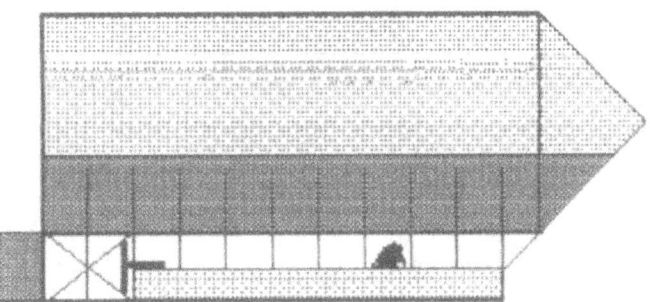

Pointed Cylinder
SCHWARTZ 1897

ENVELOPES

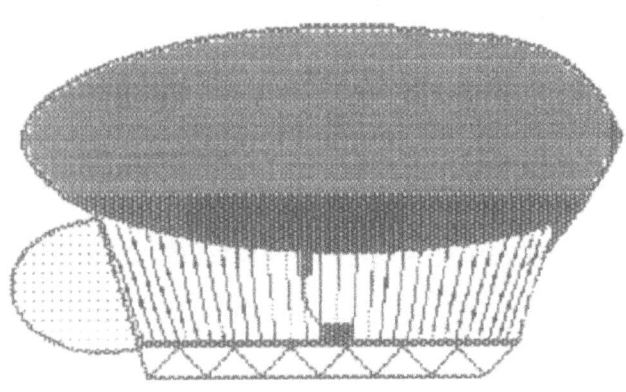

Pointed Ellipsoid
SANTOS-DUMONT 1903

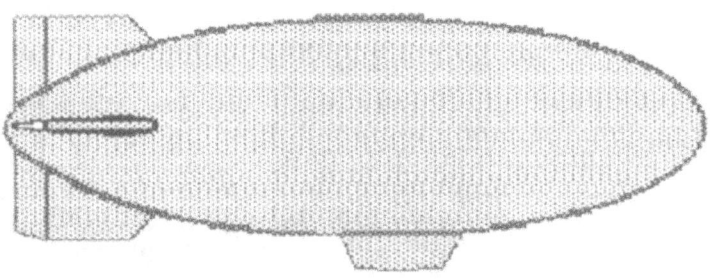

Airfoil Shaped
U.S. NAVY 1916

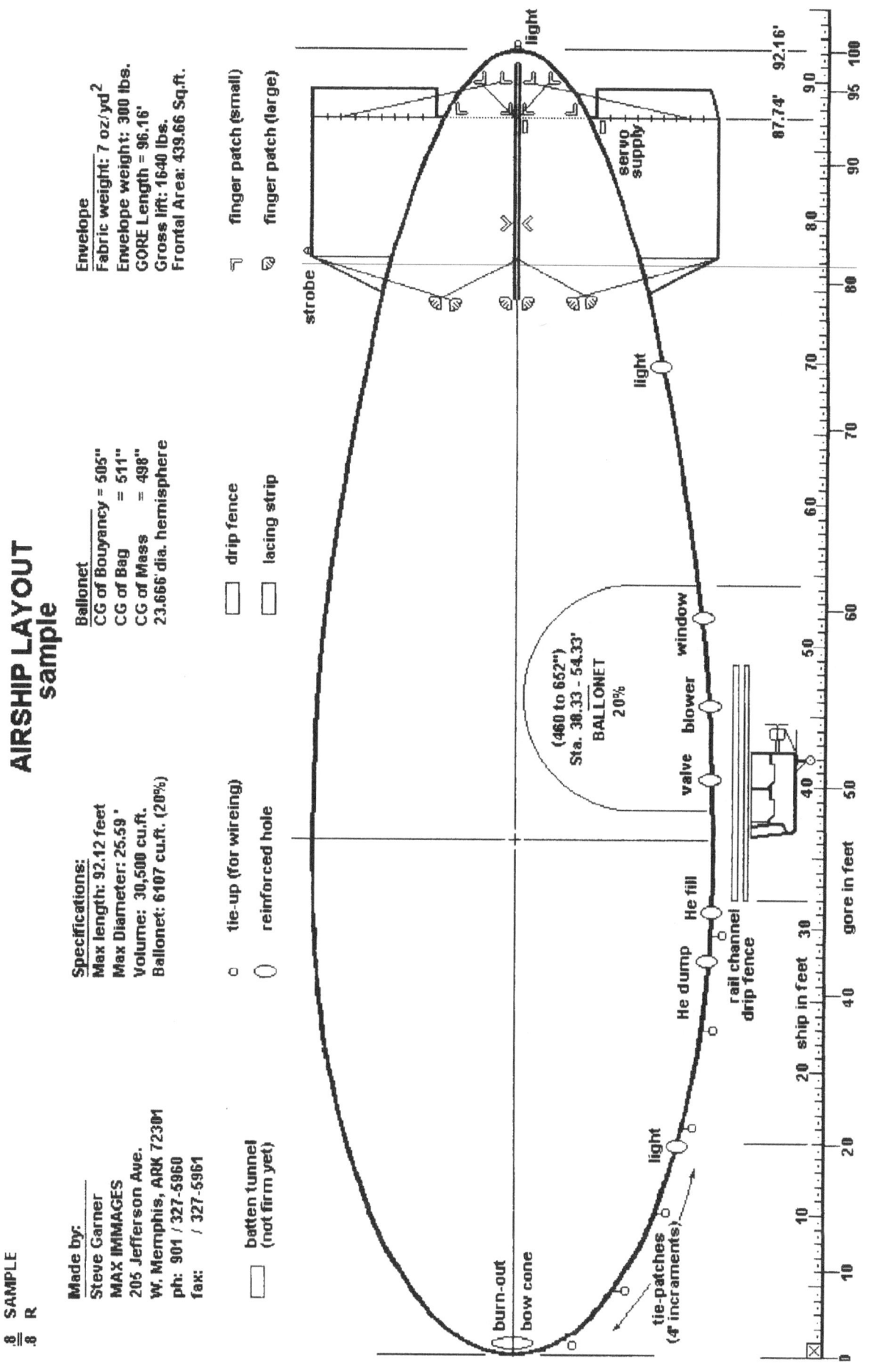

Building Small Gas Blimps

ENVELOPES
R: 5/10/96

Now comes the "big bug-a-boo". Hopefully, by now, you have done your homework and completed everything else. I doubt it, but its your mistake, and it will show up in the end.

GETTING STARTED: By now you've heard a lot of whispers on this fabric, or that fabric, and all the great deals to be had. Let me be the first to tell you, "don't". Stop now and think about this. Once you put the scissors to the fabric, you take a deep breadth and say to yourself: My God, what have I gotten into? So lets talk about a reasonable alternative:

BUYING A COMPLETED ENVELOPE: Yes they are available, and YES a good one can be made to your dimensions, at a reasonable price. When finished, it comes to you in a box, saving you a dirty, smelly, ickey job. So lets consider the fast track to getting you into the air. If you have made all of the supporting hardware, all you have to do is supply dimensions of where you want the Finger Patches, valves, access ports, scuff patches, and reinforcements. From experience, let me recommend the 3 best sources:

1. **MAX Image;** Memphis, AK;
2. RAVEN-Aerostar; Sioux Falls, SD, and
3. ILC-Dover; Fredricksberg, MD.

They are all good, but I put them in order for a reason. Maximage will sit down with you and actually refine your design from experience as a builder of small blimps. They are very competitive and their expertise will be of great benefit to you in the end. Raven builds a great quality, and has a great track record. They are competitive, but have no experience with the finished product. They will make EXACTLY what you specify, but never question possible errors. ILC is like Raven, but NOT competitive in pricing. The last two are strictly "cash & carry", engineering not included.

OK, so you still want the experience of building it yourself (it's really a 2-man job), so
Go For It!

SOME PRACTICAL ADVICE: Yes, there is no substitute for good planning. But then there is no substitute for experience either. Since it is not likely that you will ever have an assembly line of blimp envelopes, the absolute best positions for some of the attachments will never be obvious until the finished airship has been in use for a year. So lets treat this envelope as the ONLY one you'll ever make. Just make it as an airtight sphere (ellipsoid), with only one access port that you can crawl through. Do not even think about adding finger patches, scuff pads, or valve holes at this time. After the basic envelope is finished AND tested, THEN install the ballonet(s). The rest of the stuff can be added later.

Weather you stick to this plan or not, somewhere down the production sequence, all of the accessories will have to be installed with some precision. Start early, and mark on "tic" marks in 12" increments down the critical gores. Do it down the whole gore that you want as the bottom (with a fine tip permanent marker). Do the first and last 20' of all the gores with a #1 pencil. These marks don't have to be giant; just big enough for you to identify where you are at when it comes time to install the nose & fin patches.

Building Small Gas Blimps

ENVELOPE cont.
R: 12/02/97

So maybe you <u>don't</u> want a 3.8:1 x 30,000 cu.ft. envelope, but want something bigger or smaller. Well, since an airship shape is an spheroid, crank your desired volume into the standard formula: $V = 4.1888 * \text{Length} * \text{Diameter}$

Before you start transposing that formula, pick a fineness ratio (i.e. 3.8, or ?) and substitute: $D = L / 3.8$ then **$L = (4.1888 * D * 3.8) / V$**

Other deravations that may be of help are:

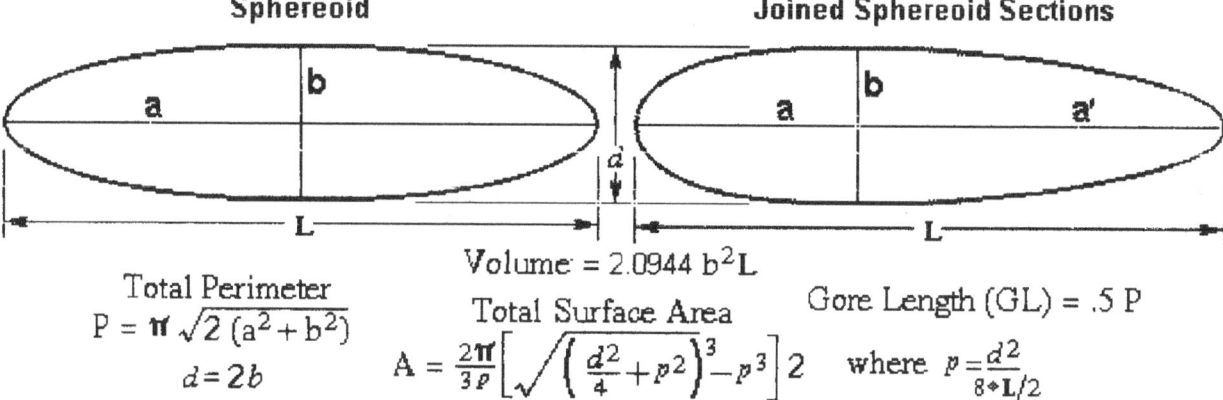

Spheroid **Joined Sphereoid Sections**

Volume = $2.0944 \, b^2 L$

Total Perimeter
$P = \pi \sqrt{2(a^2 + b^2)}$
$d = 2b$

Total Surface Area
$A = \frac{2\pi}{3p}\left[\sqrt{\left(\frac{d^2}{4} + p^2\right)^3} - p^3\right] 2$ where $p = \frac{d^2}{8*L/2}$

Gore Length (GL) = .5 P

On a **Joined Sphereoid**, it is generally assumed that $a = 1/3 \, L$.

The best profile is the one used by the U.S.Navy. It has a continuous curve over the entire length. This produces a minimum of drag from airflow separation. Here are the magic numbers that were used fron 1942 on. The big trick is to set it up in your computor, and make it spit out the numbers you need.

$$V = 0.64381 \, \pi R^2 L \quad \text{and} \quad 0.16095 \, \pi D^2 L$$
$$L = 1.25522 \, V^{1/3} (L/D)^{2/3} \sim V^{1/3} (D/L)^{2/3}$$
$$D = 1.25522 \, V^{1/3} (D/L)^{1/3} \sim V^{1/3} (D/L)^{1/3}$$
$$S = 3.80243 \, V^{2/3} (L/D)^{1/3}$$

$$y = \frac{D}{2}\sqrt{1-\left(\frac{x}{L/2}-1\right)^2} \, \left(1.19072 - 0.21263 * \frac{x}{L/2}\right)$$

Where y = the local **radius** at station x (measured from the bow);
D = the **Diameter**
L = the **Length**
x = use 100 incraments (#1 to #100)

Set "x" in a loop to repeat and print each answer.

If you can master that, take a quick check of what the maximum radius is, calculate the circumferance, and divide it by the width of your material (60"?). That will give you a ballpark figure on the number of gores it would take to make your envelope. Unfortunately, you may want an <u>even</u> number, <u>preferably</u> (but not necessarily) divisable by 4. It will make the assembly of your airship a lot easier. Once you have decided on the number of gores, rerun the formula with an additional equation to automatically print a column of the gore wudth at each station (x). That will be your gore pattern layout.

Building Small Gas Blimps

NOSE BATTENS
R: 10/18/97

So Goodyear uses battens. And they may know something that you don't. So what are they for? Maybe because they look nice, but I think they are ugly and create a lot of hull drag. They put them on because of "potential" nose deformation at higher airspeeds. "Higher" speeds? How high is high? Realistically, we are only talking 30-45 knots cruise, and 50 max. (see Engines & Props). Yes, 100 kts. is possible, but then you will have to buy some of those books listed in the bibliography. So take a look at the following chart.

Wind Miles per Hour.	Feet per Second.	$\frac{1}{2}\rho V^2$ Lbs. per Square Foot.	Head of *Water*. In.	mm.
20	29·3	1·01	·19	4·9
25	36·6	1·58	·30	7·7
30	44	2·29	·43	11
35	51·3	3·10	·59	15
40	58·6	3·98	·76	19·4
45	66	5·16	·98	25
50	73·3	6·30	1·21	30·7
55	80·6	7·58	1·46	37
60	88	9·17	1·76	44·7
65	95	10.69	2·07	52
70	102·7	12·49	2·40	61

From: **Aerodynamic Theory**

So what it is telling you is that as long as you keep the pressure above 1.5, or the speed below 38.1 kts., you've got no problem. So what if?

In The Air you want to go faster. Faster means more thrust (power), which in turn increases the drag. And to this, you want to add more drag by adding battens? How about putting the battens *inside* the envelope? It can be done fairly easily.

On The Mast poses different problems. As the wind speed increases, the internal pressure becomes critical. Above 35 kts. the envelope will stretch up to 2 feet between the nose and the gondola wheel. Above 50, all bets are off. If you lose power to the blower, so goes your pressure. I've spent **2** hours flying (with the engine running at cruise) on the mast in a 50 kt. wind. I don't expect to be that lucky again.

Forming the Battens to fit the nose contour is not a big deal if you use the best material. Yes, PVC schedule 10 white plastic tubing (used for water). Remember, the greatest amount of curve is in the first 1/3 of the 10' length. Make a simple wood fixture with the precalculated curve you want. Put a plug in one end that fits your water hose, and run HOT water thru the tubing until you get the set you want. Be sure to bevel both ends of the tubing so it doesn't scuff on the batten tunnel when you assemble it.

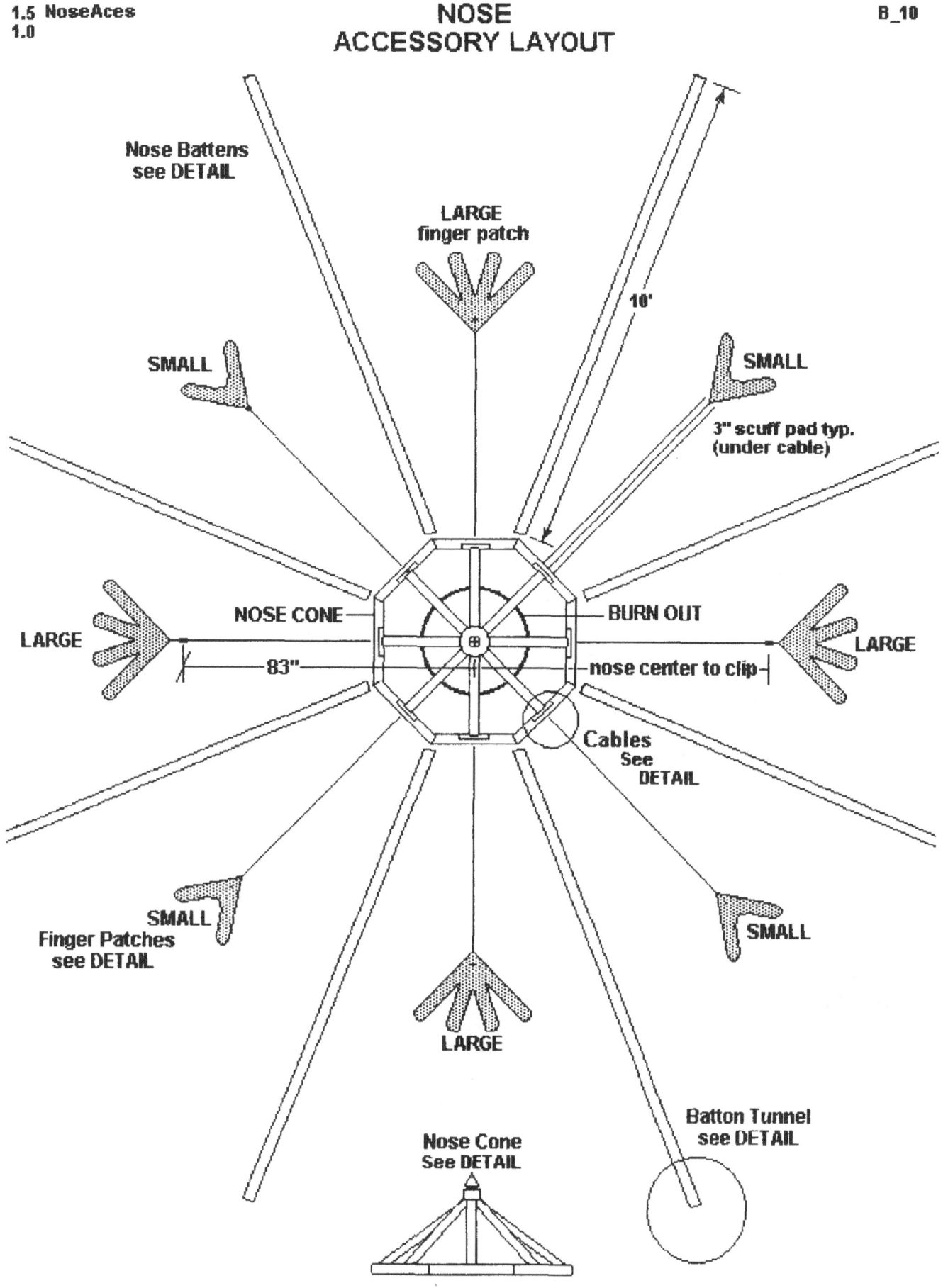

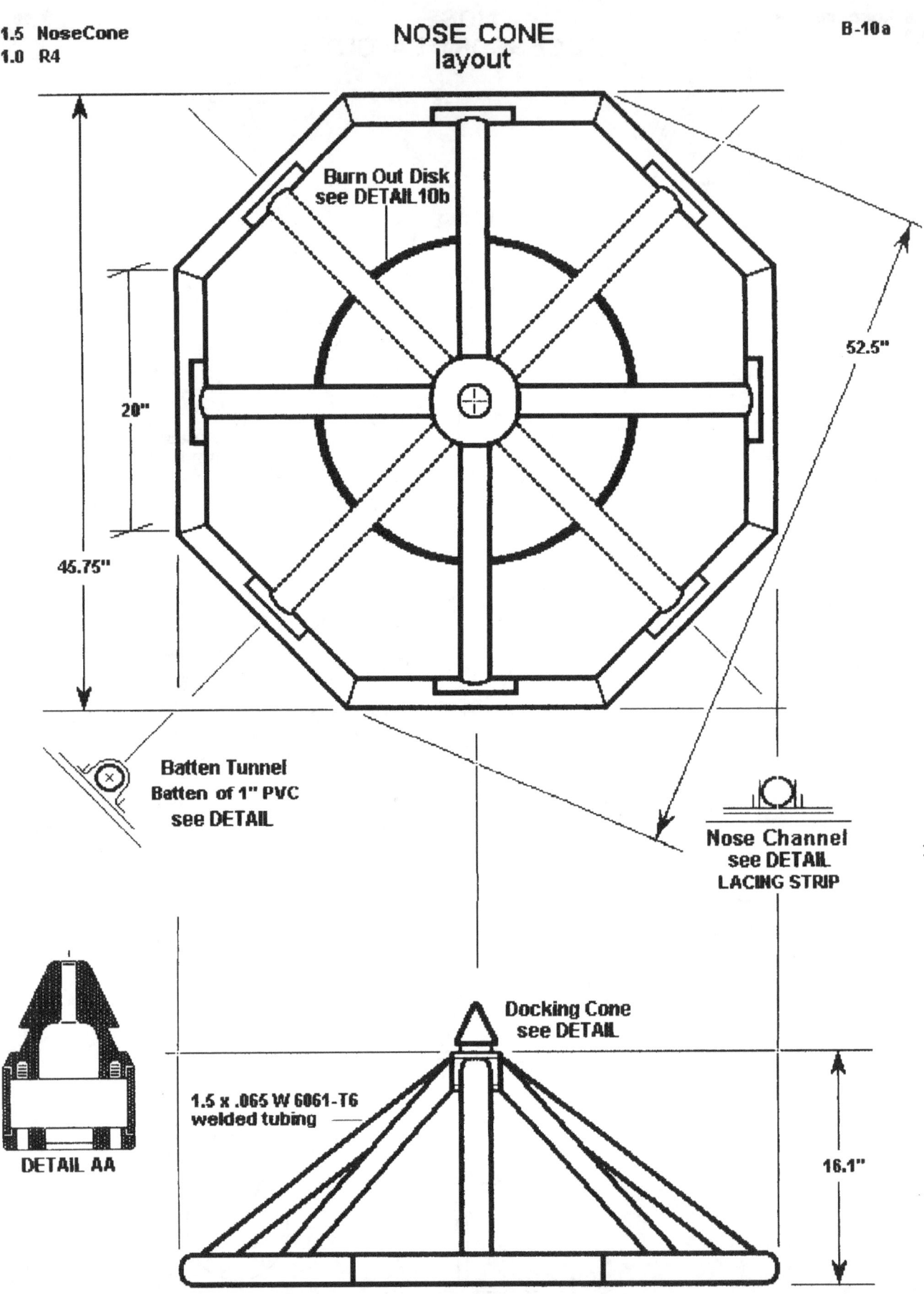

NOSE BATTEN & TUNNEL LAYOUT

1.5 NoseTunl
1.0 R2

B-10c

MATERIAL: Coated Dacron
Upper layer = 3 oz. x 7.75" wide
Bottom layer = 5 oz. x 6" wide
Use 1" Sked.40 PVC tubing in 10' lengths

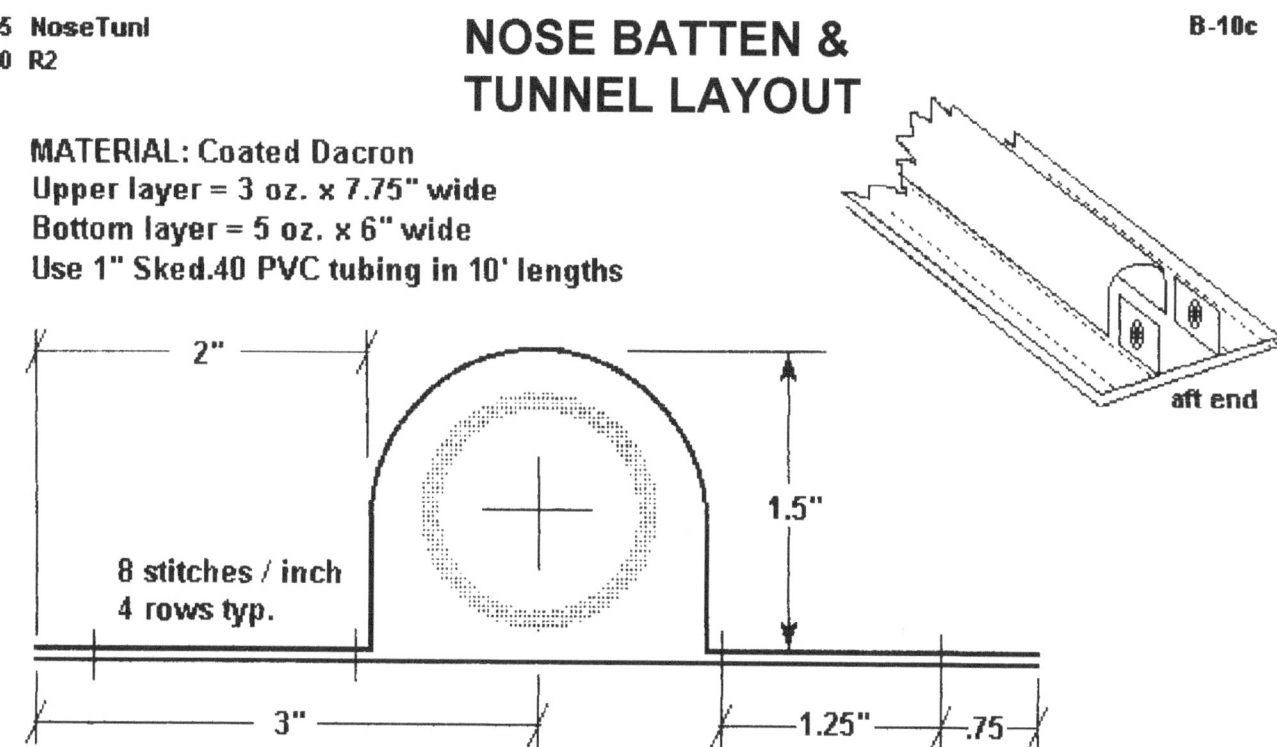

aft end

8 stitches / inch
4 rows typ.

LACING CHANNEL LAYOUT

MATERIAL: Coated Dacron
Upper layer = 3 oz. x 10" wide
Bottom layer = 5 oz. x 5" wide
Grommets, .25 I.D. every 4"
Use 1.5" x .090 wall Alum tubing in 10' lengths

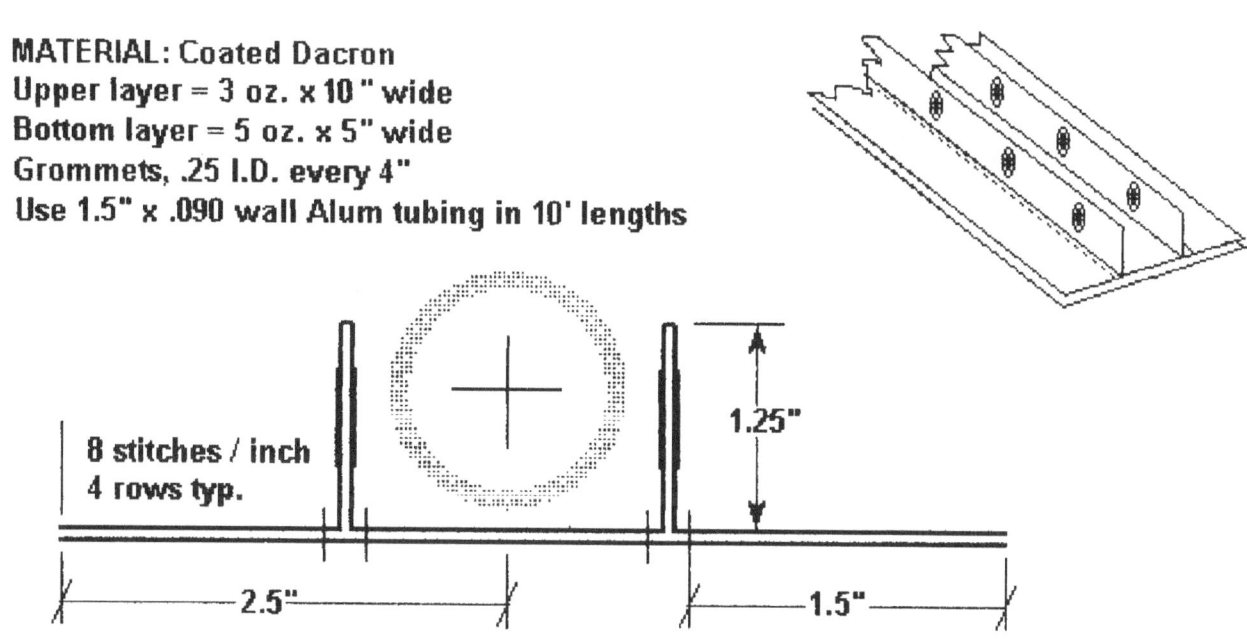

8 stitches / inch
4 rows typ.

Building Small Gas Blimps

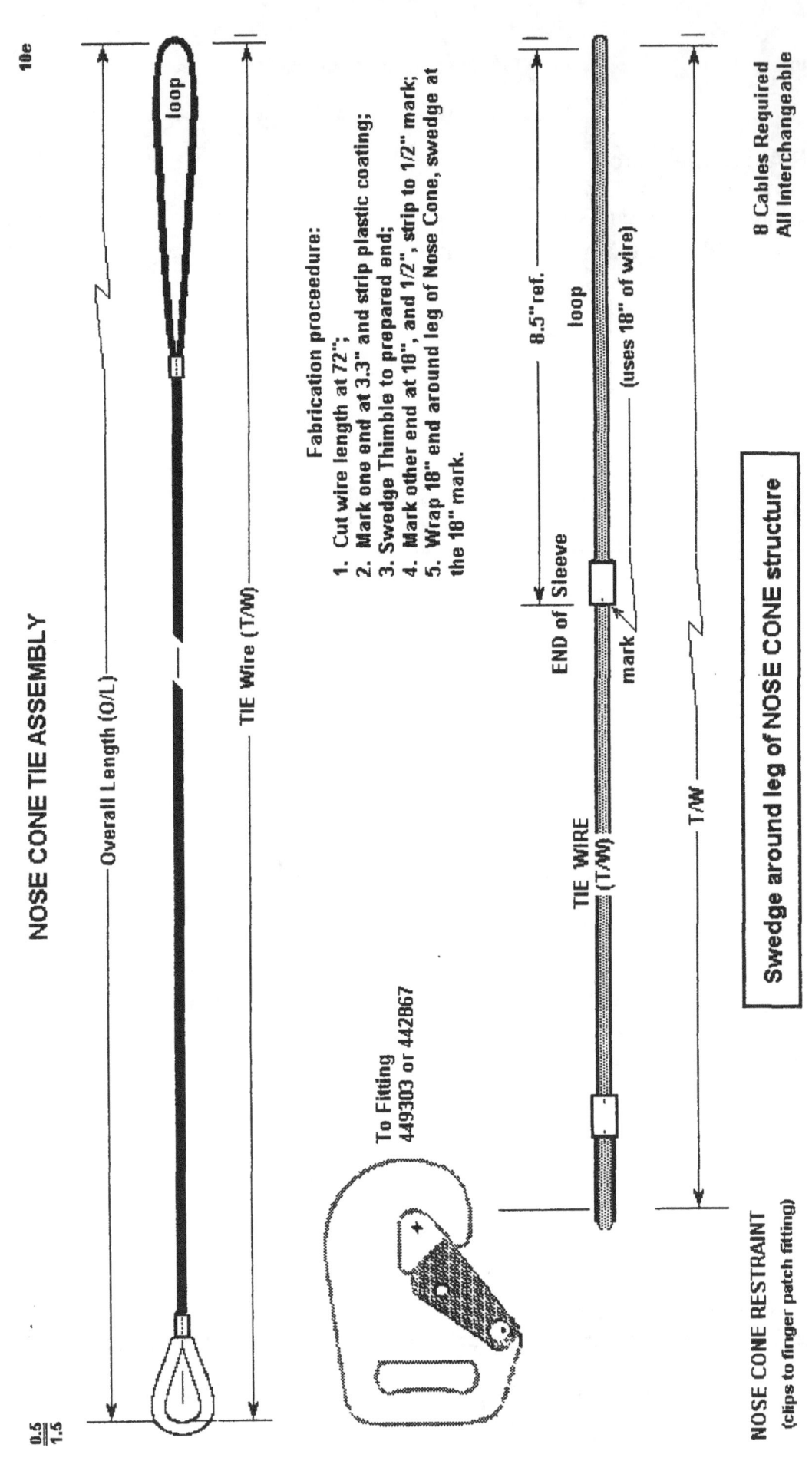

130

Building Small Gas Blimps

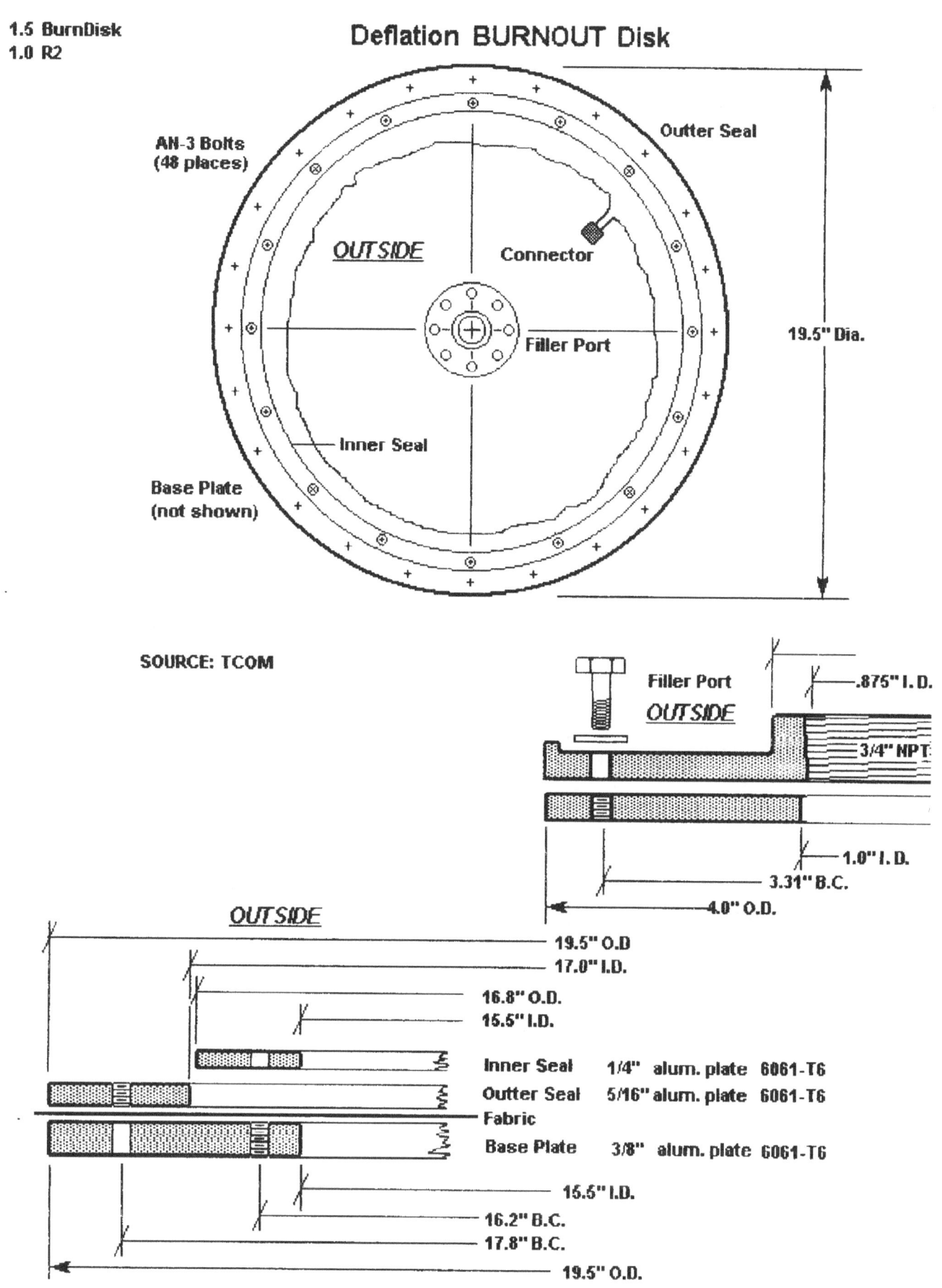

131

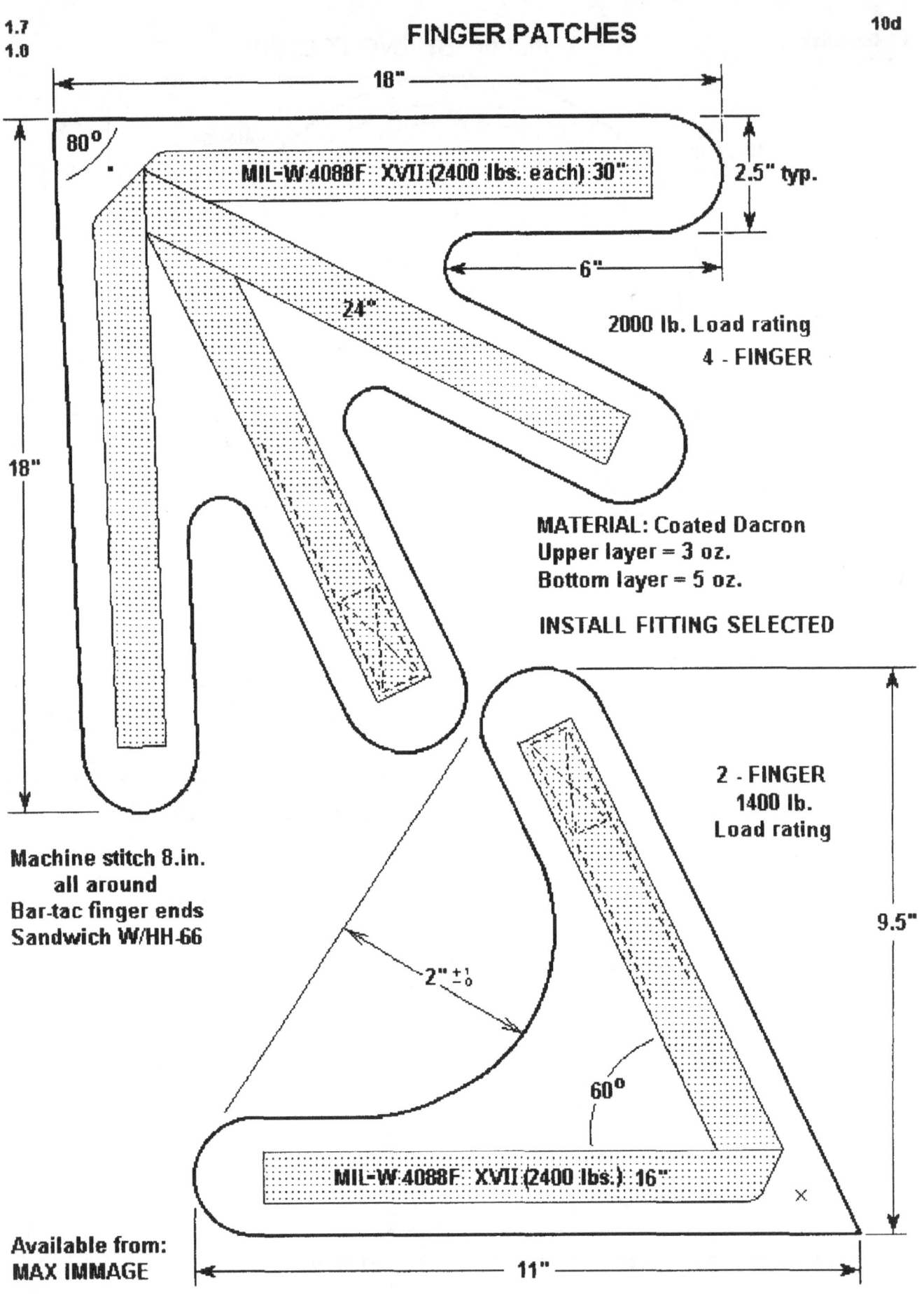

Building Small Gas Blimps

1.7 FITTINGS
1.0 R2

**FINGER PATCH
END FITTINGS**

AMSAFE part # 449303
2000 lb. rating; Wt. 1.0 oz.
Clear Nickel finished Aluminum
Anchor point: 3/8"

List price: $8.00 ea.

AMSAFE part # 442867
2000 lb. rating; Wt: 1.8 oz.
0.090 thick Chrome Plated Steel;
Anchor point: 3/8"

List price: $5.30 ea.

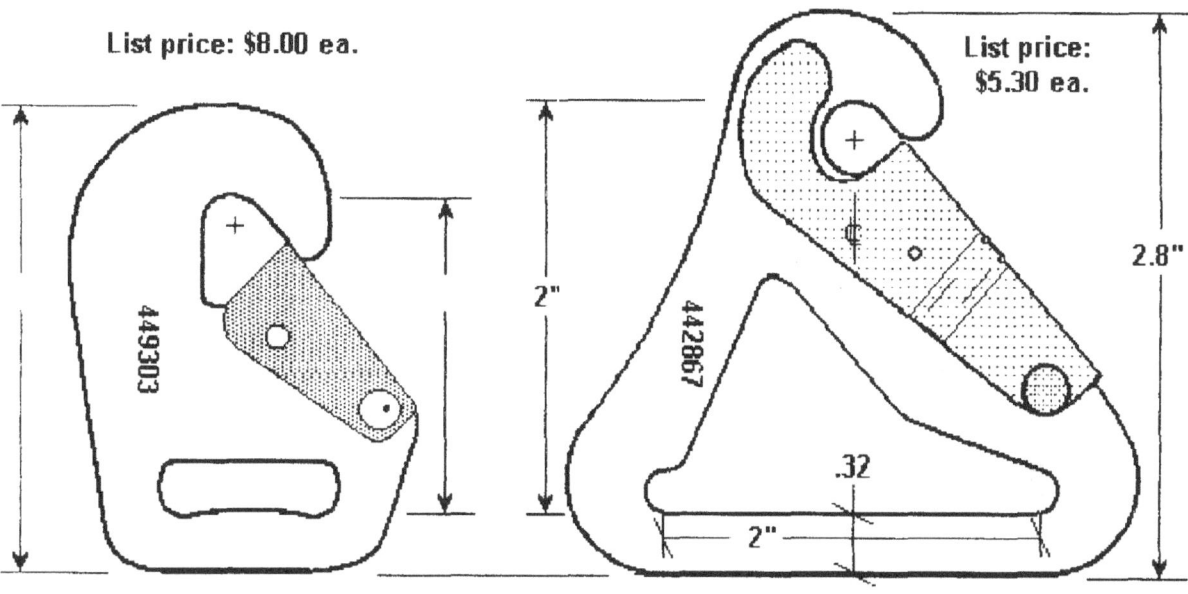

FLAT

3000 lb.; Wt.: 1.1 oz.
Chrome Plate Steel
3/8" anchor point

AMSAFE
part # 442878
$3.25 ea.

BENT

AMSAFE
part # 442879
$3.25 ea.

3000 lb.; Wt.: 1.1 oz.
Chrome Plate Steel
3/8" anchor point

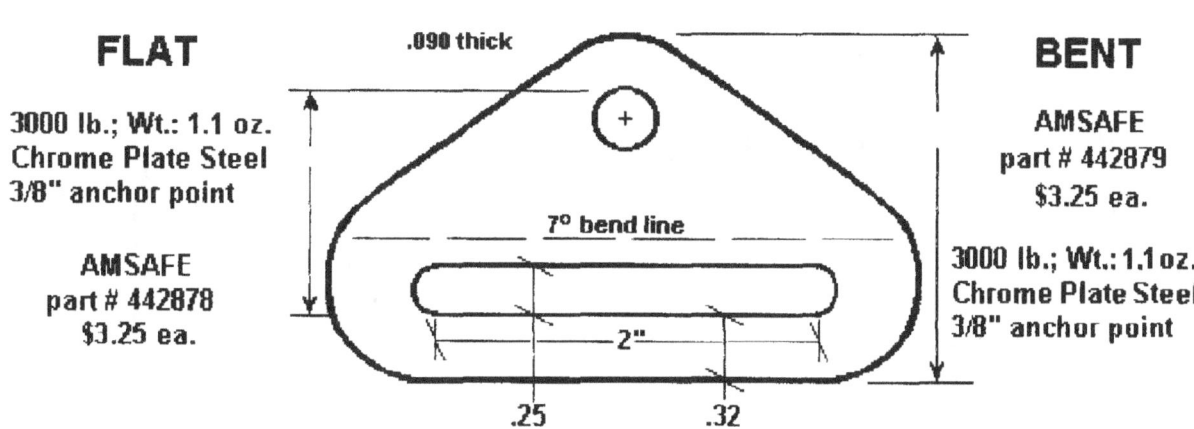

Manufactured by:
AMERICAN SAFETY EQUIPMENT
Phoenix, AZ
800 / 228 - 1567
(3 U.S. Retail Locations)

Retailer:
BELT MAKERS Inc.
1815 W. 205th. St. #304
Torrance, CA 90501
318 / 618 - 8868

Building Small Gas Blimps

CHANNEL PATCH

1.5 / 1.0 ChanPatc R2

B-13 a

SOURCE:

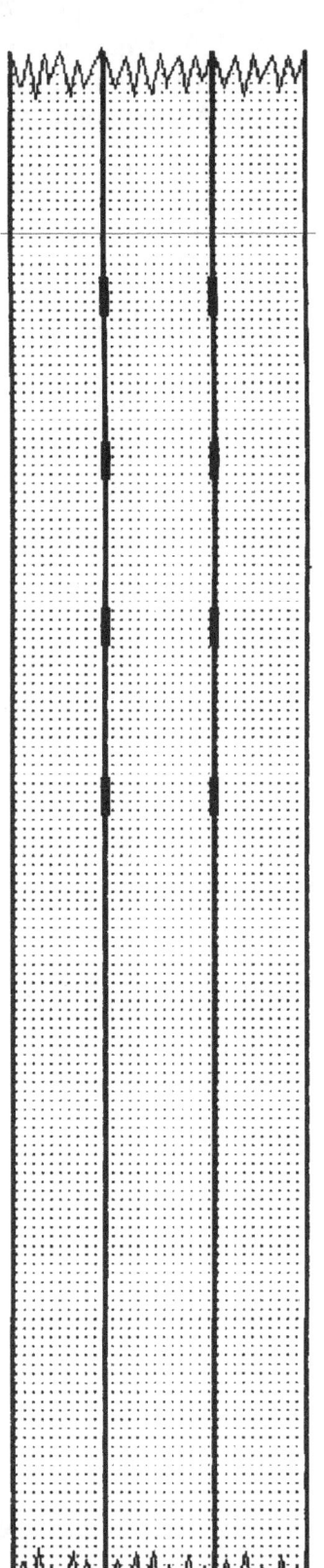
#3 Brass Grommet @ 4" spacing

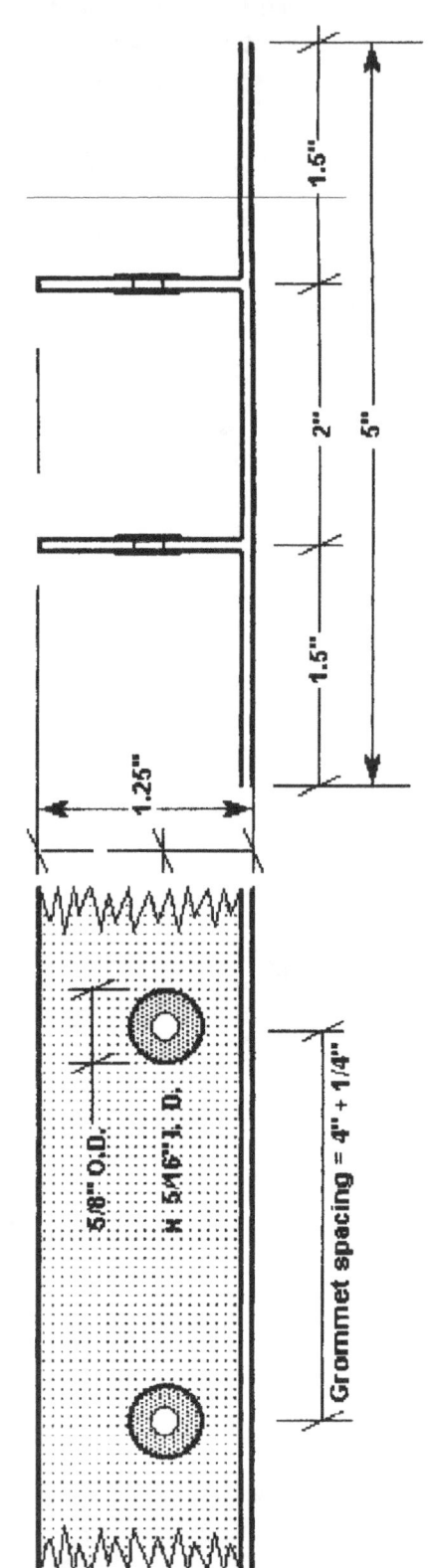
1.5"
2"
5"
1.5"
1.25"
5/8" O.D.
N 5/16" I.D.
Grommet spacing = 4" + 1/4"

134

Building Small Gas Blimps

R: 9/20/98

FABRIC & WEBBING
SUPPLIERS

AIRCRAFT SPRUCE Co.
201 Truslow Av.(POB 424)
Fullerton, CA   92632
800 / 824-1830
Great A/C source
FABRICS, Ceconite
Fin coverings
dopes and tapes

BALLY Mills
23-25 North 7th Street
Bally, PA   19503
800/845-2201
f)610/845-8013
quality products
FABRICS, polyester
Mill Spec (MS) Mfgr.
highly respected

LAMCOTEC Inc.(RECOMENDED) 413/267-4808
Munson, MA
Rick Henderson
FABRICS, polyester
Made to your specs

LOWRY Aircraft Supply
2311 E.Artesia Blvd.
Long Beach, CA   90?81
310/531-8134
hard to find items
great source
FABRICS, surplus
surplus & seconds
inspect closely

MANN-Tech Industries
225 Arlington Way
Farmingham, MA   01701
617/879-6366
well advertized
but unknown
FABRICS, polyester
Dacron & more

MICROSEAL Industries
610-T East 36th Street
Patterson, NJ   07509
201/523-0704
well known
quality work
FABRIC coatings
Polyester laminating

ORCON Corporation
1570 Atlantic Street
Union City, CA   94587
800/228-2781
exotic scrims
heat sealing
POLY FILMS
Reinforced films
great helium barriers

PERFORMANCE Textiles
3917 Liberty Road
Greensboro, NC   27406-6109
910/275-5800
Fax:275-8866
good source
FABRICS, polyester
Dacron & more
large quanities

STANDARD Brands Paints
620 Sepulveda Blvd.
Van Nuys, CA   91401
818/786-1381
most Calif. cities
ck yellow pages
FIBERGLASS
cloth & resin
Bondo, glues, paint

SNYDER Manufacturing
2900 Progress Street
Dover, OH   44622
800/837-4450
Fax/767-8070
FABRICS, polyester
In-house laminating
unknown

UNITED Textile (Unitex)
5175 Commerce Drive.
Baldwin Park, CA   90012
818/962-6281
Retail/wholesale
good source
FABRICS/hardware
WEBBING, canvas
H-66 fabric glue<-----

NOT recommended: *Gentek*

Specialty Items; see your Thomas Register. Get quality guarantee before price.

135

Building Small Gas Blimps

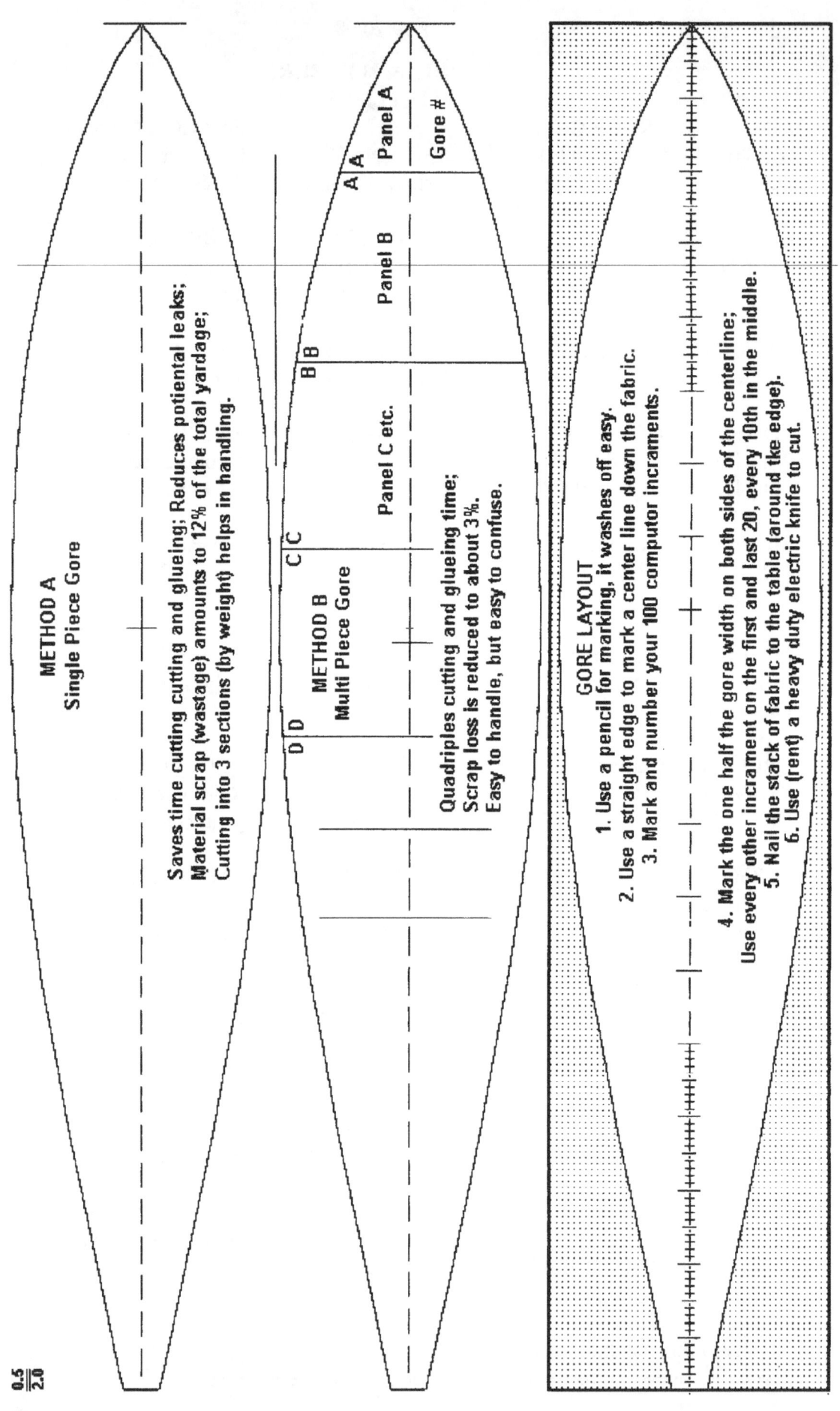

Building Small Gas Blimps

3.6:1 30,000 Cubic Feet 18 Gore

Pos. #	Long.X'	Gore L'	Rad. R'	GoreW'	Pos. #	Long.X'	Gore L'	Rad. R'	GoreW'
1	0.37	1.96	1.92	0.67	51	18.81	23.41	11.39	3.98
2	0.74	2.83	2.71	0.95	52	19.17	23.78	11.45	
3	1.11	3.53	3.31	1.15	53	19.54	23.16	11.52	
4	1.47	4.15	3.81	1.33	54	19.91	24.53	11.58	
5	1.84	4.71	4.24	1.48	55	20.28	24.89	11.64	4.06
6	2.21	5.25	4.63	1.62	56	20.64	25.28	11.69	
7	2.58	5.76	4.98	1.74	57	21.01	25.65	11.75	
8	2.95	6.25	5.31	1.85	58	21.38	26.02	11.81	
9	3.32	6.73	5.61	1.96	59	21.75	26.39	11.85	
10	3.69	7.19	5.89	2.06	60	22.12	26.77	11.91	4.16
11	4.06	7.65	6.15	2.15	61	22.49	27.14	11.95	
12	4.42	8.09	6.41	2.24	62	22.86	27.51	12.01	
13	4.79	8.53	6.64	2.31	63	23.22	27.88	12.04	
14	5.16	8.97	6.87	2.39	64	23.59	28.25	12.09	
15	5.53	9.39	7.08	2.47	65	23.96	28.62	12.13	4.23
16	5.91	9.82	7.29		66	24.33	28.99	12.17	
17	6.27	10.23	7.49		67	24.71	29.37	12.21	
18	6.64	10.73	7.68		68	25.07	29.74	12.25	
19	7.01	11.06	7.87		69	25.44	30.11	12.28	
20	7.37	11.46	8.03	2.81	70	25.61	30.48	12.32	4.29
21	7.74	11.87	8.21		71	26.17	30.85	12.35	
22	8.11	12.28	8.36		72	26.54	31.21	12.38	
23	8.48	12.68	8.52		73	26.91	31.59	12.41	
24	8.85	13.07	8.67		74	27.26	31.22	12.44	
25	9.22	13.46	8.82	3.08	75	27.65	32.33	12.47	4.35
26	9.58	12.95	8.96		76	28.02	32.69	12.51	
27	9.95	14.26	9.11		77	28.39	33.07	12.52	
28	10.32	14.65	9.23		78	28.75	33.44	12.55	
29	10.69	15.04	9.36		79	28.12	33.81	12.57	
30	11.06	15.43	9.48	3.31	80	29.49	34.17	12.59	4.39
31	11.43	15.82	9.61		81	29.86	34.54	12.62	
32	11.81	16.21	9.72		82	30.23	34.91	12.63	
33	12.17	16.59	9.83		83	30.61	35.28	12.65	
34	12.53	16.97	9.94		84	30.97	35.65	12.67	
35	12.91	17.36	10.05	3.51	85	31.33	36.02	12.69	4.43
36	13.27	17.74	10.15		86	31.71	36.39	12.71	
37	13.64	18.12	10.25		87	32.07	36.76	12.72	
38	14.01	18.51	10.35		88	32.44	37.13	12.73	
39	14.38	18.88	10.44		89	32.81	37.86	12.74	
40	14.75	19.27	10.54	3.68	90	33.18	37.86	12.75	4.48
41	15.11	19.64	10.63		91	33.55	38.23	12.76	
42	15.48	20.02	10.71		92	33.91	38.59	12.77	
43	15.85	24.41	10.81		93	34.28	38.97	12.78	
44	16.22	20.78	10.88		94	34.65	39.34	12.78	
45	16.59	21.16	10.96	3.38	95	35.02	39.71	12.79	4.46
46	16.96	21.53	11.04		96	35.39	39.71	12.79	
47	17.53	21.91	11.11		97	35.76	40.44	12.79	
48	17.69	22.28	11.18		98	36.13	40.81	12.79	
49	18.06	22.66	11.25		99	36.51	41.18	12.79	
50	18.43	23.03	11.32	4.02	100	36.86	41.55	12.79	4.46

Chart 1 of 2

Building Small Gas Blimps

3.6:1 A/F   30,000 Cubic Feet   18 Gore

Pos. #	Long.X'	Gore L'	Rad.R'	GoreW'	Pos. #	Long.X'	Gore L'	Rad.R'	GoreW'
101	37.42	42.11	12.79	4.46	151	65.06	69.89	10.39	3.63
102	37.97	42.66	12.79		52	65.62	70.45	10.29	
103	38.52	43.21	12.79		53	66.17	71.01	10.21	
104	39.08	43.76	12.78		54	66.72	71.57	10.11	
105	39.63	44.32	12.77		55	67.28	72.13	10.01	
106	40.18	44.87	12.76	4.45	56	67.83	72.69	9.901	3.45
107	40.73	45.42	12.75		57	68.38	73.25	9.801	
108	41.29	45.97	12.73		58	68.94	73.82	9.701	
109	41.84	46.53	12.72		59	69.49	74.38	9.59	
110	42.39	47.08	12.71		160	70.04	74.94	9.48	
111	42.95	47.63	12.68	4.42	61	70.59	75.51	9.37	3.27
12	43.51	48.19	12.65		62	71.15	76.07	9.26	
13	44.05	48.74	12.63		63	71.71	76.64	9.15	
14	44.61	49.29	12.61		64	72.25	77.21	9.03	
15	45.16	49.85	12.57		65	73.81	77.77	8.91	
16	45.71	50.41	12.54	4.37	66	73.36	78.33	8.79	3.07
17	46.26	50.96	12.51		67	73.91	78.89	8.67	
18	46.82	51.51	12.47		68	74.47	79.47	8.54	
19	47.37	52.06	12.44		69	75.02	80.03	8.41	
120	47.92	52.62	12.41		170	75.57	80.61	8.28	
121	48.48	53.17	12.36	4.31	71	76.12	81.17	8.15	2.84
22	49.03	53.73	12.32		72	76.68	81.74	8.01	
23	49.58	54.28	12.28		73	77.23	82.31	7.67	
24	50.14	54.84	12.23		74	77.78	82.88	7.73	
25	50.69	55.39	12.19		75	78.34	83.45	7.58	
26	51.24	55.95	12.14	4.24	76	78.88	84.03	7.43	2.59
27	51.79	56.51	12.09		77	79.44	84.61	7.28	2.54
28	52.35	57.06	12.04		78	79.99	85.18	7.12	2.48
29	52.91	57.61	11.98		79	80.55	85.75	6.96	2.43
130	53.45	58.17	11.93		180	81.11	86.33	6.79	2.37
131	54.01	58.73	11.87	4.14	81	81.65	86.91	6.62	2.31
32	54.56	59.28	11.81		82	82.21	87.41	6.45	2.25
33	55.11	59.84	11.75		83	82.76	88.07	6.27	2.19
34	55.66	60.37	11.69		84	83.31	88.65	6.08	2.12
35	56.22	60.95	11.63		85	83.87	89.24	5.89	2.06
36	56.77	61.51	11.56	4.03	86	84.42	89.83	5.69	1.99
37	57.32	62.06	11.51		87	84.97	90.42	5.48	1.91
38	57.88	62.62	11.43		88	85.52	91.01	5.27	1.84
39	58.43	63.18	11.36		89	86.57	91.61	5.04	1.76
140	58.98	63.74	11.29		190	86.63	92.21	4.81	1.68
141	59.54	64.29	11.21	3.91	91	87.18	92.81	4.56	1.59
42	60.08	64.85	11.14		92	87.74	93.42	4.31	1.51
43	60.64	65.41	11.06		93	88.29	93.88	4.02	1.41
44	61.19	65.97	10.98		94	88.84	94.84	3.72	1.29
45	61.75	66.53	10.91		95	89.41	95.31	3.41	1.19
46	62.31	67.09	10.82	3.78	96	89.95	95.97	3.04	1.06
47	62.85	67.65	10.74		97	90.51	96.66	2.63	0.92
48	63.41	68.21	10.65		98	91.05	97.39	2.15	0.75
49	63.96	68.76	10.57		99	91.81	98.23	1.52	0.53
150	64.51	69.32	10.48		200	92.16	99.85	0	0

Building Small Gas Blimps

Envelopes cont.

R: 4/12/98

Inflation sleeve: Hopefully, you thought ahead and engineered in a way for a person to get inside the inflated envelope and install a ballonet(s) It should be a sleeve at least 12" in diameter and long enough to tie closed (18"). The position of choice is on the envelope bottom where the valves are installed. A standard Burnout Ring works best. An access sleeve in the tail cone also works well, except that it is impossible to get into when the envelope is inflated (out of reach).

Air Inflation / Deflation: Materials needed (For pressure check, adding reinforcements, positioning fins)

        1 ea. Blower, 10" centrifugal, 1.5" W.P.(test inflation deflation)
        50' 4" Perforated Flexible PVC pipe (construction drainage)
        1 ea. Helimec pressure gage, 0-10" W.P.
        10' 1/4" flex tubing (from pressure gage to envelope)
        5' Bungee cord (for affixing envelope access tube to blower)

Inflate to 1.5" WP and monitor leak rate over 24 hours Any noticeable decrease warrants a complete inspection. If you are satisfied with the holding pressure, increase the pressure to 2 5" for another 24 hours, and repeat the inspection process. Repeat at 3.5". A final at 5" should be made for 5 minutes (or 6 for 3) to test the burst safety factor.

Cold Air Blowers: CAUTION; <u>use only electric powered centrifugal blowers</u>. DO NOT use a 2-cycle engine driven fan; they suck exhaust (carbon monoxide) into the envelope. NOT a good place to go when installing the ballonet(s) or looking for pinholes.

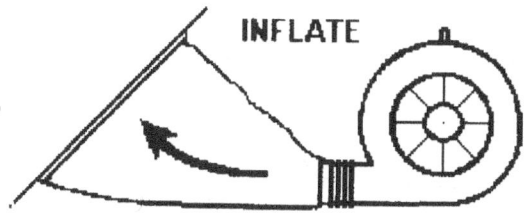
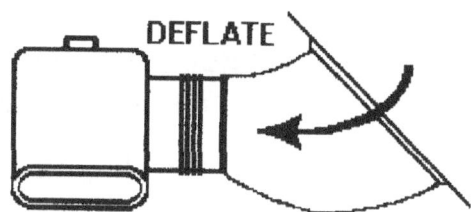

The only active source of good quality centrifugal blowers is: *AIR PLUS MANUFACTURING* of Santee, CA. See the "Appendices" section for choices. The smaller ones are good to 1.5" (2.5" when new) of water pressure. But only the largest will reach 5". Note that the deflation cycle requires an intake adapter to accomodate the port neck.

The smaller ones ARE good for maintaining the 1" to 1.5" of ballonet pressure when the airship is on the mast. Most blimps, however use specially designed units with varying degrees of success.

Building Small Gas Blimps

Air Plus Manufacturing
9932 Prospect Avenue
Suite 139
Santee, California 92071

phone: 619.596.4808
toll-free: 1.888.999.4808
fax: 619.596.4877
www.air-plus.com

Black Box High Pressure Blower

Performance Specifications

Motor:
  1 H.P. 3450 RPM   110/220 volts   50/60 Hz.
  12 amps at 110 volts
  6 amps at 220 volts
  Single speed Motor

This blower requires a dedicated 20 amp (110 volt) circuit to operate properly. If an extension cord is used – it must be of the following size:
  Up to 50 ft. in length 14-3 gauge cord
  50–100 ft. in length 12-3 gauge cord
  Longer length cords should not be used.

Wheel:
  9.125 in. X 3.00 in. High RPM Design. Forward incline blade. Wheel is single inlet.

Switch:
  Off/On rotary switch

Output:
  Maximum output is 2500 CFM at free air. (0" W.C.)
  Tested pressure is 4.3" W.C. on a 14.5' x 12.5' x 17' high bounce unit. (Castle style)

Black Box Blowers must have back pressure to operate without overheating.

Shipping Information:
  Box dimensions: 25 ½ in. X 17 ½ in. X 25 in. (LxWxH)
  Shipping weight: 52 lbs.
  Actual blower weight: 45 lbs.

Part # BB10

SUPERIOR AIR MOVERS FOR INFLATABLES, RESTORATION AND COOLING

Building Small Gas Blimps

Envelopes cont.
R: 11/3/97

Cutting preparation

Obviously, I haven't talked you into purchasing a completed envelope from a manufacturer, or you wouldn't be reading this. So let me try something else. How about taking all your inspected fabric to a professional cutter? He has all the tools and expertise to eliminate this section. The yellow pages of your phone book will give you a list of canvas fabricators and commercial cutters. They may even let you participate.

OR, Materials you will need are: (besides a 100' LONG clean room, with adequate lighting)

- 100'   Table(s) 4' wide plywood (don't try working on the floor)
- 1 ea. Electric knife, heavy duty vertical (cut 20 layers of 5-7 oz. fabric)
- 1 ea. Straight edge, 10' x 1.5" steel (for long/accurate centerlines)
- 1 ea. Surveyors Tape, 100' in feet & tenths (or metric, if programmed)
- 1 ea. Wood molding, 10' x 3/4 half round (for drawing curves)
- 10 ea. Pencils, #1 (for all cut and finger patch positions)
- 3 ea. Pens, fine tip black, fabric permanent (for critical positions)
- 3 ea. Extension cords, 50' x 10 amp. (keep everything within reach)

Hopefully, you fixed all the pinholes at the light-table inspection, but now is the time to make a last check of any suspicious areas. Do it now as you won't have time when the serious gluing starts. Try not to place any patches within 1 1/4" of the longitudinal gore seams (cut another gore if you have too). Special care will have to be made for these during air inflation.

Gluing preparation:

Sorry, this is something you are going to have to do yourself. Set your pace to do one seam a day. Go for cleanliness, accuracy and quality, NOT speed. Materials needed: (besides a Large temperature controlled, well vented room)

- 1 ea.   Blower Fan (for removing toxic fumes)
- 100'   Table(s) 4' wide plywood (throw it away when done)
- 1 case   Glue/cement, H-66 (6 Gal.)
- 1 ea.   Heat Gun, heavy duty (setting glue)
- 20 rolls Paper Masking Tape, 3/4" wide (keep glue line neat)
- 2 Gal.   Denatured Alcohol * (flashing surface before gluing)
- 50 ea.   Paint Brushes, 1 1/2" wide (discard when hard)
- 1 ea.   Degreasing dispenser (wetting rag with *)
- 1 ea.   Breathing Mask, disposable cartridge (USE IT)
- 20 ea.   Disposable charcoal cartridges (good, but no guarantee)
- 1 pr.   Kneepads, Industrial/disposable (they'll be full of glue)
- 6 ea.   Wallpaper Rollers, 1" wood wheel (rolling each glued seam)
- 3 ea.   Kitchen knives, 3/4" flat blade (delaminating bad seams)
- 1 ball   Waxed cord (getting distances on an inflated curve)
- 20 ea.   2" precut hole patches (for pin holes & suspicious imperfections)
- 24 ea.   Large Finger Patches (for nose & fin attachments)
- 48 ea.   Small Finger Patches (for nose & fin attachments)
- 20 ea.   Clamps, 2" openong, spring loaded (holding fabric on table

* MEK is better, but extremely hazardous

Building Small Gas Blimps

Envelopes cont.

ASSEMBLY PROCEDURE
R: 12/04/97

Surface preparation: The quality of your work will be evident if your seams are all wavy, inconsistent in lay, and the bonding pulled apart to re-lay the piece. All of this can be virtually eliminated by carefully marking each seam position. The recommended way is to mark the 1" seam line with a silver pencil. However, you will quickly find out that #1 pencil is more visible.

The next step is to clean the waxy film off the area (with a damp solvent cloth) you are going to apply the glue. It is NOT a step to overlook; an uncleaned seam will pull apart eventually. Cleaning specifically includes you NOT putting your hands on the area you just cleaned; and if you are going to lunch, clean it again before applying glue.

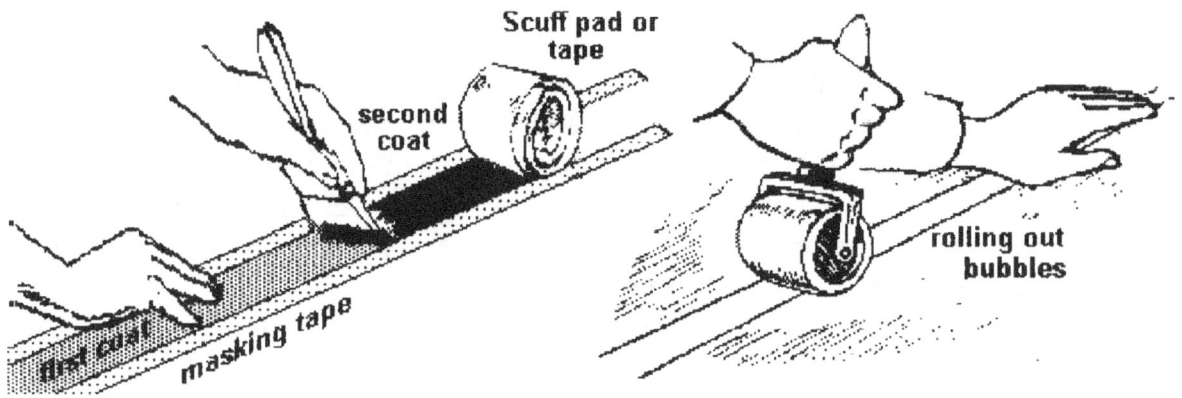

Adhesive application: Apply the adhesive with a paint brush in two coats, letting the first dry before applying the second. Understand that you are doing this to BOTH surfaces, as evenly as you possibly can. Wait about 15 to 30 seconds (depends on temperature & humidity) for both surfaces to get "tacky". Then carefully join the surfaces exactly to your marks. If you hesitate and try to pull it apart to reposition it, you have most likely pulled the vinyl coating off one surface (and you will have a gas leak). If you did such a boo-boo, at least reglue the pieces and be liberal with the glue over the newly exposed area.

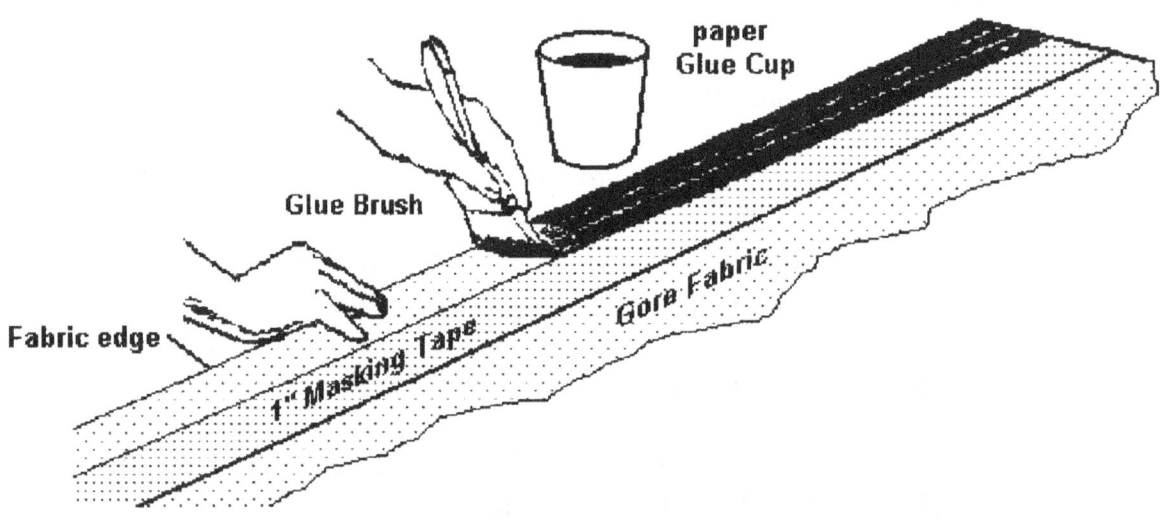

Building Small Gas Blimps

Envelopes cont.

Bonding: Each glued strip/section must immediately be smoothed with a plastic squeegee, rollered firmly, and allowed to set for at least 4 hours before moving. Special care should taken to eliminate trapped air bubbles and wrinkles. It can be rerolled as many times as needed in the first 12 hours.

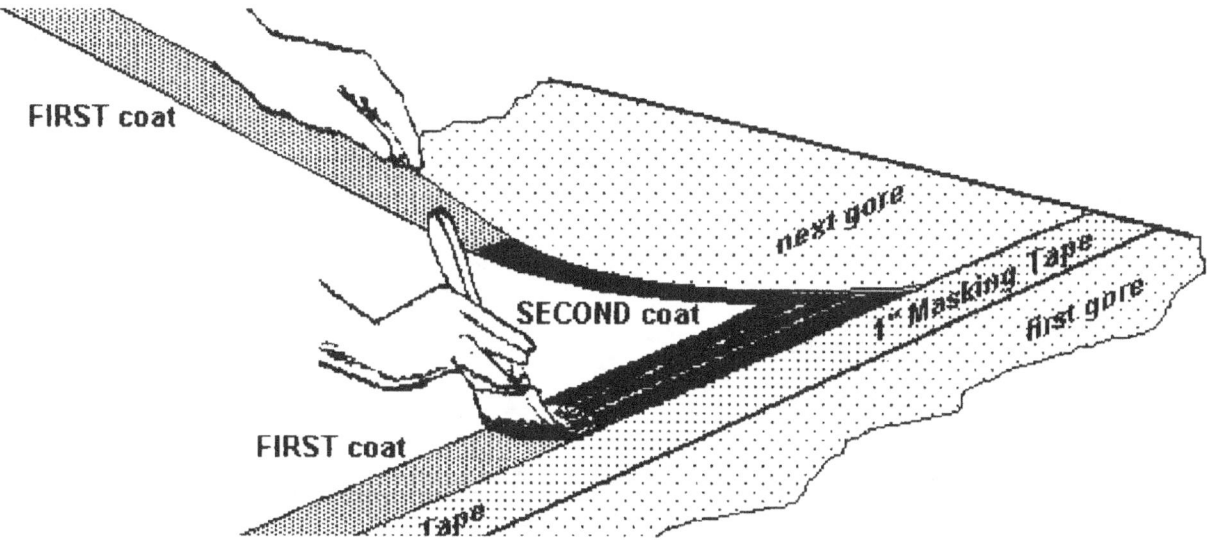

Curing: It is considered "set" after 4 hours and can be moved as long as there is no tension on the seam. In no case should an attempt be made to crease or pack it up for at least 12 hours (the seam reaches 90% strength). Do NOT attempt an air inflation test until the finished envelope has cured at least 48 hours.

Inspection: The adhesive bonding process should result in a part that is free of lumps, wrinkles, and uniformly attached over the whole bond area.

Repairs: If the bonded joint (or patch) is found to be misalign, use a putty knife dipped in MEK* to slide carefully between the surfaces as soon as possible after initial bonding. The sooner, the better. After 12 hours You'll have to leave it and patch over it if possible. Believe me, it does happen. * Remember, MEK is extremely hazardous.

Warning: Remember that I said clean both surfaces with a damp cloth, meaning just flashing the surface with Alcohol (or MEK). To spill a large quantify (teaspoon or more) onto your envelope can be a disaster. To get it off, blot it quickly. If you miss it, it will dissolve the envelope coating; and turn the area to "mush". The area affected will most probably leak without great care, and will always leave an evil looking spot.

DEFLATION
R:12/05/97

There are several considerations for deflating a blimp. They may be because you want to put in storage for the Winter, make some serious repairs or modifications, or an emergency. Obviously, loosing several thousand dollars of helium takes some thought, but it may be a better trade off than loosing something more valuable. If you have to make the sacrifice, at least minimize the damage. A very slow (24 hours) deflation can be made with a well placed bullet, from the bottom and out the top, at the greatest diameter. Eventually EVERY airship sustains (at least) one hit. If it is loosing an unusual amount of gas, patch the hole on the bottom first, then rent a cherry-picker (man-lift) immediately to patch the exit point. I put that in passing, now lets get in to CONTROLLED deflation's.

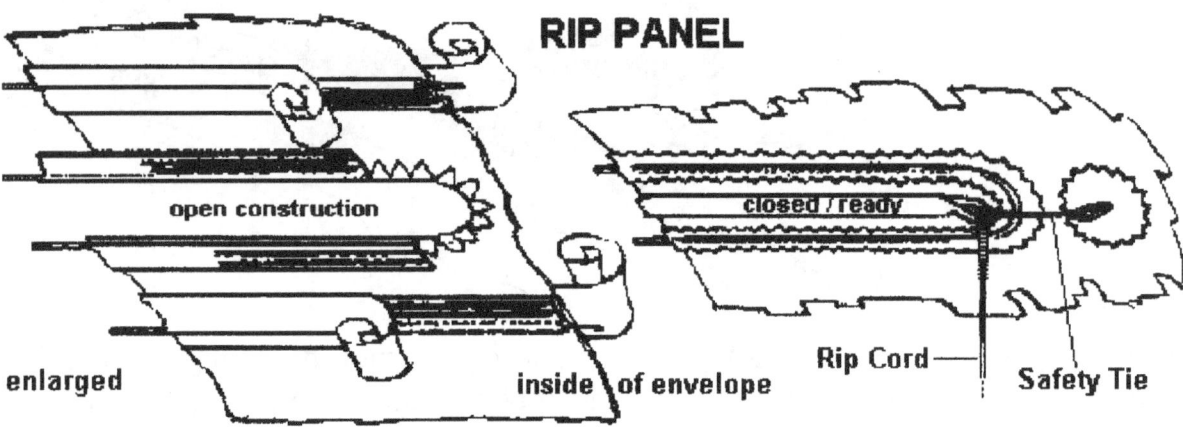

The Rip Panel is the tried and tested method. It requires careful installation as the envelope is made, and very careful repair to reseal it. It also meets the FAA requirement for an emergency, on, or off the mast. The pull-lanyard must be accessible to the pilot, and to the ground crew (standing on the ground). The position and design are important basic considerations. Position it on the top centerline (helium goes up), aft of the widest point (so the gas bubble leaves the nose last), and the rip panel kept narrow (6-12"), so the wind (airflow) doesn't just replace the gas with air. Since almost all (99.9%) airships are lost at the mast, the lanyard can be easily attached to the mast at the approach of severe weather.

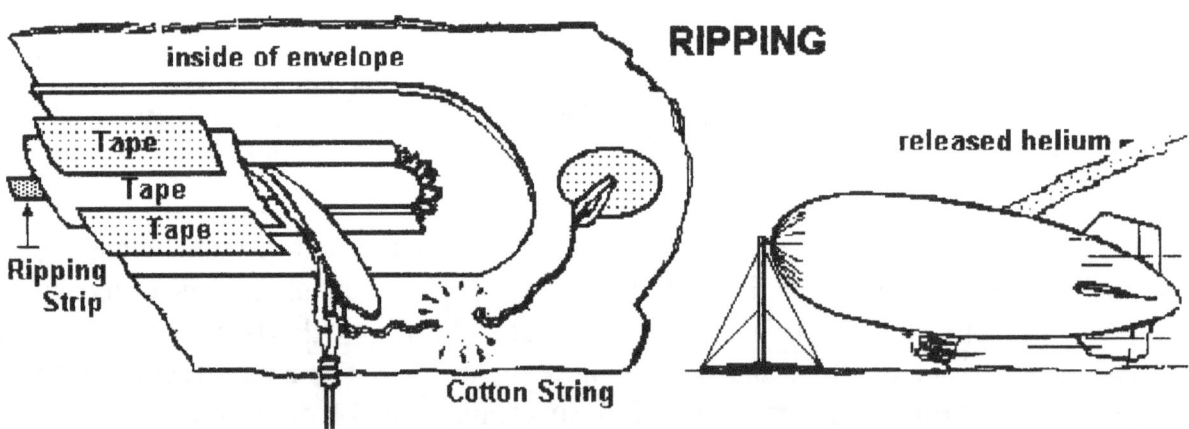

DEFLATION cont.
R:12/05/97

The Burn-Out is something relatively new. It incorporates a heat element into a large replaceable port. It works, is relatively effective, and reinstalls easily. You can be very creative on where you put it (nose, tail, top), but its one drawback is that it needs peak electrical power to activate it. Bottom line, it takes a good battery, must bypass the master switch, and the activation be safetied from buttonitis. People just love to push buttons at night. Approaching severe weather, however, means that a crewman must stay in it.

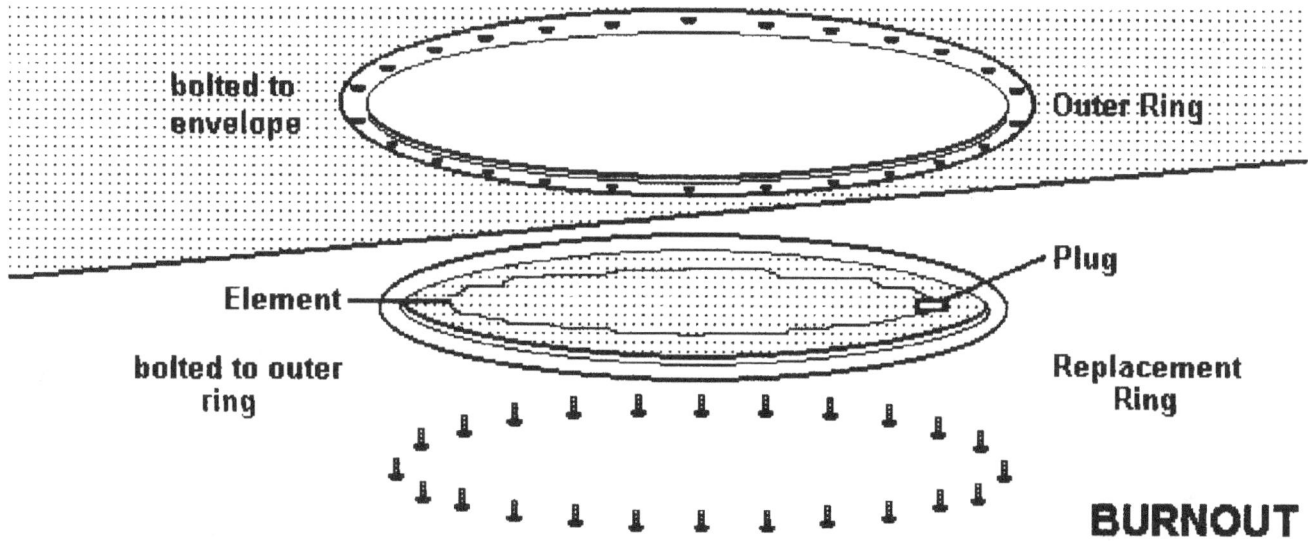

BURNOUT

Opening the Helium Valve (and/or other ports) will give you a nice slow (2-hours) deflation. Understand that you must disconnect the gondola and roll the envelope to get the valve port on top. Easier said than done.

The Drip Fence has been around for more than 100 years. It was handed down from the old gas ballooning gurus that found out rain ran down the sides of curved surfaces an dripped on your head. All it takes is a simple "T" shaped length of barrier strip. Glue it right to the envelope after you know the exact position of the gondola. Allow enough distance all around where you expect to walk, and especially include electronic and rust areas. Use an arrow design in front of the windshield so water rolls around the gondola. Make sure the rear propeller area is protected also. Bad enough to have to work on the engine in bad weather, but worse to be miserable with water dripping down the back of your neck.

ENVELOPE cont.

INTERNAL CATENARYS

Yes the Goodyear blimp uses them, but they (GZ-20's) are 214,000 (214K) cu.ft. lifting a 10,000 pound gondola and 8 people inside. And your blimp is how big? I definitely do not recommend catenaries for a volume of less than 60K., and certainly not 30K.

Lets explain their purpose: They are a compromising attempt to help support a heavy gondola with the internal envelope pressure. Yes, the gondola weight will distort the profile of any fabric structure. But with the stability of the new fabrics and low gondola weights, external finger patches are sufficient for a small blimp.

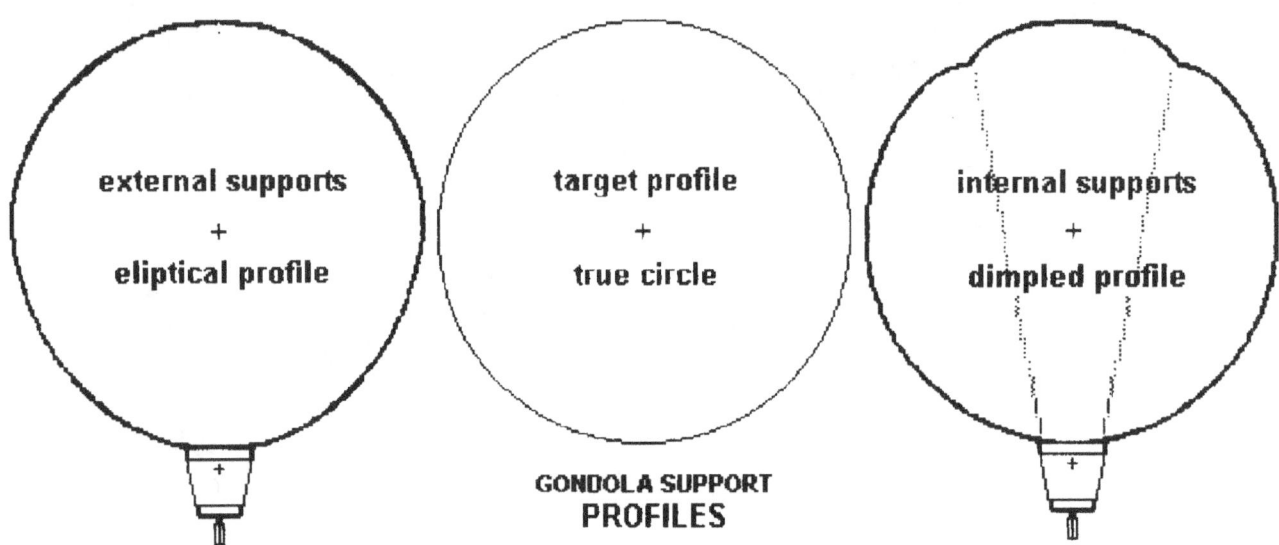

GONDOLA SUPPORT PROFILES

But if you think you've GOT to have them, here are a few things to think about:
    You are going to have to be creative on ballonet placement,
        It can't conflict with the support cable positions;
    The support cable connections are going to be real sources for gas leaks,
        More cables, more sources.
    The curtain & rigging will add a minimum of 100 pounds to your weight.
        More weight, less useful load;
    Rigging to keep the cable tensions equal is very hard and time consuming,
        It is not likely that it will be right on one inflation;
    In the end you have traded distortion on the bottom of the envelope for two distortions on the top.

INTERNAL CATENARIES
Excerpts from U.S.Navy K-Ship Training Manual

BACKGROUND: In response to many inquires on the technique of supporting a heavy gondola (car) from the top of the blimp envelope, the following offers one approach to the solution. The K-Ship gondola had an Empty Weight of 8080 lbs. The addition of crew and equipment (Useful Load) brought the total suspended weight to 13000 lbs.

LOAD DISTRIBUTION: This load was supported by four catenary curtains located in the top of the envelope for attachment of the inside suspension system. These curtains were cemented and sewed into the seams between the top five gores of the ship, and extended from panel 19 to 60 (138', almost half the gore length of 86 panels). See Fig. 1.

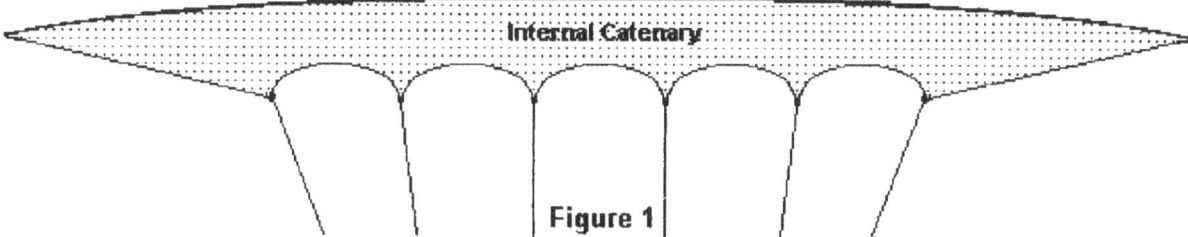

Figure 1

There were six catenary curves, or concentration points on each curtain for the attachment of suspension cables. An aluminum alloy ring, with steel bushing inserts, was lashed securely into each point. The suspension cables were attached to these bushings by means of shackles. The lower edges of the curtains were reinforced by 3/16" steel cables. Those cables also helped secure the rings in the concentration points. See Fig. 2. At the ends of each curtain the cables bridled out and were anchored to finger patches on the envelope. See Fig. 3.

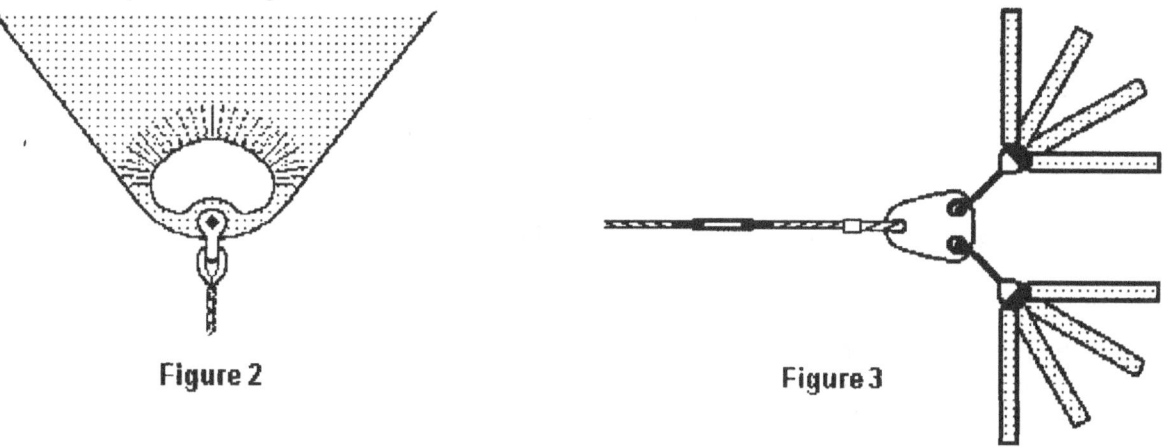

Figure 2   **Figure 3**

INSIDE SUSPENSION: The inside suspension of the car consists of a system of cables attached at one end to fittings around the top of the car framework and at the other end to the catenary curtains at the top of the envelope. The six cables attached to each of the outer curtains connected to five suspension points on each side of the car and to one point on the rear end of the car. The six cables which attached to each of the inner curtains bridled together approximately on the ship's axis, each with the corresponding cable from the other inner curtain, and from these bridles dropped to the car. The aft cables of this latter system continued from the ship's axis as a single cable which was attached ti the one point on the after end of the car. See Fig. 5.

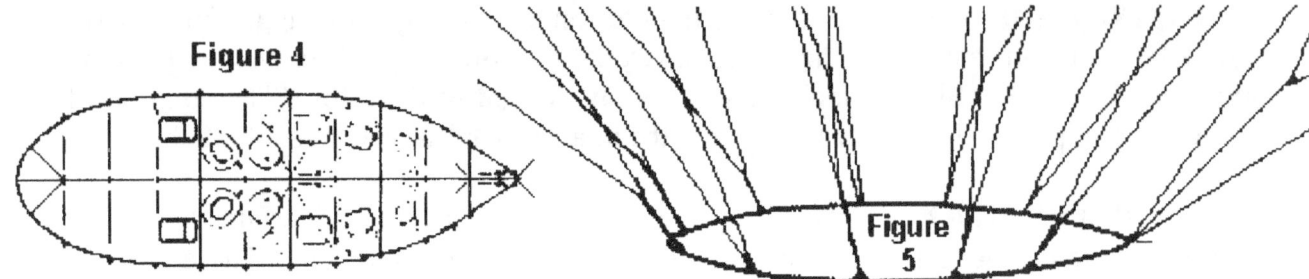

Figure 4

Figure 5

Shackle and link fittings gave universal connections between the cables and the ten suspension points on the car. Turnbuckles were provided in each cable just above the car to give sufficient adjustment in tensioning the cables. These turnbuckles are made accessible by fabric sleeves extending into the gas space, with a gas tight fitting where the cable enters the gas compartment. See Fig. 6.

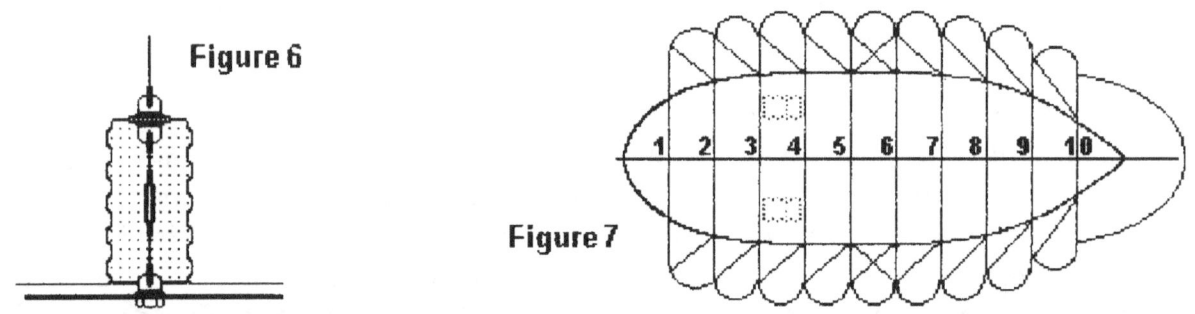

Figure 6

Figure 7

CABLE TENSIONS: Following are the car loading and suspension tensions calculated and used for the K-Ships at 1.5" WP. Any considerable changes in lift necessarily required a recalculation and adjustment. Although it is not possible to obtain the exact tensions, they should be approached as closely as possible. Tensions on opposite suspensions should be made reasonably equal (within 10%).

Figure 8

Frame#	1	2	3	4	5	6	7	8	9	10	11
Cable Inner System: (pounds)											
Tension	965	-	485	-	570	-	580	-	505	-	1010
Cable Outer System: (pounds)											
Tension	530	-	520	-	600	-	605	-	535	-	550
Load Distribution: (pounds)											
Tension	-	150	850	1000	750	450	800	900	650	150	-

ASSOCIATED SUPPORT: The above system was designed to support the entire weight. It was not to be confused with the Outside Catenary System which was designed to overcome transverse loads fore & aft (from thrust loads), and side loads (from yaw and gusts). Basically it kept the car in alignment and in position. See Fig. 7.

ENVELOPE cont.
R:12/04/97

BANNERS; I would like to discourage you from this fixation, but the temptation to make some side money will be too great. But let me list some reasons why I don't like them:

    1. It makes too much aerodynamic drag; and more at higher speed as it flutters;
    2. They are a hassle to make, and a hassle to get them to lay right;
    3. They are a hassle to change, and a hassle to get them to look right;
    4. The FAA will take a dim view at any message on an experimental aircraft.

Let me tell you this; Don't use sticky-back *Vinyl* on the envelope for a message if you ever expect to change it. The goo never comes off. If the message owner pays for the envelope, that's different. Painting a message on eliminates the first 3 objections. But no matter how you approach it, #4 will always be watching.

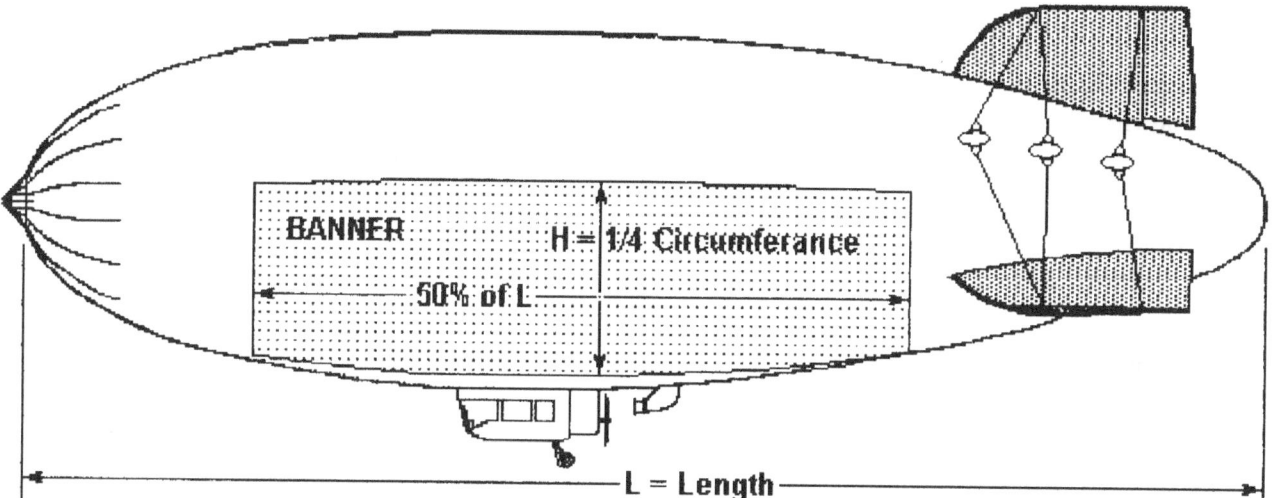

Note how the banner is offset in the drawing. The lower position allows it to be better seen better at 1000' above the ground. The tie-tabs should keep it under tension, and more will be needed on the forward end to keep it from pealing in the slipstream.

Building Small Gas Blimps

INFLATION
R:12/05/97

I have already mentioned several times that helium lifts 65 pounds pet 1000 cu.ft. For practical purposes, I'm happy to leave it at that. But if you are one of those ringers that is going to call me at 2 o'clock in the morning to say "It ain't so", go ahead and use the technical treatises in the referance section. But lets keep it simple, this ain't rocket science.

Getting Ready to place that order with the gas supplier? Weather you need 6000 (6K) or 60000 cu.ft.(60K), it pays to shop around for prices (by grade*). But be ready to answer some questions, like: How much do you need?
>(of course, cost ranges $10 to $100/1000)

When do you want it? (be flexible to his schedule,
>don't pick odd hours, weekends, or holidays)

How far out of town are you? Nearest cross roads?
>(better be specific if out in the boonies)

How do you want it? Bulk, Palletized, or Cylinders?
>(30K is 97 cylinders, ea. pallet 25; you need a fork lift)

Can I get my truck there without getting stuck?
>(containers are *very* heavy, any surprises will cost you)

How long will you need my truck and driver?
>(demurrage charge $20/hour or 100/day)

How do you plan to pay for it? (don't feel hurt,
>but your check may not be good enough)

If you only have an ultralight blimp of less than 10K (30 bottles) you may want to go get it yourself from the supplier. Remember that those bottles weigh 100+ pounds each and are back breakers to handle. I've used 50 in a day by myself, but I wouldn't wish that on anyone. A good trailer is the key.

Getting Set to Helium inflate requires that you are well organized. Your envelope is all laid out, with the nose reinforcements, pressure system and access ports all installed and tested. Fins attached and positioned. Your sand bags filled to maximum and tied securely, your net and sand bags are positioned. Hoses and connections ready. Crew is all there and fully briefed on the sequence of events, each individuals duties, what constitutes an emergency, AND signals for emergency stop. Be sure you have some hamburgers on the way because everybody is there until the gondola is attached, and that could be after sunset.

CAUTION: A simple gas diffuser is needed for an initial inflation. 2000 psi coming out of a filling nozzle will cut fabric like a knife.

GO, position gas supply as close to the filling port as possible. When the envelope is about 30% full, things will happen rapidly. Move fast, but NO RUNNING (yes the excitement is overwhelming). Pay attention to detail, Murphy's Law is waiting, and usually at the fin attachments.

* You want "the cheapest 99.9900%

Building Small Gas Blimps

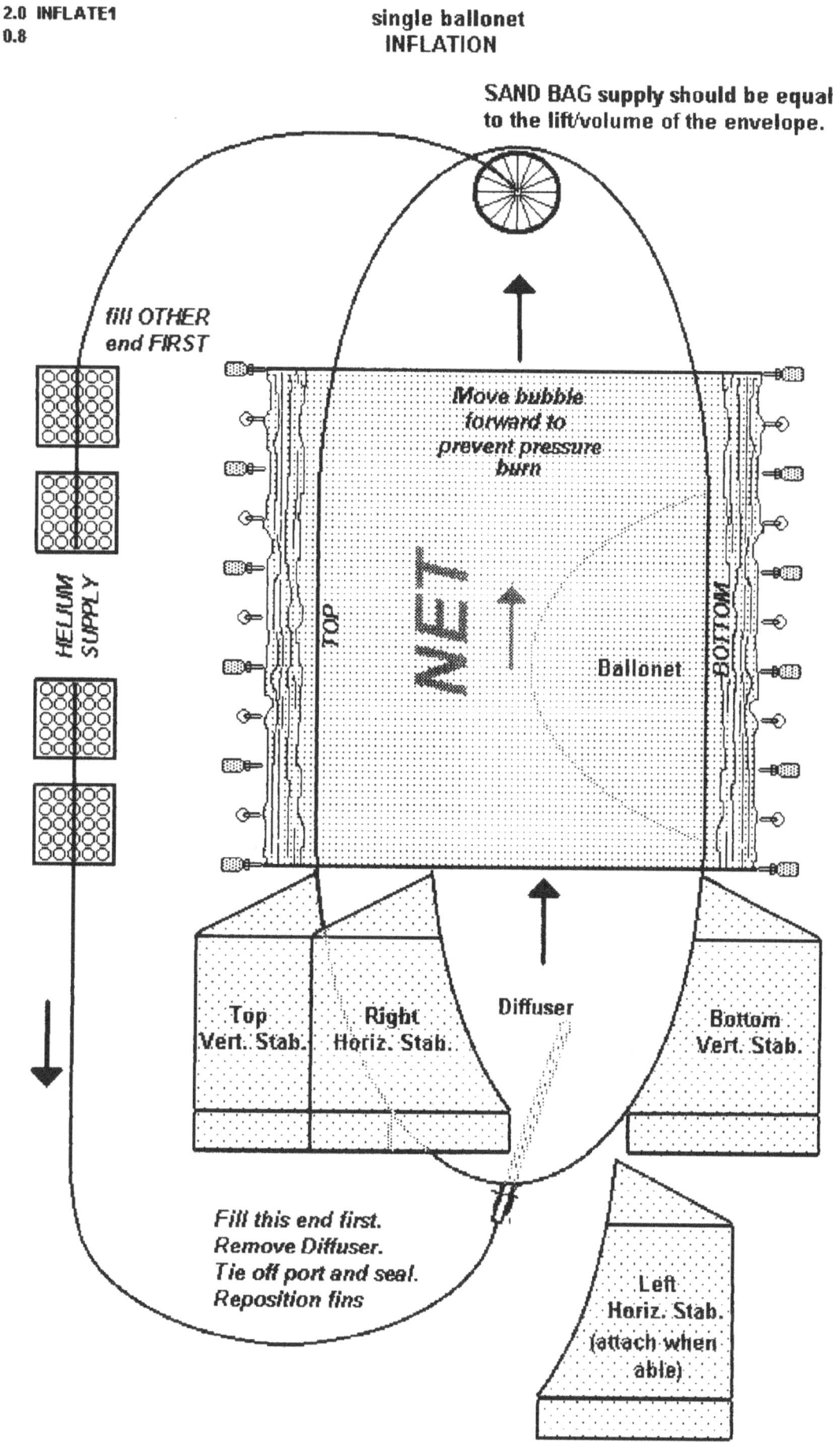

151

BIBLIOGRAPHY
R: 11/03/97

BOOKS:

Aerodynamic Theory Vol. VI	William F. Durand		194?
Dover Publications	New York, NY	629.	
Airship Aerodynamics	U.S. Government		1941
War Department	Washington, DC	629.141	
Airship Design	Chs. P. Burgess		1927
Ronald Press Co.	New York, NY	629.15/B912a	
Science of Flight, Vol. 1	P.H. Sumner		1926
Crosby/Lockwood	London	629.13/S956	
Ultralight Aircraft,	Michael Markowski		1982
U/L Publications,	Hummerstown, PA,	629.133.34	
Machinerys Handbook	Erik Oberg	621.02	1995
Principals of Aerostatics	U.S. Army Air Corps		1933
War Department	Washington, DC	629.141	

MAGAZINES:

Airship	Airship Association	(01149499-3515)	(quarterly)
4 Hamilton Place;	London, W1V 0VQ England		
Airship Homebuilder	Jessie Blens, author	(334/8976132)	(periodically)
Rt.2, Box 53-4;	Elba, AL 36323		
Gas Bag Journal	N. Zealand LTA Institute	(01164	(quarterly)
(subscribe from : ABAC; P.O. Box 90864; San Diego, CA 92169)			
Ultralight Magazine	E.A.A. Association,		(monthly)
P.O. Box 229;	Hales Corners, WI 53130		
Ultralight Flight	U.S.U.A Association	(423/629-5375)	(bi-monthly)
1085 Bailey Ave;	Chattanooga, TN 37404		

CATALOGS:

AeroCrafter Supply Co.	$20	Specialty aircraft parts	
940-G Adams Street; Benicia,	CA 94510		707 / 747-1509
Aircraft Spruce & Company		Specialty aircraft parts	
201 Truslow Av.(POB 424) Fullerton, CA 92632			800 / 824-1830
Chief Aircraft Supply		Aircraft parts	
1301 Brookside Blvd. Grants Pass, OR 97526			800 / 447-3408
Dwyer Instruments		Helium/gas pressure gauges	
102 Highway 212	Michigan City, IN 46361		219 / 879-8000
McMaster-Carr Supply Co.		Specialty parts & tools	
9630 Norwalk Blvd. Santa Fe Spr, 90670			310 / 695-2449
WAG Aero Supply		Aircraft parts	
1216 North Road; Lyons, WI 53148			800 / 558-6868
W.W. Grainger (nationwide)		Specialty equipment	
every US city over 100,000 population			800 /

Building Small Gas Blimps

Bibliography cont.

GOVT. PUBLICATIONS:
Most of these are available for you to review at your local FAA-FSDO office. They can also tell you where to obtain copies for your personal files.

Federal Aviation Regulations

Part 21	Certification	Part 23	Standards
Part 33	Engines	Part 35	Propellers
Part 43	Maintenance	Part 45	Markings
Part 47	Registration	Part 61	Pilots/Instructors
Part 91	Flight Rules	Part 103	Ultralights
Part 830	Accidents		

Advisory Circulars (AC's)

20-32	Carbon monoxide	20-43	Fuel control
20-36	Parts	23-9	Flight loads
43-3	Testing	43-13	Inspection
65-12	Powerplants	65-15	Airframes

The following is a must for your personal reference files:

FAA-P-8110-2   Airship Design Criteria

MANUALS:

Aero Fabric Coverings	Poly-Fiber Inc.; Riverside,	CA
Aircraft Batteries	Gill/Teledyne Products; Redlands,	CA
Airship Flight Manual 138S	Aerotek Airships (now U.S.L.T.A.)	OR
Ducted Fans	Hovey Enterprises; Newhall,	CA
Engines, L-2000 series	Limbach Flugmotoren;	Germany
Engines, Konig-430	Konig Motorenbau, Berlin	Germany
Engines, Rotax-3/4/500	Bombardier Gunskirchen	Austria
Ground Handling Airships	Aerotek Airships (now U.S.L.T.A.)	OR
Parachute Manual	Dan Pointer Santa Barbara	CA
Pilot Training Manual	Aerotek Airships (now U.S.L.T.A.)	OR

TECHNICAL REPORTS:

More than 3000 highly technical reports have been written on all aspects of airhips. The best places to find them are in the following order:

* 1. Assoc. of Balloon & Airship Constructors	San Diego,	CA
2. National Air & Space Library	Washington,	DC
3. Akron Public Library (LTAS collection)	Akron,	OH
4. Royal Aeronautical Society	London	England
5. Daniel Guggenheim Research Center	Akron,	OH
6. Univ. of Texas (Rosendahl collection)	N. Dallas	TX

Unknown/unverified/unseen

1. Goodyear archives	Akron,	OH
2. Lockheed archives (Goodyear?)	Lancaster,	CA

* http://abac.archivale.com and http://www.archivale.com/catalog

Building Small Gas Blimps

AER-D-13-MW Bureau of Aeronautics May, 1934
Technical Information Branch

Design Memorandum No. 169

Airship Fin and Rudder Loads

By

Charles P. Burgess.

In the nonrigid airships built for the Navy during and shortly after WW-1, no attempt was made to calculate the required or actual strength of the tail surfaces. The areas were based on coefficients derived from experience; and it was assumed that if the structural weight was about .75 lb./sq.ft..., disposed in accordance with good engineering judgment, the strength would be adequate. The original fins of D-Class nonrigid airships gave the first indication that this simple method of design might be inadequate. They were replaced by surfaces designed for a mean ultimate load of 10 lb./sq.ft... These surfaces gave long service without trouble, but were considered to be unnecessarily strong and heavy.

It is understood that verbal information was given by British liaison officers in this country during WW-1 that the British considered the tail surfaces for all classes of airships to be amply strong if designed for working loads of 1 lbs./sq.ft.. over the greater part of the area, increased to 2 lbs./sq.ft.. around the outer boundary of the fins and forward part of movable surfaces, with a safety factor of four.

In 1918, information was received that the surfaces of Italian M-Class semirigid airships were designed to the following criteria:
Fixed surfaces, 5 kg./sq.m. with safety factor of
5 = 5.1 lbs./sq.ft. ultimate load.
Movable surfaces, 12 kg./sq.m. with safety factor of
6 = 14.7 lbs./sq.ft. ultimate load.

The pressure distribution experiments carried out by the National Advisory Committee for Aeronautics on the tail surfaces of the Navy's nonrigid airship C-7 in 1022 indicated that the mean working load on the tail surfaces is not likely to exceed .4q, with maximums of 1.6q on the leading edges and overhanging balancing surfaces.

The **SHENANDOAH** was designed for a limiting safe speed of 50 knots at 6000' altitude. In this condition, q = 7.0 lbs./sq.ft.. The surfaces were designed for a mean ultimate load of 5 lbs./sq.ft. This is equivalent to 2.5 lbs./sq.ft., or .36q with a safety factor of 2.0. The top fin failed while the ship was riding to a high mooring mast

Building Small Gas Blimps

AER-D-13-MW　　　　　　　Bureau of Aeronautics　　　　　　　May, 1934
　　　　　　　　　　　　Technical Information Branch

in a heavy gale; and the outer boundary girders of other fins showed weakness in flight; but the surfaces were intact even after the ship broke in the air and crashed to the ground.

The surfaces of the **LOS ANGELES** were designed for a mean load of 5.0 lbs./sq.ft., or .33q, with safety factors of 2.0 to 2.5.

In the **AKRON** and **MACON**, the mean designed loading was 6.7 lbs./sq.ft. or .42q, on safety factors of 2 to 3.

The Airship Design Competition, 1927, called for the following loadings, based on a speed of 70 kts., to be increased as the square of the speed for higher speeds, with minimum safety factors of 2.5:

Forward third of fins,	22.5 lbs./sq.ft. =	1.5q
Remainder of fins	7.5 " " =	0.5q
Control surfaces	15.0 " " =	1.0q
Balancing surfaces	45.0 " " =	3.0q

Some designers protested against these loadings as unduly severe; and in the Airship Design Competition, 1928, no tail surface loadings were specified. It was left to the designers to provide surfaces of sufficient strength to meet the aerodynamic conditions specified for the general strength criteria of the airship as a whole.

It is not known what fin and rudder loadings, but assumed for the **GRAF ZEPPELIN** or the **R-100** and **R101**. The Airship Stressing Panel did not specify loadings, but assumed certain maneuvers for which the loadings were to be determined by wind tunnel experiments.

Modern nonrigid airship practice is exemplified by the metal framed fins of the Navy's J and K airships, both of which have given trouble in flight. The recent failure of the K-Class fins was so serious as to indicate the need for considerable reinforcement.

Conclusions and Recommendations.

The mean loading of 4q corresponds to the total fin and rudder force occurring at about 6 degrees angle of attack at the fins with full contrary rudder or elevator. Larger angles of attack are probably quite common in rough air, but at reduced speed owing to increased resistance. In exceptional cases, a violent gust may momentarily increase the air speed; and more severe assumptions than have commonly been made may be desirable to take account of such cases. On the other hand, the recent troubles which have occurred in flight have been indicative of local rather than general weakness. The local forces near the leading edge may well be considerably in excess of

AER-D-13-MW   Bureau of Aeronautics   May, 1934
                 Technical Information Branch

the design assumptions. Further theoretical and experimental investigations of general and local fin loads are recommended.

Exceptionally large fins, as on the **MACON**, should hold the ship to a steadier course with less angle of attack, and therefore less intensity of load than occurs with smaller fins; but the total load transferred to the ship's structure is probably increased by unusually large fins.

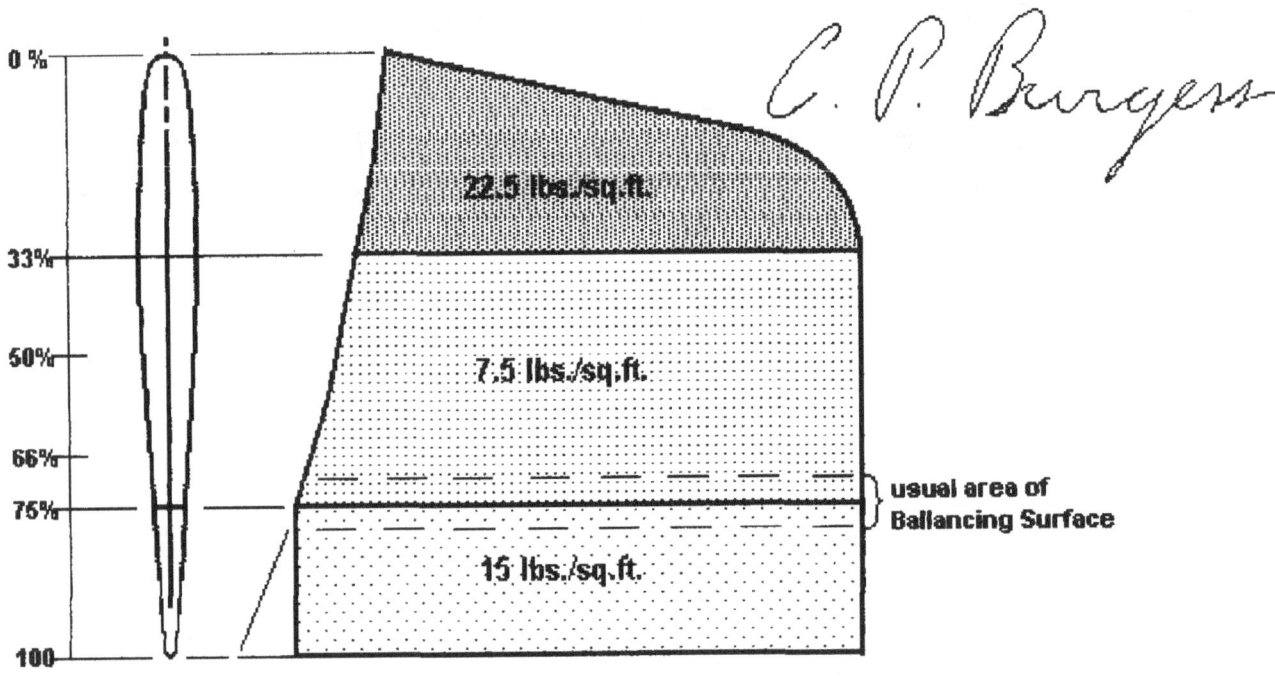

Building Small Gas Blimps

R: 12/03/97

Gas AIRSHIP Parts
SUPPLIERS

Supplier	Contact	Notes
AEROSTAR-Raven P.O. Box 1007 Sioux Falls, SD	605/336-2750 Fax/	ENVELOPE Mfgr. quality work
AMERICAN BLIMP Co. 1900 NE 25th Ave. Hillsboro, OR 97124	503/693-1611 Lightship Jim Thiele, Pres.	BLIMP Mfgr. A-60/A150 **status LIMITED**
BLIMP WORKS Route 2, Box 86 Statesville, NC 28677	704/876-6848 Fax/876-1251 Tracy Barnes, Pres.	BLIMP Mfgr. Whispership **status LIMITED**
BOULDER Blimp Co. 2840-E Wilderness Pl. Boulder, CO 80301	303/664-1122 Fax/449-2074 Frank Rider, Pres.	BLIMP Mfgr. Inflatable specialist **status LIMITED**
COMPOSITE Services P.O. Box 144 Eugene, OR 97440	541/344-5323 Bruce Blake, Pres.	BLIMP Engineer FAA-DER quality work
ILC-Dover; P.O. Box 266 Frederica, DE 19946	302/335-3911 quality work	ENVELOPE Mfgr. Retail fabrics
MAX IMAGE 205 Jefferson St. W. Memphus, ARK 72301	901/327-5960 Fax:327-5961 Steve Garner, Pres.	ENVELOPE Mfgr. reasonable prices good engineer
STEFAN LTA, Inc. 4825 Overkill Dr. Ft. Collins, CO 80626	970/223-9107 Fax/482-8907 Karl Stefan, Pres.	BLIMP Engineer FAA-DER **status LIMITED**
T-COM Aerostats 7115 Thomas Edison Drive Columbia, MD 21046-2113	410/312-2386 Weeksville based good quality control	FABRIC, Helium VALVES Burn-out rings
U.S. Lighter Than Air 1600 Valley River Rd. Eugene, OR 97401	503/683-4983 see N-Wave for parts Ray Olma, Pres.	BLIMP Mfgr. 138-S **status UNKNOWN**
WORLD WIDE Aeros 22900 Ventura Blvd. Woodland Hills, CA 91364	818/876-9144 Fax/876-9145 Igor Pasternak, Pres.	BLIMP Mfgr. **poor work** **NOT RECOMMENDED**

Catalog Section

ENGINES

HIRTH AIRCRAFT ENGINES

Aircraft Spruce & Specialty Company proudly announces its distributorship of Hirth aircraft engines, worldwide. Hirth aircraft engines, built in Germany, utilize some of the world's most advanced powerplant technology in their new two-stroke engine designs. Advances such as the Nikasil process, use of hypereutectic alloys, exhaust port design and high-strength chrome-moly steel crankshafts have made this new breed of engines vastly superior to older engine designs. These new advances have improved performance, reliability and safety while also yielding better economy, lower noise and much longer life. The compact size of the Hirth engines make for easy installations and the light weight will provide shorter takeoff, better climb and faster cruise. The 12 volt, 11 amp alternator provides standard electrical service and the electric starter delivers fast, easy starting. The exciting new F30 engine (four cylinder, two stroke) develops 110 HP at 5700 RPM, consumes only 7 GPH at full power and weighs only 84 lbs. complete. The Hirth F30 will certainly be the clear choice for light aircraft builders in need of efficient, quiet yet powerful engines. All Hirth engines come complete with carb(s), exhaust manifold(s), electric starter, alternator, rectifier, starter relay, air filter(s), fuel pump, ignition system, manuals, and one year warranty. A complete line of accessories such as reduction drives, tuned exhaust systems, engine mounts and dual ignitions are available.

HIRTH AIRCRAFT ENGINE PRICES AS OF 1-1-91

Model	Displacement	WT. (Lbs.)	HP	Part No.	Price
F22	383 cc	50	25	08-04800	$2047.00
263	383 cc	57	28	08-04810	$1830.00
F23	521 cc	60	50	08-04820	$1921.00
2703	521 cc	74	53	08-04830	$2217.00
F30	1042 cc	80	110	08-04840	$4224.00

HIRTH ENGINE INFORMATION PACK   P/N 08-04850 $8.00

THESE FEATURES MAKE A HIRTH ENGINE THE BEST POWER PLANT CHOICE

1. New technology design which creates more reliability and efficiency.
2. Nikasil cylinders which are extremely long lasting and almost totally sieze-proof.
3. One full year warranty covering parts & labor.
4. 1000 hour recommended TBO. Chances are that even after 1000 hours you will not have to do anything to cylinders.
5. No rotary valve. Simple, efficient port valve-ing with no extra moving parts.
6. Nikasil cylinders allow MUCH better heat dissipation in cylinders, reducing overheat worries (NO STEEL OR CHROME SLEEVES!).

ROTAX ENGINES

ROTAX 277 ENGINE

The Rotax 277 is the ideal engine for lighter weight ultralight aircraft. It delivers an amazing 28 HP at 6000 RPM and features shock mounted main bearings and a balanced crankshaft for smooth operation. High horsepower, compact design, reliability and light weight are important features of the Rotax 27 engine. It comes complete with a Bing 36MM (slide valve) carburetor, fuel pump, factory tuned exhaust, tool kit and owner's manual.

P/N 08-00100 (no Drive) $1194.00
P/N 08-00110 (with Gear Drive) $1952.00

ROTAX 503 ENGINE

The Rotax 503 is one of the highest horsepower ultralight/homebuilt aircraft engines available, producing 46 HP at 6250 RPM. Ideally suited for powering heavier 2-place ultralights, small homebuilts or standard ultralights from high altitude bases. It comes complete with a Bing 36mm (slide valve) carburetor, fuel pump, factory tuned exhaust, tool kit and owner's manual. Wt. 67 lbs.

P/N 08-00400 (no Drive) $1487.00
P/N 08-00410 (with Gear Drive) $2433.00

ROTAX 582 ENGINE

The new Rotax 582 2-cycle engine is a twin cylinder, liquid cooled engine which delivers 66 HP and improved mid-range torque and throttle response over the model 532. The 582 also offers the improved reliability and reduced maintenance of dual electronic ignition. It features an oil injection system which automatically adjusts fuel-to-oil ratio according to power setting, enabling owners to fill fuel tanks with straight gasoline. The new 582 comes complete with dual Bing carburetors, fuel pump, special tuned exhaust, spark plugs, tool kit, mounting studs and an operator's manual.

P/N 08-04570 (no Drive) $2689.00
P/N 08-04580 (with Gear Drive) $2978.00

ROTAX 912 4-CYCLE ENGINE

The new Rotax 912 engine is a 4-cylinder, 4-stroke opposed cylinder engine which delivers 79 HP at 5500 RPM. It includes CDI doubleignition with opposite side angle dual spark plugs in each cylinder. Weight-to-horsepower ratio is approximately .64 HP per pound. The Rotax 912 features built-in electric starter and a heavy duty gear reduction box. The engine weighs 123 lbs. including starter, carburetors, fuel pump, air filter, and oil system, and the exhaust system weighs 9 lbs.

P/N 08-04590 $7329.00

Other Rotax Models:

ROTAX 447 ENGINE (40HP)

P/N 08-00300 (no Drive) $1443.00
P/N 08-00310 (with Gear Drive) $1988.00

ROTAX 532 ENGINE

P/N 08-00450 (no Drive) $2389.00
P/N 08-00460 (with Gear Drive) $2694.00

Rotax 532 cooling system - call for price

ROTAX ENGINE ORDER FORM

Aircraft type _____
Engine Type _____
Cooling System _____
Carburetion (Dual or single) _____
Tractor _____ Or Pusher _____
Ratio of Reduction Drive _____
Style of Reduction (Up or Down) _____
Prop Bolt Mounting Pattern _____
Exhaust System Type _____

Optional Powerplant Equipment
Air Filter _____
Rotax Electric Start _____
ADS 5100 Electric Start _____
ADS 5200 Electric Start _____
Muffler Silencer Kit _____
Intake Silencer Kit _____
Remote Cable Choke Kit _____
Engine Primer System _____

HOW TO SELECT CORRECT ROTAX PART NUMBERS

To order an engine, please determine the correct part number by using the information below. Please state both engine and gear box part numbers when ordering.

1) Rotax Engine — R
2) Engine Model — 277, 377, 447, 503, 532, 582
3) Engine Type — FA - Free Air / FC - Fan Cooled / LC - Liquid Cooled
4) Carburetion — SC - Single Carb / DC - Dual Carb
5) Exhaust System — EX - Eipper / SM - Side Mount / ST - Straight / FI - Fisher

E.G...If you would want to order a Rotax 503 Fan Cooled, Dual Carb, with Straight Exhaust, the Part Number would be:

R503FCDCST — ROTAX / MODEL / TYPE / CARBURETION / EXHAUST TYPE

We carry all Rotax gear boxes, exhaust systems, and other accessories as well as KFM engines and accessories. Contact us for your best price on all ultralight and ARV powerplant parts requirements. Please verify current Rotax prices before ordering.

175
Prices Subject to Change Without Notice

AIRCRAFT SPRUCE & SPECIALTY CO.
P.O. BOX 424, FULLERTON, CA 92632

ORDER TOLL FREE (800) 824-1930
Or (714) 870-7551 • Fax (714) 871-7289

Homebuilts

Options for Hirth Engines

Hirth's most popular engine, the 2706 produces more horsepower and torque per pound than any other engine in its class.

TIFFEN, Ohio—Recreational Power Engineering has released several new products for use with Hirth aircraft engines.

The G-27 gearbox has been replaced with the new G-50 gearbox. The G-50 is a much stronger unit with an improved vibration dampener. All 2700 series engines ordered with a gearbox will now be supplied with G-50s.

A new double ball joint exhaust system is now available as an option on new engines and can be retrofitted to most of the existing Hirths in the field. Hirth has developed an improved intake manifold system for the 2706 65 hp engine. When equipped with the new manifolds, the 2706 will produce 42 percent more torque in the 4,500 rpm range and 14 percent more torque at full power. Besides the increase in performance, two side benefits of the new intake system are lower idle rpms and easier control of mid-range EGT temperatures.

Initial testing of older 2706 installations with the new intake have produced some very impressive results. Two plus degrees of additional pitch had to be added to the ground adjustable props in order to hold peak hp rpm to the recommended 6200. Climb performance increased substantially, but most impressive was that the cruise rpm drops 400 to 500 rpm. All new 2706s will be equipped with the new manifolds. An upgrade kit to convert older 2706s over to the new intake system is available from all authorized Hirth dealers.

For more information, call (419) 585-7002, or write: Hirth Recreational Power Engineering, 5479 E. Co. Rd. 38, Tiffin, OH 44883.

Catalog Section

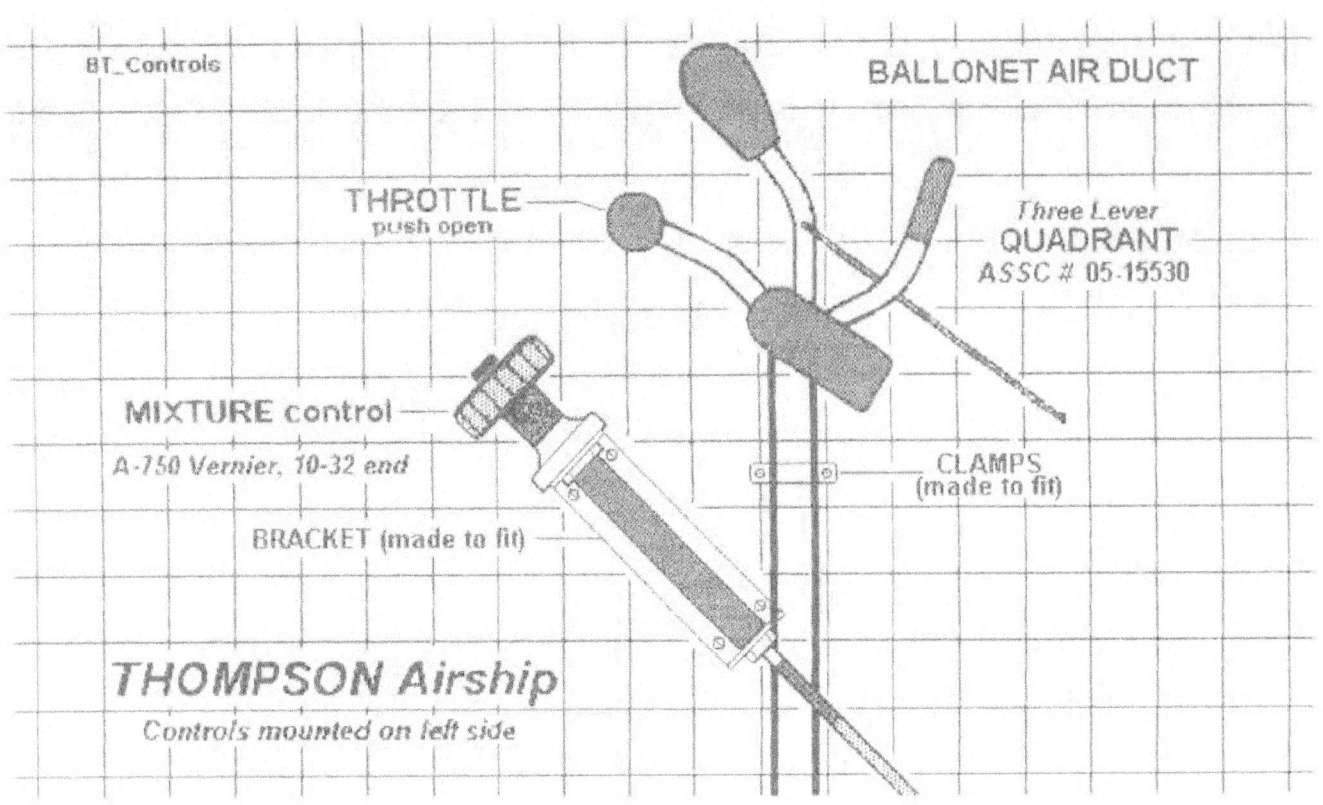

PRECISION THROTTLE QUADRANTS

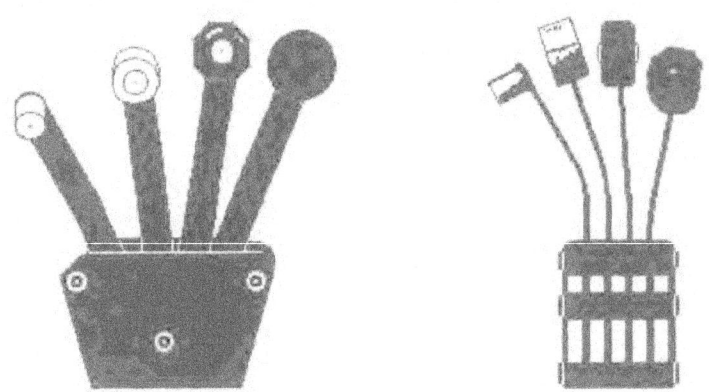

These quadrants are produced on a CNC mill for optimum precision and are 6061T6 aluminum anodized to a black finish. Available in single to four levers, these durable quadrants feature throttle (black) prop (blue), mixture (red), and alternate air (aluminum gray) respectively. A friction device offers adjustment to suit the pilot. The control cable end of the levers are left undrilled to allow builder to select hole location. Use with A-920 or A-930 push-pull controls. Ideal for any experimetal aircraft.

AIRCRAFT SPRUCE prices subject to change without notice.
Single Lever Quadrant (black knob)   P/N 05-15510 $147.00
Two Lever Quadrant (black, red knobs)   P/N 05-15520 $162.00
Three Lever Quadrant (black, red, blue knobs)   P/N 05-15530 $177.00
Four Lever Quadrant (black, red, blue, gray knobs) P/N 05-15540 $192.00
Quadrants with five to eight levers available on special order.

PHENOLIC PULLEYS

Part Number	Cable Size (In.)	A Dia.	B Dia.	C Dia.	D (Bore) Dia.	E	Allowable Load Limit (Lbs.)	Price Each
AN210-1A	1/16, 5/64, 3/32	1.250	.972	.625	.1900	.297	185	4.92
AN210-2A	1/16, 5/64, 3/32	2.500	2.222	.777	.1900	.297	500	5.84
AN210-3A	1/8, 5/32, 3/16	2.000	1.510	.901	.2500	.484	450	5.97
AN210-4A	1/8, 5/32, 3/16	3.500	3.010	.901	.2500	.484	1200	7.54
AN210-1B	1/16, 5/64, 3/32	1.250	.972	.625	.1900	.297	185	6.87
AN210-2B	1/16, 5/64, 3/32	2.500	2.222	.777	.1900	.297	500	7.82
AN210-3B	1/8, 5/32, 3/16	2.000	1.510	.901	.2500	.484	450	7.71
AN210-4B	1/8, 5/32, 3/16	3.500	3.010	.901	.2500	.484	1200	8.42
MS20219-1	1/16, 3/32	1.312	1.000	.423	.2500	.438	480	6.96
MS20219-2	1/16, 3/32	1.750	1.438	.423	.2500	.438	480	7.92
MS20219-3	1/16, 3/32	1.750	1.438	.769	.6250	.438	480	7.16
MS20219-4	1/16, 3/32	2.625	2.312	.423	.2500	.438	920	8.30
MS20220-1	1/8, 5/32, 3/16	1.755	1.255	.475	.3125	.625	500	8.42
MS20220-2	1/8, 5/32, 3/16	3.005	2.505	.475	.3125	.625	1680	13.49
MS20220-3	1/8, 5/32, 3/16	4.255	3.755	.475	.3125	.625	2500	14.48
MS20220-4	1/8, 5/32, 3/16	5.505	5.005	.475	.3125	.625	2500	15.86
A-223	1/16, 3/32, 1/8	1.000	.688	---	.2500	.281	---	1.45
A-224	1/16, 3/32, 1/8	1.750	1.375	---	.2500	.250	---	1.88
A-138 (NAS)	1/16, 3/32, 1/8	1.250	.875	---	.2500	.265	---	2.13
A-123 (NAS)	1/16, 3/32, 1/8	2.000	1.500	---	.2500	.265	---	2.71
A-124 (NAS)	1/16, 3/32	2.500	2.000	---	.2500	.265	---	2.85
A-130 (NAS)	1/8, 5/32	2.000	1.375	---	.3750	.438	---	2.95
A-120 (NAS)	5/32, 3/16	3.000	2.375	---	.3750	.438	---	3.55
BB-120-1	3/16	3.000	2.375	---	.2500	.440	---	9.39
BB-128-1	3/16	2.500	1.875	---	.2500	.440	---	9.12

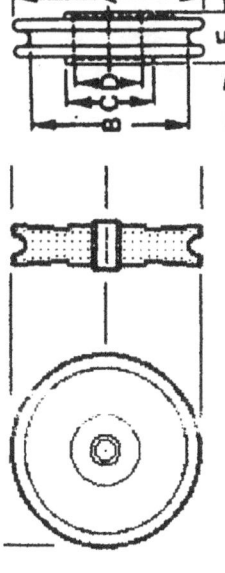

ALUMINUM PULLEYS

Part No	Price Each
MS20220A1	10.68
MS20220A2	18.62
MS20220A3	25.43
MS20220A4	29.27

Dimensions of these aluminum pulleys are same as equivalent size of MS20220 Phenolic Pulleys in table

ATCO SURPLUS PULLEYS

Obtained in a special purchase, these quality pulleys are offered at an excellent price. Quantity limited.

P/N	Cable Size	Bore	Price Each
05-	1/8", 5/32", 3/16"	.250	$1.35

Less 10% for 12, 15% for 25, 20% for 100 or more. May be assorted. No discount on surplus pulleys

Prices Subject to Change Without Notice

ORDER TOLL FREE (800) 824-1930
Or (714) 870-7551 • Fax (714) 871-7289

AIRCRAFT SPRUCE & SPECIALTY CO.
P.O. BOX 424, FULLERTON, CA 92632

Catalog Section

AeroCrafter™
NEW! HOMEBUILT AIRCRAFT SOURCEBOOK

304 page SECOND EDITION
OVER 500 PHOTOS & 3-VIEW PLANS

SAVE TIME & MONEY! For less than the cost of one or two Manufacturer's Info Packages, get this Huge, Illustrated "Bible of Homebuilt Aircraft" and...

- **SEE** every type of aircraft available - Fixed Wing, Ultralights, Rotorcraft, Gliders, Motorgliders, Sea Planes, Amphibians, Powered Parachutes.
- **DISCOVER** the fascinating world of Kit Planes that you can build!
- **COMPARE** the Specs, Performance, Costs, etc. on designs in Composite, Fiberglass, Metal, Wood, Fabric.
- **FULL COLOR BUILDER'S GALLERY** See and read the unbiased accounts of planes that people just like you have built and flown.
- **ALREADY STARTED BUILDING?** Save time and money finding the best sources for everything you'll need. ENGINES, HARDWARE, ELECTRONICS, TOOLS, COMPONENTS, ACCESSORIES, BOOKS, INSURANCE, FAA ADVISORIES, SCHOOLS, ASSOCIATIONS, SERVICES & MORE. Over 1200 Suppliers and Services for everything you'll need!

ORDER TODAY! (707) 747-1509

$20.00 plus $4.50 S/H ($6.50 Canada). CA res. add $1.45 tax.
Call 707 747-1509 Visa/Master Card or send Check or M.O. to:
AeroCrafter ™ 940 Adams St., #G(2), Benicia, CA 94510

Same day shipping via Priority Mail.

Catalog Section

SYNTHETIC WEBBING

For service contact our closest representative

UNITED TEXTILE & SUPPLY CO.

Los Angeles, California 90026
(213) 483-9600
Toll Free Numbers:
National (800) 421-8506
California (800) 252-0009

UNITEX-SOUTHWEST DIVISION
609 N. Great Southwest Parkway
Arlington, Texas 76011
Phone (817) 265-5301
Toll Free Numbers:
National (800) 433-5000
Texas (800) 772-5311

UNITEX-EASTERN DIVISION
501 Roosevelt Avenue
Pawtucket, Rhode Island 02862
Phone (401) 723-6500
Toll Free Number:
National (800) 556-7254

TIMOTHY BARANOUSKAS
48614 Tonopah Drive
Fremont, California 94538
(415) 657-0469

TIMOTHY DeBOE
206 Lions Club Road
Greenville, South Carolina 29611
(Terr. CA, VA, S.C. & N.C.)
(803) 246-1951

HANK O'MARA
4054 140th S.E.
Bellevue, Washington
(206) 746-9851

DOUG REED
RFD Rt. 10 Box 173
Cookeville, Tennessee 38501
(615) 526-2556

UNITED TEXTILE and SUPPLY CO.
DIVISION OF ELIZABETH WEBBING MILLS CO., INC.

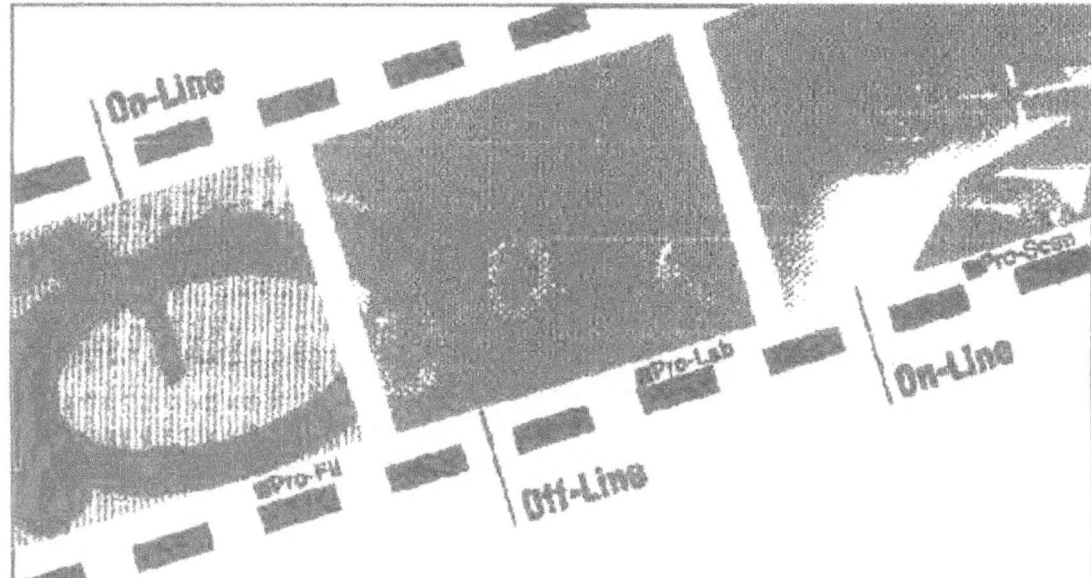

NEW PRODUCTS!

APPLIANCE/TOOL MOTORS

INDUSTRIAL/COMMERCIAL VACUUM MOTORS

7.2 INCH DIAMETER BYPASS VACUUM MOTORS/BLOWERS

AMETEK New

E47185 LR31393

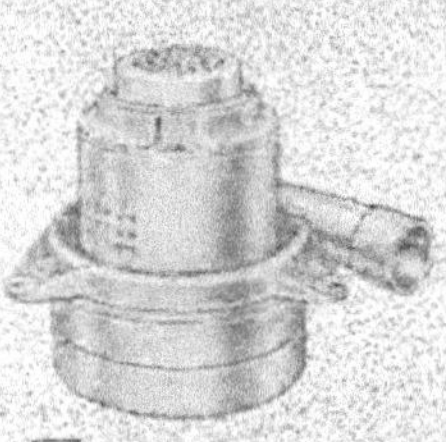

A No. 2M188 Tangential Discharge

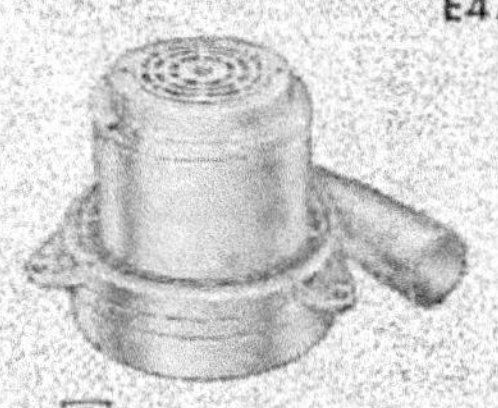

B No. 2M174 Tangential Discharge

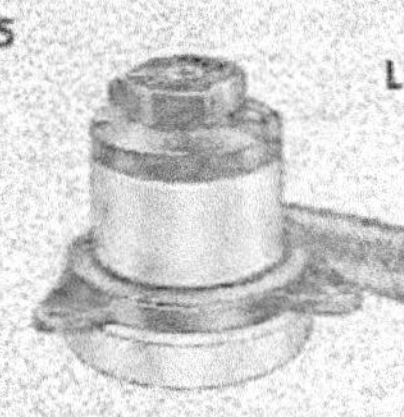

C No. 2M416 Tangential Discharge

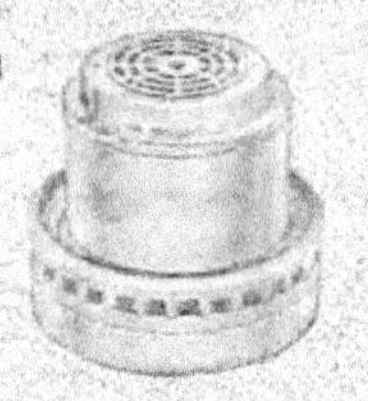

D No. 2M173 Peripheral Discharge

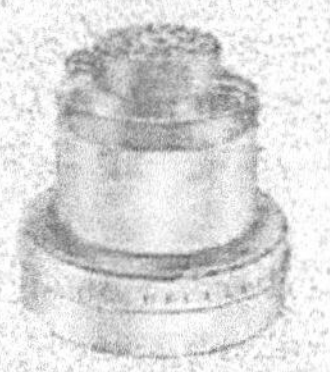

E No. 2M414 Peripheral Discharge

- Tangential or peripheral air discharge
- Copper windings

Typical Applications: Central vacuums, steam carpet cleaners, carwash vacuums, and commercial/industrial vacuum systems.

NOTE: Thru flow vacuum motors are to be used in dry applications only. Bypass motors should not come in contact with foam, liquids, or moisture laden air. Applications should be designed to protect the motor's fan system, housing, and electric components.

Bearings: Ball
Mounting: All position
Enclosure: Open
Thermal Protection: None
Insulation Class: A
Ambient: 40°C
Duty: Intermittent
Average Life: 700 hours
Brand: Ametek

FOR REPLACEMENT BRUSH MECHANISMS, SEE PAGE 187

Style	Blower Stages	Volts 50/60 Hz	Max. Amps	Overall Ht.	Vacuum (in H2O Sealed)	CFM (2" Orifice)	Max. Air Watts	Special Features*	Ametek Model		Stock No.	List	Each	Shpg. Wt.
A	3	120	13.8	7 59/64"	134.0	92.0	403	1, 3	116103-00		2M188	$256.00	$221.75	10.0
A	3	120	13.8	8 59/64	134.0	92.0	403	1, 3, 4	116161-00		4M884	269.00	232.75	10.0
A	3	120	12.6	7 29/32	134.0	94.0	438	1	116119-00		2M202	230.00	199.00	11.0
C	3	120	12.7	8 5/8	134.0	94.0	406	2, 3	117507-13		4M885	216.00	187.00	11.0
C	3	120	12.7	8 5/8	134.0	94.0	406	2	117507-00		2M419	196.00	170.00	9.0
D	3	120	14.7	7 13/16	131.4	105.0	447	1, 3	116118-00		4M891	250.00	216.50	10.0
E	3	120	12.0	8 5/8	134.5	92.0	395	2	117511-00		2M415	205.00	177.50	9.0
A	3	120	14.0	8 5/8	145.9	102.5	530	1	117500-12	New	3HV25	331.00	199.00	11.0
C	3	220	6.2	8 5/8	131.0	86.0	392	2	117741-00		4M882	212.00	183.50	11.0
A	3	220	5.5	8 59/64	106.0	87.0	306	1, 3	116136-00		4M883	281.00	243.25	10.0
A	3	240	7.0	8 5/8	145.6	103.2	511	1	117502-12	New	3HV26	239.00	204.50	11.0
A	2	120	12.7	6 15/16	110.1	105.4	392	1, 3	115937		2M178	235.00	203.50	9.5
A	2	220	5.5	6 15/16	95.0	98.9	308	1, 3	115950		4M888	261.00	226.00	9.0
B	2	120	13.0	6 11/16	110.0	102.7	384	1	115334		2M174	202.00	174.75	8.9
B	2	240	5.9	6 9/16	109.6	101.2	356	1	115684		4M887	242.00	209.50	9.3
C	2	120	12.8	7 11/16	123.0	110.0	455	2, 3	117467-13		4M889	204.00	176.75	10.0
C	2	120	12.8	7 11/16	123.0	110.0	455	2	117467-00		2M418	185.00	160.00	8.3
C	2	120	12.8	7 11/16	123.0	110.0	437	2, 3	117465-13		2M417	207.00	179.25	7.7
C	2	120	12.8	7 11/16	123.0	110.0	437	2	117465-00		2M416	186.00	161.00	7.5
C	2	240	6.1	7 11/16	114.8	108.1	440	2, 3	117157-13		4M886	202.00	174.75	10.0
C	2	100	14.3	7 9/16	114.9	104.1	415	2, 3	117917-13		4M890	206.00	178.25	9.3
D	2	240	5.5	6 9/16	91.9	106.8	312	1	115519		4M892	234.00	202.75	9.3
D	2	220	5.5	6 13/16	88.7	104.8	301	1, 3	115963		4M893	252.00	218.00	9.3
D	2	120	13.0	6 13/16	106.1	116.4	401	1, 3	115962		2M187	230.00	199.00	8.5
D	2	120	13.0	6 9/16	110.0	102.7	384	1	115330		2M173	195.00	168.75	8.5
E	2	120	12.9	7 9/16	113.3	104.0	407	2	117508-00		2M414	166.00	143.50	8.5
E	2	120	10.9	7 1/4	109.1	97.8	386	2	117560-01		4M894	228.00	197.50	9.2

(*) Special Features: (1) Metal mounting brackets; (2) Thermoset brackets; (3) Air sealed bearing protection; (4) 2-inch inlet tube

CAUTION: Not for fans in unattended areas.

Refer to page 5 for UL507 Standard, proper thermal protection, and other motor selection information.

7.5 INCH DIAMETER INDUSTRIAL/COMMERCIAL VACUUM MOTORS/BLOWERS

Applications: Commercial vacuums, dental evacuators, hopper loaders, and material handling and transfer systems.

Features: Permanently lubricated ball bearings, peripheral discharge, long life design, and copper windings.

Bearings: Ball
Mounting: All position

Enclosure: Open
Thermal Protection: None
Insulation Class: A (except Nos. 4M880 and 4M878 are Class B)
Ambient: 40°C
Duty: Intermittent
Average Life: 1000 hours
Brand: Ametek

AMETEK

E47185 LR31393

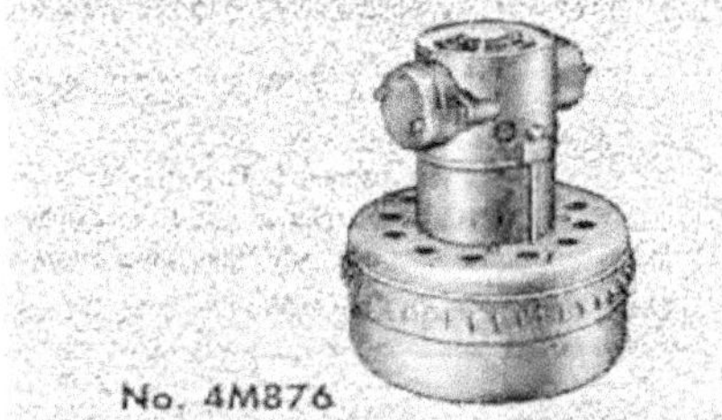

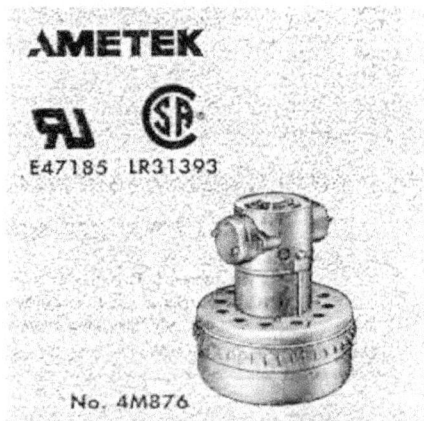

No. 4M876

Blower Stages	Volts 50/60 Hz	Max. Amps	Over-all Ht.	Vacuum (In. H2O Sealed)	CFM (2" Orifice)	Max. Air Watts	Spec. Feat.*	Ametek Model	Stock No.	List	Each	Shpg. Wt.
3	120	11.6	9 7/16"	88.2	126.0	360	—	114787	2M179	$405.00	$350.25	12.0
3	220	5.2	9 7/16	84.2	122.0	364	—	114789	4M877	497.00	429.75	13.0
3	42‡	30.0	9 7/16	87.9	126.0	360	—	115419	4M876	589.00	509.50	13.0
2	120	10.6	8 5/16	73.4	131.0	329	—	114786	4M881	401.00	347.00	11.0
2	220	4.8	8 5/16	70.2	146.0	329	—	114788	4M879	475.00	411.00	12.0
2	115	10.0	10 1/2	69.6	132.0	315	1	114586	4M880	1455.00	1260.00	16.0
2	230	5.0	10 1/2	61.2	123.0	299	1	114589	4M878	1537.00	1330.00	11.0

(*) Special Features: (1) Hazardous duty Class I Group D; Class II Groups E, F, and G. (‡) DC Volts.

Catalog Section

APPLIANCE/TOOL MOTORS — NEW PRODUCTS!

BRUSHLESS BLOWERS

5.7 INCH DIAMETER BRUSHLESS BLOWERS

- 50/60 Hz with continuous, long life operation
- Tangential discharge
- Aluminum construction
- Centrifugal fan system
- Brushless DC motor drive
- Low noise to power ratio
- Adjustable air performance
- Compact size
- Air intake and exhaust tube
- Mechanical control features a built-in 4-turn potentiometer
- Electronic control requires a remote 0-10 volt DC signal (not included)

Typical Applications: Business machines, computer peripherals, material handling, air samplers/evacuators, packaging equipment, photographic equipment, and medical equipment.

Bearings: Ball
Mounting: All position
Enclosure: Open
Thermal Protection: Auto
Insulation Class: A
Ambient: 40°C
Brand: Ametek

AMETEK
UL E99403
CSA LR43448
No. 4M961

Watts	Blower Stages	Volts 60 Hz	Max. Amps	Overall Height	Vacuum (In. H₂O Sealed)	CFM (2" Orifice)	Control	Ametek Model		Stock No.	List	Each	Shpg. Wt.
250	1	120	4.6	5⅞	34.0	72.0	Mechanical	116634-01		4M965	$619.00	$402.00	6.0
250	1	120	4.7	5⅞	34.0	72.0	Electronic	116640-01		4M968	623.00	404.50	5.0
250	1	120	4.8	6¼	28.0	129.0	Mechanical	116637-03		4M966	628.00	408.00	5.5
250	1	120	4.8	6¼	28.0	120.0	Electronic	116643-01		4M967	633.00	411.25	4.9
250	2	120	4.8	7⅛	49.0	90.0	Mechanical	116636-03		4M963	648.00	421.25	7.0
250	2	120	4.8	7⅛	49.0	90.0	Electronic	116642-01		4M964	653.00	424.00	5.3
250	3	120	5.2	7⅞	75.0	50.0	Mechanical	116632-06		4M961	654.00	424.75	6.5
250	3	120	5.2	7⅞	75.0	50.0	Electronic	116638-08		4M962	658.00	427.25	6.0
400	1	240	14.0	5⅞	26.0	74.0	Mechanical	117634-01	New	3HV28	692.00	449.50	5.0
400	1	240	14.0	5⅞	26.0	74.0	Electronic	117640-01	New	3HV33	696.00	452.50	5.0
400	1	240	14.0	6¼	25.0	115.0	Mechanical	117637-01	New	3HV31	702.00	456.25	5.0
400	1	240	14.0	6¼	25.0	115.0	Electronic	117643-02	New	3HV36	708.00	460.00	5.0
400	2	240	14.0	7⅛	39.0	100.0	Mechanical	117636-01	New	3HV30	725.00	471.25	5.0
400	2	240	14.0	7⅛	39.0	100.0	Electronic	117642-01	New	3HV35	730.00	474.25	5.0
400	3	240	14.0	7⅞	50.0	86.0	Mechanical	117635-03	New	3HV29	724.00	470.50	5.0
400	3	240	14.0	7⅞	50.0	86.0	Electronic	117641-01	New	3HV34	748.00	486.25	5.0
400	3	240	14.0	7⅞	60.0	58.0	Mechanical	117632-02	New	3HV27	731.00	475.00	5.0
400	3	240	14.0	7⅞	60.0	58.0	Electronic	117638-03	New	3HV32	736.00	478.00	5.0
800	1	120	9.0	6⅛	36.0	140.0	Mechanical	119102-01	New	3HV44	918.00	597.00	5.0
800	1	120	9.0	6⅛	36.0	140.0	Electronic	119101-01	New	3HV43	921.00	599.00	5.0
800	2	120	9.0	7¼	56.0	107.0	Mechanical	119104-01	New	3HV46	924.00	601.00	5.0
800	2	120	9.0	7¼	56.0	107.0	Electronic	119103-01	New	3HV45	927.00	603.00	5.0
800	3	120	9.0	7⅞	85.0	68.0	Mechanical	117418-01	New	3HV24	949.00	617.00	5.0
800	3	120	9.0	7⅞	85.0	68.0	Electronic	117417-01	New	3HV23	949.00	617.00	5.0
1200	1	240	9.0	6¼	46.0	176.0	Mechanical	119152-02	New	3HV48	960.00	624.00	5.0
1200	1	240	9.0	6¼	46.0	176.0	Electronic	119151-01	New	3HV47	933.00	606.50	5.0
1200	2	240	9.0	7¼	80.0	137.0	Mechanical	119154-01	New	3HV50	965.00	627.00	5.0
1200	2	240	9.0	7¼	80.0	137.0	Electronic	119153-01	New	3HV49	936.00	608.50	5.0
1200	3	240	14.0	7⅞	118.0	85.0	Mechanical	117416-01	New	3HV22	997.00	648.00	5.0
1200	3	240	14.0	7⅞	118.0	85.0	Electronic	117415-01	New	3HV21	997.00	648.00	5.0

VISIT THE GRAINGER WEB SITE TO PLACE YOUR ORDER!

http://www.grainger.com

Link up with Grainger's Web site and tap into a fast, convenient way to access up-to-date information.

Our online catalog allows you to search and display products as well as place orders online... day or night.

DESK SPACE AT A PREMIUM?

USE THE GRAINGER CD-ROM, ELECTRONIC CATALOG TO REVIEW THOUSANDS OF PRODUCTS!

Our Grainger Electronic Catalog is a state of the art CD-ROM selection guide which features products and information from our General Catalog.

The perfect reference tool when you need product and technical information on the road. Contact your local branch for more information.

GRAINGER — BUSINESS TO BUSINESS SALES

Catalog Section

RUNNING RIGGING

SeaFit Economy Polyester Yacht Braid — New!

Description: Traditional double-braid polyester line with a filament cover and braided core
Stretch: 3.5% @ 15% of breaking strength
Optimum Use: Non-critical applications that do not require low stretch or good abrasion resistance
Benefits: Most economical double braid available
Comments: Offers some of the benefits of more expensive braids without a high price tag, but will not last as long as higher quality line. Moderate stretch characteristics. All white, with no tracer.

Dia.	Avg. Brk. Strength	Model	ShWt 100 ft.	List/ft
1/4"	1,800	367468	4	$.44
5/16"	2,400	367476	5	.64
3/8"	3,700	367490	6	.72
7/16"	5,000	367484	8	1.00
1/2"	7,000	367500	10	1.20

Sta-Set X Polyester Yacht Braid

Description: Patented all-polyester braid with parallel core construction, and a filament/spun cover for improved handling and greater abrasion resistance
Stretch: 1.6% @ 15% of breaking strength
Optimum Use: Cruising boat halyards and sheets, club racer running rigging
Benefits: 1/3 less stretch than Sta-Set at only a small price premium
Comments: In many applications, Sta-Set X can be used in lieu of expensive racing braids, especially when sized conservatively. Soft cover improves grip in the hand or on a winch. Identified by two fine red tracers in the cover forming the characteristic "X".

The most cost-effective way to reduce stretch!

Dia.	Avg. Brk. Strength	Model	ShWt 100 ft.	List/ft
3/16"	1,600	221341	3	$.34
1/4"	2,700	198689	4	.52
5/16"	4,400	198697	6	.71
3/8"	5,500	198705	7	.89
7/16"	7,400	198713	8	1.16
1/2"	9,600	198721	10	1.44
5/8"	15,000	198739	17	2.28

Color Coded Sta-Set X Polyester Braid

White Sta-Set X with red, blue, or green "flecks" for easy identification. Same qualities as white Sta-Set X.

Dia	Avg. Brk. Strength	Model Red	Blue	Green	ShWt 100 ft.	List/ft
3/16"	1,600	276891	188864	188914	3	$.35
1/4"	2,700	276883	188870	152134	4	.54
5/16"	4,400	267296	152159	152140	6	.75
3/8"	5,500	152118	188880	152126	7	.93
7/16"	7,400	276875	188898	188920	8	1.17
1/2"	9,600	276867	188906	188930	10	1.48

Sta-Set Polyester Yacht Braid

Description: Strong and durable polyester double braid with a braided core and continuous filament braided cover
Stretch: 2.4% at 15% of breaking strength
Optimum Uses: All cruising boat applications, racing boat control lines
Benefits: Cost-effective, long lasting, lots of color choices for simplified sailing
Comments: The mainstay of the New England Ropes yachting ropes. A great balance of stretch, strength, suppleness, and cost. Identified by a fine red tracer

The long-lasting way to improve performance!

Dia	Avg. Brk. Strength	Model	ShWt 100 ft.	List/ft
3/16"	1,200	121996	3	$.34
1/4"	2,000	121988	4	.49
5/16"	3,000	122028	5	.70
3/8"	4,400	122010	6	.87
7/16"	6,000	122044	8	1.07
1/2"	8,200	121970	9	1.40
9/16"	11,000	122051	16	1.86
5/8"	14,000	122036	17	2.22
3/4"	20,000	122002	20	2.78

Color Coded Sta-Set Polyester Braid

White Sta-Set with red, blue or green "flecks" to make identifying each line quick and easy. Same qualities as white Sta-Set.

Dia	Avg. Brk. Strength	Model Red	Blue	Green	ShWt 100 ft.	List/ft
3/16"	1,200	121657	121533	121590	3	$.36
1/4"	2,000	121640	121525	121582	4	.53
5/16"	3,000	121673	121558	121616	5	.73
3/8"	4,400	121665	121541	121608	6	.91
7/16"	6,000	121681	121566	121624	8	1.17
1/2"	8,200	121632	121517	121574	9	1.48

Solid Color Sta-Set Polyester Braid

In blue, green or black with a bold white tracer and fine red tracer or red with bold white tracer. Same qualities as white Sta-Set.

Dia	Avg. Brk. Strength	Model Red	Blue	Green	Black	ShWt 100 ft.	List/ft
3/16"	1,200	243659	243642	243667	352971	3	$.38
1/4"	2,000	243683	243675	243691	352989	4	.57
5/16"	3,000	243717	243709	243725	352997	5	.81
3/8"	4,400	243741	243733	243758	315028	6	1.00
7/16"	6,000	243774	243766	243782	353003	8	1.30
1/2"	8,200	243808	243790	243816	315036	9	1.63

NEW ENGLAND ROPES

Catalog Section

RIGGING WIRE AND TOOLS

Loos & Co. Marine Wire Rope
The unique chemistry of this top-quality rigging cable has been developed and perfected for sailboat applications. Each strand is preformed and polished. This process eliminates the tendency of the wire to collect carbon particles and become brittle. Exceptionally strong Loos wire rope is available in several different configurations and materials.

1 x 19 Brite-Stay™ Stainless Wire
Ideal for standing rigging applications.
Stiff, one strand cable made of 19 wires twisted together.

Type 302/304 Commercial Grade Stainless Steel
Standard pack is a 250' spool. Sold per foot.

Cable Dia.	Brk. Strength	Model	LOOS#	ShWt/ft.	List
3/32"	1,200 lb.	119305	SC09419LYR	.2	$.39
1/8"	2,100 lb.	119289	SC12519LYR	.2	.62
5/32"	3,300 lb.	119339	SC15619LYR	.2	.74
3/16"	4,700 lb.	119297	SC18819LYR	.2	1.05
7/32"	6,300 lb.	479808	SC21919LYR	.2	1.21
1/4"	8,200 lb.	479790	SC25019LYR	.2	1.60
9/32"	10,300 lb.	479782	SC28119LYR	.2	1.94
5/16"	12,500 lb.	479774	SC31319LYR	.2	2.39
3/8"	17,500 lb.	479766	SC37519LYR	.2	3.33

Type 316 Extra Corrosion-Resistant Stainless Steel
Standard pack is a 250' spool. Sold per foot.

Cable Dia.	Brk. Strength	Model	LOOS#	ShWt/ft.	List
3/16"	4,000 lb.	188948	SZ18819LYR	.2	$1.39
1/4"	6,900 lb.	188955	SZ25019LYR	.2	1.99
9/32"	8,700 lb.	188963	SZ28119LYR	.2	2.08
5/16"	10,600 lb.	188971	SZ31319LYR	.2	3.15
3/8"	14,800 lb.	188989	SZ37519LYR	.2	4.09

7 x 19 Flexible Stainless Wire
Ideal for running rigging applications
Strong, medium flexibility cable commonly used for wire halyards. Each of the seven strands consists of 19 wires. Type 302/304 domestic stainless. Standard Pack is 250' spool. Sold per foot.

Wire Dia.	Brk. Strength	Model	LOOS#	ShWt/ft.	List
1/16"	700 lb.	258121	SC06379	.2	$.44
3/32"	1,000 lb.	119214	SC09479	.2	.95
1/8"	1,760 lb.	119198	SC12579	.2	.99
5/32"	2,400 lb.	119248	SC15679	.2	1.30
3/16"	3,700 lb.	119206	SC18879	.2	1.35
7/32"	5,000 lb.	173361	SC1979	.2	1.70
1/4"	6,400 lb.	299537	SC25079	.2	1.80

 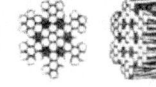 

7 x 7 Lanyard Wire
1/16" diameter wire only. Similar construction to lifeline wire, but no vinyl cover. Available on 250' spool. Sold per foot.

Model 119354 **LOOS#** SC06377
ShWt .2 List $.44/Ft.

7 x 19 Galvanized Wire
Standard pack is 500' spool. Sold per foot.

Dia.	Strength	Model	LOOS#	ShWt	List
3/16"	4,200 lb.	163925	GC18879	.2	$.60
1/4"	7,000 lb.	254500	GC19579	.2	.65

Loos Professional Swaging Tool
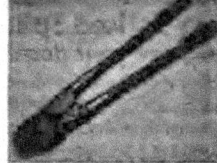
A powerful seven-hole swaging tool featuring adjustable jaws made of the finest grade hardened steel. The adjustment allows you to compensate for tool wear over time. Compresses oval sleeves for wire sizes: 1/32", 3/64", 1/16", 3/32", 1/8", 5/32", and 3/16". 6.5 lb.; 26" long.
Model 323816 LOOS#NO.1SC
ShWt 15 List $196.28

Landman Swaging Tool
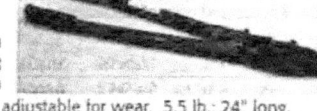
Here is a quality tool that covers a wide range of sleeve sizes. It swages seven sizes of oval and stop sleeves from 1/16" to 3/16". Not adjustable for wear. 5.5 lb.; 24" long.
Model 410530 LANDM#HSC-600
ShWt 6 List $191.03

S.F.Tool Hand Swaging Tools

Swage your own oval sleeves with this cadmium-plated steel tool. Pressure is transmitted to the copper sleeve by tightening two bolts on the tool with a wrench. A must for every tool box.

Fits Wire	Model	S&F#	ShWt	List
1/16", 3/32", 1/8"	129817	NO2	2	$20.85
1/8", 5/32", 3/16", 1/4"	129825	NO3	4	44.35

Boatman's Wire Cutters
The shearing action of the precision blades gives a clean, sharp cut to steel cable, rather than spreading and squashing the way some cutters do. These cutters are comparable in quality to models you've seen costing two or three times as much!

Cuts Wire	Model	LANDM#	ShWt	List
to 1/4"	206037	WR-6	5	$87.66
to 3/8"	206011	WR-10	7	124.26
to 9/16"	206029	WR-14	10	150.33

Professional Wire Cutters

These excellent cutters will handle wire rigging to 3/16" diameter. Plastic-coated cushion-grip handles make it easier on your hands when you're cutting lots of wire at a time.
Model 413203 LANDM#RC-8
ShWt 4 List $45.50

Wire Cutters

These wire cutters are made of heat-treated steel and will cut up to 3/16" wire. Protected against rust.
Model 129833 S&F#316
ShWt 2 List $17.95

Catalog Section

SAILBOAT HARDWARE

Avibank Stainless Fastpins

Model	Dia	Length	AVIBK#	ShWt	List
102533	3/16"	1/2"	3D0.50	.5	$4.41
102558	3/16"	1"	3D1.00	.5	4.41
102566	3/16"	1 1/2"	3D1.50	.5	5.13
102616	1/4"	1/2"	4D0.50	.5	5.23
102632	1/4"	1"	4D1.00	.5	6.51
102640	1/4"	1 1/2"	4D1.50	.5	7.10
102657	1/4"	2"	4D2.00	.5	7.90
102723	5/16"	1"	5D1.00	.5	6.37
102731	5/16"	1 1/2"	5D1.50	.5	7.33
102749	5/16"	2"	5D2.00	.5	7.33
102798	3/8"	1"	6D1.00	.5	6.51
102806	3/8"	1 1/2"	6D1.50	.5	7.90
102814	3/8"	2"	6D2.00	.5	7.90
102822	3/8"	2 1/2"	6D2.50	.5	9.04

Avibank Ball-Lok Quick-Release Pins

Model	Dia	Usable Length	AVIBK#	ShWt	List
102574	3/16"	1/2"	3M0.50	.5	$13.42
102582	3/16"	3/4"	3M0.75	.5	13.42
102590	3/16"	1"	3M1.00	.5	13.42
102665	1/4"	1/2"	4M0.50	.5	14.02
102681	1/4"	1"	4M1.00	.5	14.02
102699	1/4"	1 1/2"	4M1.50	.5	16.72
102707	1/4"	2"	4M2.00	.5	16.72
102756	5/16"	3/4"	5M0.75	.5	14.58
102764	5/16"	1"	5M1.00	.5	14.58
274357	5/16"	1 1/2"	5M1.50	.5	16.79
274365	5/16"	2"	5M2.00	.5	16.79
273920	5/16"	2 1/2"	5M2.50	.5	20.27
102830	3/8"	1"	6M1.00	.5	15.33
102848	3/8"	1 1/2"	6M1.50	.5	17.35
102855	3/8"	2"	6M2.00	.5	17.35
102863	3/8"	2 1/2"	6M2.50	.5	20.27

Monel Seizing Wire

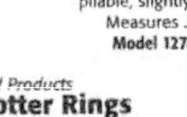

Top-quality, non-corrosive, pliable, slightly magnetic. Measures .037" x 30'.
Model 127854 ShWt .5 List $8.95

S&J Products Cotter Rings

Stainless

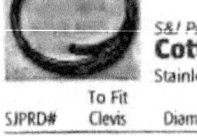

SJPRD#	To Fit Clevis	Diam.	Qty./Pk.	Model	ShWt	List
R1	3/16"	3/8"	10	272286	.5	$1.99
R2	1/4"	7/16"	6	272294	.5	1.99
R3	5/16"	3/4"	4	272302	.5	1.99
R4	3/8"	7/8"	4	272310	.5	1.99
R5	1/2"	1 1/8"	4	272245	.5	1.99
R6	5/8"	1 3/16"	4	272252	.5	1.99

Cotter Pins and Rings Mini-Kit

Contains 4 popular sizes of cotter pins and 5 sizes of cotter rings. All stainless steel. 60 pins and 49 rings total.
Model 243139 SJPRO#970006
ShWt 3 List $15.25

Johnson Clevis Pin Kits

Assorted sizes of stainless-steel clevis and cotter pins.
Small Boat Kit—46 pieces. 3/16"–5/16".
Model 420679 JOHNS#37-504
ShWt 1 List $21.40
Big Boat Kit—36 pieces. 5/16"–1/2".
Model 420661 JOHNS#37-505
ShWt 2 List $35.00

S&J Products Clevis Pin Assortment

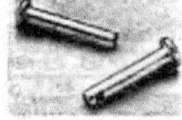

Contains common clevis pins used on boats from 8 to 27' long. 15 clevis pins from 1/4" to 3/8" diameter, and the cotter pins and rings to match.
Model 489757 SJPRO#970023
ShWt 2 List $25.21

S&J Products Clevis Pins

Stainless steel. Poly bagged.

Model	SJPRD#	Diam.	Length	Head	#Per Pkg.	ShWt	List
445553	CL22	3/16"	13/32"	5/16"	2	.5	$3.04
304220	CL23	3/16"	1/2"	5/16"	2	.5	3.04
242727	CL1	3/16"	3/4"	5/16"	2	.5	3.04
243188	CL2	3/16"	7/8"	5/16"	2	.5	3.04
242735	CL3	3/16"	1"	5/16"	2	.5	3.04
242743	CL4	3/16"	1 1/8"	5/16"	2	.5	3.04
448367	CL29	1/4"	7/16"	3/8"	2	.5	3.67
304212	CL30	1/4"	1/2"	3/8"	2	.5	3.67
242750	CL5	1/4"	3/4"	3/8"	2	.5	3.67
242768	CL6	1/4"	7/8"	3/8"	2	.5	3.67
242776	CL7	1/4"	1"	3/8"	2	.5	3.67
242784	CL8	1/4"	1 1/8"	3/8"	2	.5	3.67
242792	CL9	1/4"	1 1/4"	3/8"	2	.5	3.67
334961	CL34	5/16"	9/16"	7/16"	2	.5	3.04
334979	CL36	5/16"	11/16"	7/16"	2	.5	3.04
242800	CL10	5/16"	3/4"	7/16"	1	.5	3.04
242818	CL11	5/16"	1"	7/16"	1	.5	3.04
242826	CL12	5/16"	1 1/4"	7/16"	1	.5	3.04
242834	CL13	5/16"	1 1/2"	7/16"	1	.5	3.04
242842	CL14	5/16"	2"	7/16"	1	.5	3.04
334987	CL40	3/8"	3/4"	1/2"	1	.5	3.67
242859	CL15	3/8"	1"	1/2"	1	.5	3.67
242867	CL16	3/8"	1 1/2"	1/2"	1	.5	3.67
242875	CL17	3/8"	2"	1/2"	1	.5	3.67
304238	CL43	7/16"	3/4"	7/8"	1	.5	4.20
242883	CL18	1/2"	1"	5/8"	1	.5	4.20
334995	CL48	1/2"	1 3/32"	5/8"	1	.5	4.20
335000	CL49	1/2"	1 1/4"	5/8"	1	.5	4.20
242891	CL19	1/2"	1 1/2"	5/8"	1	.5	4.20
242909	CL20	1/2"	2"	5/8"	1	.5	4.20

S&J Products Cotter Pins

Stainless

Model	SJPRD#	Diam.	Length	#Per Pkg	ShWt	List
257931	C1	1/16"	1/2"	24	.5	$1.99
283895	C2	1/16"	3/4"	18	.5	1.99
257956	C3	1/16"	1"	18	.5	1.99
257964	C4	1/16"	1 1/2"	10	.5	1.99
257972	C5	1/16"	2"	6	.5	1.99
257980	C6	3/32"	5/8"	10	.5	1.99
257998	C7	3/32"	3/4"	10	.5	1.99
258004	C8	3/32"	1"	8	.5	1.99
258012	C9	3/32"	1 1/2"	6	.5	1.99
258020	C10	3/32"	2"	3	.5	1.99
258038	C11	1/8"	3/4"	7	.5	1.99
258046	C12	1/8"	1"	6	.5	1.99
258053	C13	1/8"	1 1/4"	4	.5	1.99
258061	C14	1/8"	1 1/2"	4	.5	1.99
258079	C15	1/8"	1 3/4"	4	.5	1.99
258087	C16	1/8"	2"	3	.5	1.99
258095	C17	1/8"	2 1/2"	3	.5	1.99
258103	C18	5/32"	1 1/2"	3	.5	1.99
258111	C19	5/32"	2"	3	.5	1.99
258129	C20	5/32"	2 1/2"	2	.5	1.99

Cotter Ring Mini-Kit

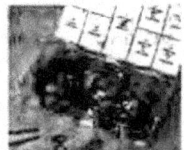

Contains 30 common cotter rings used on boats. Stainless.
Model 489799 SJPRO#970039
ShWt 2 List $4.75

Cotter Pin Mini-Kit

An assortment of 44 handy stainless steel cotter pins for most jobs on board.
Model 489781 SJPRO#970038
ShWt 2 List $5.21

CLEVIS PINS AND COTTER PINS

ANTI-COLLISION LIGHTING SYSTEM

POSITION LIGHTS AND ANTI-COLLISION LIGHT DISTRIBUTION PATTERNS REQUIREMENTS

Write for a complete catalog of products
WHELEN ENGINEERING
Route 145, Winthrop Road
Chester, CT 06412-0684
Tel: (860) 526-5904 Fax: (860) 526-4078

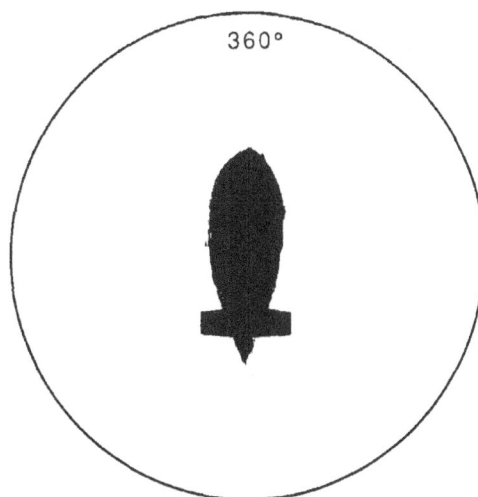

FIG. 5. An approved anti-collision strobe light system must project light 360 around the aircraft's vertical axis. One or more strobe lights can be used.

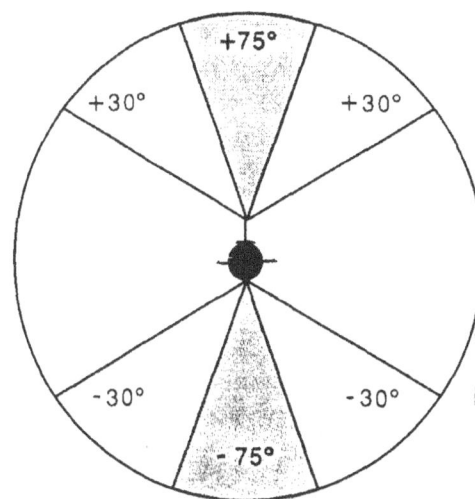

FIG. 6. An approved anti-collision strobe light system must project light + or - 30 above and below the horizontal plane of the aircraft. One or more strobe lights can be used. The + or - 75 projected light is required since July 18, 1977.

FIG. 7. Approved light pattern in the horizontal plane. The anti-collision wingtip mounted lights must converge within 1200 feet directly in front and rear of the aircraft on center line. If the wingtip strobe light convergence is greater than 1200 ft. in back of the aircraft, a 3rd light is necessary.

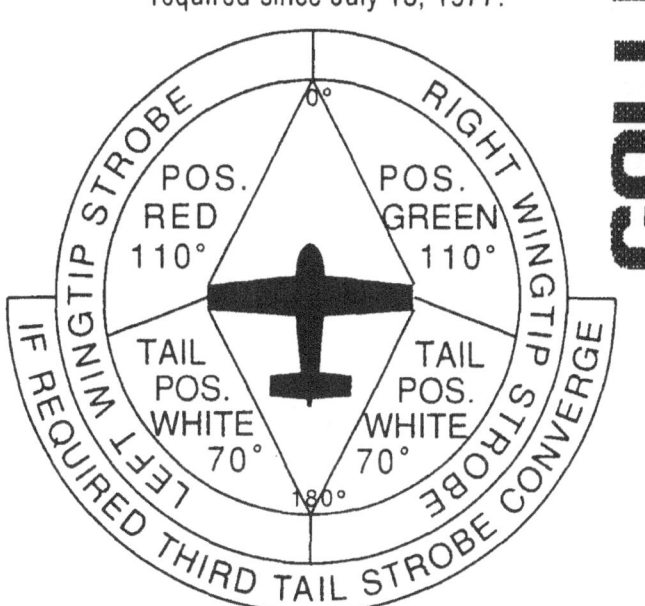

Catalog Section

RIGGING ACCESSORIES

Copper Oval Sleeves
Used to make eyes in 7x19 wire rope. Zinc-plated copper.

Size	Model	LANDM#	ShWt	List
1/16"	115022	S-11-002	.25	$.25
3/32"	115063	S-11-003	.25	.49
1/8"	115048	S-11-004	.25	.51
5/32"	115071	S-11-005	.25	.80
3/16"	115055	S-11-006	.25	1.44
7/32"	185025	670135	.25	1.70
1/4"	115030	S-11-008	.25	2.02

Copper Stop Sleeves
Circular sleeves that put a slug of metal on a wire for halyard locks.

Size	Model	LANDM#	ShWt	List
1/16"	494260	670610	.25	$.37
3/32"	494278	S-60-003	.25	.54
1/8"	494286	S-60-004	.25	.61
5/32"	494294	S-60-005	.25	.69
3/16"	494302	S-60-006	.25	.83
1/4"	494310	S-60-008	.5	2.29

Loos Rig Tension Gauges
Allows accurate and quick adjustment of wire rigging. Tension is read in pounds. Anodized aluminum. Professional model has heavy-duty spring and "self-holding" capability.

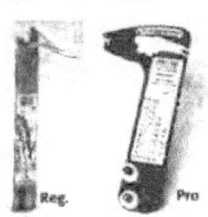
Reg. / Pro

Style	For Wire	Model	LOOS#	ShWt	List
Reg.	3/32"-5/32"	179358	91A	1	$45.87
Reg.	3/16"-9/32"	179366	90F	1	48.32
Pro	3/32"-5/32"	289027	PT-1	2	66.58
Pro	3/16"-1/4"	289019	PT-2	2	69.00

Stainless / Galvanized Iron

Wire-Rope Clamps

Material	Size	Model	ShWt	List
Stainless	1/8"	272088	.5	$3.50
	5/32"-3/16"	272096	.5	3.95
	7/32"-1/4"	272104	1	4.75
	9/32"-5/16"	272112	1	6.35
	3/8"	480194	1	9.40
Galvanized	1/8"	135483	.5	$.58
	3/16"	135491	.5	.69
	1/4"	135475	.5	.75
	5/16"	534354	1	.89
	3/8"	534362	2	1.10

Nylon "O" Rings
Injection-molded nylon. May be used for hanging gear or light webbing assemblies. Not suitable for use in critical applications. ShWt .5.

Stock Dia.	Inside Dia.	Model	RONST#	List
3/16"	3/4"	283848	PNP52C	$.64
5/16"	1 1/4"	283830	PNP52B	.86

Single-Shank Stainless Ball Stop
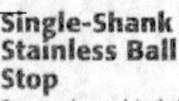
Commonly used in halyards. Type 303 stainless steel.

Size	Model	LOOS#	ShWt	List
1/16"	111213	MS 664-C2	.5	$.49
3/32"	111221	MS 664-C3	.5	.73
1/8"	111239	MS 664-C4	.5	.85
5/32"	111247	MS 664-C5	.5	1.29
3/16"	111254	MS 664-C6	.5	1.69
7/32"	111260	MS 664-C7	.5	4.49

Stainless Thimbles

Size	Model	WHARD#	ShWt	List
1/16"	126789	645.01	.5	$.39
3/32"	265066	645.02	.5	.60
1/8"	126763	645.03	.5	.70
5/32"	126805	645.04	.5	.58
3/16"	126771	645.05	.5	.90
1/4"	126813	645.06	.5	1.20
5/16"	126797	645.08	.5	1.97
3/8"	126821	645.10	.5	4.99

Size	Model	SEDOG#	ShWt	List
1/2"	135285	170012	.5	5.40
5/8"	373043	170016	.5	6.65
3/4"	373050	170020	.5	14.65

Schaefer Halyard Exit Plates

Lessens halyard chafe without the expense of a sheave box. Stamped stainless steel. ShWt 1.

Halyard Size	Style	Cutout Size	Model	SCHFR#	List
to 1/2"	Flat	2 1/4"x13/16"	149492	34-46	$7.15
to 3/4"	Flat	3 1/4"x1"	163857	34-48	17.15
to 3/4"	Curved	3 1/4"x1"	285470	34-49	18.20

Seke Jamstopper

Clamps around any rope or wire to act as a stopper or dead end. Use it on the forestay to eliminate overhoisting, or to stop jib hanks from fouling on the turnbuckle. Use it on a halyard to keep the halyard shackle from jamming in the mast sheave box. Fits 1/4" rope or wire, but can be drilled to fit up to 3/8".

Model 254617 SEKE#JAMSTOPPER
ShWt 1 List $7.95

Ronstan Rope End Stoppers

Makes lines up to 1/4" easy to grasp and prevents them from running free. 1 1/8" ball has countersunk hole to hide bitter-end knot. Also used to prevent halyards from jamming in sheave boxes. A 1/8" oval swage will fit inside the countersunk portion.

Color	Model	RONST#
Green	545749	RF1315GRN
Black	545723	RF1315BLK
Red	545756	RF1315R
Blue	545731	RF1315BLU

ShWt .5 List $1.99

Stainless Steel "D" Rings

Polished heliarc-welded stainless steel rings. ShWt .5.

Wire Dia.	Inside Width	Model	COLET#	List
3/16"	1"	534214	PD1810	$5.20
1/4"	1 1/2"	534222	PD2515	6.32
1/4"	2"	111252	PD2520	7.37

Stainless Steel "O" Rings
Heliarc welded and polished. Rings will tend to distort before breaking, providing indication of overload. ShWt .25.

Wire Dia.	Inside Dia.	Min. Break Strength	Model	COLET#	List
1/8"	3/4"	2028	534149	RR1207	$3.41
3/16"	1"	4564	111609	RR1810	5.50
3/16"	1 1/4"	4564	534156	RR1812	5.81
1/4"	1"	8246	111641	RR2510	6.31
1/4"	1 1/4"	8246	534164	RR2512	6.89
1/4"	1 1/2"	8246	534172	RR2515	7.37
5/16"	1 1/2"	12844	534180	RR3115	8.41
5/16"	2"	12844	111716	RR3120	10.52
3/8"	2"	18456	534198	RR3720	13.59
7/16"	2 1/2"	25028	534206	RR4125	19.02

Stainless Steel Rectangle Loops
Heliarc welded, in the dimensions shown. Can be used with eyestraps to attach webbing to surfaces, or sewn to fabric as a webbing pass-thru. ShWt .5.

Wire Dia.	Inside Dim.	Model	COLET#	List
3/16"	1/2" x 1"	534255	LS1810	$4.25
3/16"	1/2" x 2"	534263	LS1820	5.20

Stainless Steel Triangle Loops
Use for backstay adjusters or other applications where you need to connect three blocks together. Can also be used to terminate webbing assemblies. Heliarc welded, in the dimensions shown. ShWt .5.

Wire Dia.	Inside Width	Model	COLET#	List
3/16"	1"	534248	PT1810	$6.33
1/4"	2"	534230	PT2520	7.37

Stainless Steel "S" Hooks
Polished stainless steel. USA.

Wire Dia.	Inside Width	Model	COLET#	ShWt	List
1/8"	1/2"	534271	SH1250	.5	$1.35
3/16"	5/8"	534289	SH1806	.5	1.85

Stainless Steel Quick Links
Very strong 316 stainless ovals that can be used to join line, chain, shackles, or other assemblies. Barrel bolt closes easily with fingers, but should be tightened with a wrench for ultimate security. Each ShWt .5.

Wire Dia.	Length	Min. Break Strength	Model	SEDOG#	List
1/8"	1 1/4"	2420	599282	153003	$4.10
1/4"	2 1/4"	7150	599290	153006	7.60
5/16"	2 15/16"	12000	599308	153008	13.00
3/8"	3 1/2"	19800	599316	153010	18.95

SAILBOAT HARDWARE

Catalog Section

SNAP HOOKS & SHACKLES

 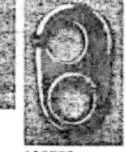 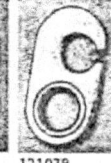

Ronstan
Sister Clips
Sold individually. ShWt .5

Size	Material	Dimensions	MWL	Model	RONST#	List
Small	Nylon	1 5/8" long x 11/32" I.D.	—	286940	PNP16	$1.03
Small	Stainless	15/16" Long x 1/4" I.D.	150	110793	RF536	2.10
Medium	Stainless	1 9/16" Long x 3/8" I.D.	550	121079	262	4.15

Stainless Tangs

Each ShWt .5.
Flat

Width	Length	SCHFR#	Model	List
1/2"	3"	85-13	252379	$5.25
1/2"	4"	85-14	251769	5.55
1/2"	6"	85-16	252387	6.65

West Marine
Safety Carabiner for Harness Tethers

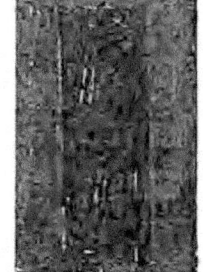

This is a very good quality snap that is very difficult to cause to release accidentally, yet allows one hand attachment and detachment. Stainless steel with a locking barrel. About 4" long with a 5/8" opening. Safe Working Load: 1500 lb. Use with locking clip to keep webbing from moving away from end of snap.

Model 528745 WHARD#575.10 ShWt 1 List $14.34

West Marine
Genius Carabiners
They even fit over large mooring eyes!

These stainless carabiners feature a really ingenious innovation. The gate opens at an angle to allow "fatter" rings to fit into the carabiner.

Length	SWL	Model	WHARD#	ShWt	List
3 1/8"	500 lb.	317057	590-08	.5	$7.95
4"	750 lb.	317065	590-10	.5	11.35
5"	1000 lb.	317073	590-12	.5	13.95

Wichard
Stainless Snap Hooks

Drop forged for great strength. Asymmetric, with eye

Length	SWL	Model	WICHD#	ShWt	List
2 3/8"	1100 lb.	273912	2323	1	$12.70
3 5/32"	2640 lb.	274332	2325	1	16.70
4"	3630 lb.	274340	2326	1	22.45

Stainless Carabiners

With eye

Length	SWL	Model	WHARD#	ShWt	List
2"	250 lb.	115980	557.05	.5	$4.95
2 3/8"	250 lb.	115998	557.06	.5	6.25
2 3/4"	400 lb.	116004	557.07	.5	6.95
3 1/8"	500 lb.	116012	557.08	.5	7.95
4"	750 lb.	116020	557.10	.5	11.25
5"	1000 lb.	116038	557.12	.5	14.25

Without eye

Length	SWL	Model	WHARD#	ShWt	List
2"	250 lb.	115923	555.05	.5	$4.50
2 3/8"	250 lb.	115931	555.06	.5	5.70
2 3/4"	400 lb.	115949	555.07	.5	6.95
3 1/8"	500 lb.	115956	555.08	.5	7.50
4"	750 lb.	115964	555.10	.5	10.50
5"	1000 lb.	115972	555.12	.5	13.50

Wichard
Stainless Steel Shackles

Wichard shackles are known for their beautiful quality and smooth contours. They are drop forged from stainless steel. The forging process ensures that the shape, unlike cast shackles, remains intact at loads which far exceed the deformation point. The quality is legendary, and deservedly so. Each ShWt .5.

Bow Shackles—captive pin

Model	Pin Dia.	Inside Width	Inside Length	Bkg. Load	WICHD#	List
116178	3/16"	5/16"	11/16"	3310	1442	$10.15
116186	1/4"	1/2"	1 1/8"	4630	1443	13.55
116194	5/16"	5/8"	1 1/2"	8160	1444	19.80
116202	3/8"	13/16"	1 3/4"	11465	1445	31.70

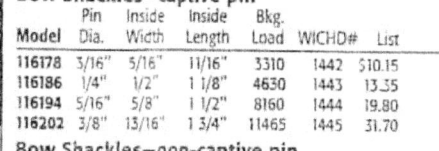

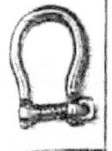

Bow Shackles—non-captive pin

Model	Pin Dia.	Inside Width	Inside Length	Bkg. Load	WICHD#	List
179721	5/32"	5/16"	23/32"	1750	1241	$5.95
179739	3/16"	13/32"	29/32"	2200	1242	7.75
179747	1/4"	15/32"	1 3/32"	3525	1243	8.60
179754	5/16"	5/8"	1 15/32"	5950	1244	13.60
179762	3/8"	25/32"	1 27/32"	9480	1245	22.50

Keypin Shackles—captive pin

Model	Pin Dia.	Inside Width	Inside Length	Bkg. Load	WICHD#	List
116301	3/16"	1/2"	1 1/2"	2645	1432	$10.80
116319	1/4"	5/8"	1 3/4"	3745	1433	14.45
116327	5/16"	7/8"	2 3/8"	5500	1434	26.20

Keypin w/Bar Shackles—captive pin

Model	Pin Dia.	Inside Width	Inside Length	Bkg. Load	WICHD#	List
116335	3/16"	1/2"	1 1/2"	2645	81432	$15.30
116343	1/4"	5/8"	1 3/4"	3745	81433	19.60
116350	5/16"	7/8"	2 3/8"	5500	81434	31.70

D Shackles—captive pin

Model	Pin Dia.	Inside Width	Inside Length	Bkg. Load	WICHD#	List
116111	5/32"	5/16"	1/2"	2200	1401	$7.60
116129	3/16"	13/32"	5/8"	3300	1402	9.25
116137	1/4"	15/32"	25/32"	4850	1403	11.85
116145	5/16"	5/8"	1 1/32"	8150	1404	18.10
116152	3/8"	25/32"	1 5/16"	11500	1405	28.40
116160	1/2"	15/32"	1 17/32"	15400	1406	43.20

D Shackles—non-captive pin

Model	Pin Dia.	Inside Width	Inside Length	Bkg. Load	WICHD#	List
179671	5/32"	5/16"	17/32"	1765	1201	$5.00
179689	3/16"	13/32"	5/8"	2200	1202	5.75
179697	1/4"	15/32"	25/32"	3525	1203	7.45
179705	5/16"	5/8"	1 1/32"	5950	1204	11.75
179713	3/8"	25/32"	1 5/16"	9500	1205	19.25

Long D Shackles—captive pin

Model	Pin Dia.	Inside Width	Inside Length	Bkg. Load	WICHD#	List
116251	5/32"	5/16"	1 7/32"	2200	1411	$9.25
116269	3/16"	3/8"	1 9/16"	3300	1412	10.90
116277	1/4"	1/2"	2"	4850	1413	14.80
116285	5/16"	5/8"	2 1/2"	8150	1414	23.45
116293	3/8"	13/16"	3 1/8"	11500	1415	37.75

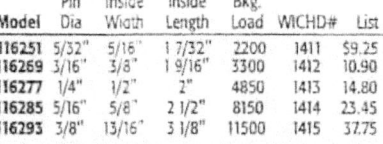

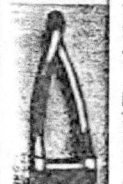

Twisted D Shackles—captive pin

Model	Pin Dia.	Inside Width	Inside Length	Bkg. Load	WICHD#	List
116210	3/16"	5/16"	1 1/2"	3310	1422	$11.95
116228	1/4"	1/2"	1 3/4"	4630	1423	16.10
116236	5/16"	5/8"	2 1/8"	8160	1424	24.75
116244	3/8"	13/16"	3"	11465	1425	39.85

Harken
Stamped D Shackles ShWt .5.

Model	Pin Dia.	Inside Width	Inside Length	Bkg. Load	HARKN#	List
114223	3/16"	7/32"	1/2"	1500	072	$2.20
196329	1/4"	5/8"	11/16"	2000	138	3.65

Schaefer
Stamped D Shackles ShWt .5.

Model	Pin Dia.	Inside Width	Inside Length	Bkg. Load	SCHFR#	List
138412	3/16"	27/64"	39/64"	1000	93-32	$5.50
138404	1/4"	7/16"	5/8"	1250	93-31	5.20
138420	1/4"	1/2"	13/16"	1750	93-33	4.70
138396	1/4"	11/16"	1"	1750	93-21	5.90

Catalog Section

Page 24 SEAT BELTS & HARNESSES
HARNESS INSTALLATION KITS

Because We Know Safety Is A Priority In Your Airplane!

Although your aircraft wasn't manufactured with shoulder restraint systems, we have developed restraint systems for many popular aircraft. These systems were developed in cooperation with the FAA. *Each H-600 Series is fully FAA/STC/PMA approved* for your installation and comes in a full array of colors. Our line of seat belts and harnesses are considered the best by professionals. Kits shown include one standard seat belt with 3-bar slide (retain your present end fitting.), one "Y" style shoulder harness (with attaching hardware), and installation instructions.

Our line of seat belts and harnesses is constructed with quality materials and careful attention to detail. Each installation kit shown includes one belt, one harness, attaching hardware and installation instructions. The standard seat belt with 3-bar slide and the Y-style shoulder harness included is shown and described further on the next pages.

Choose from the following kits for your FAA/STC/PMA approved installation!

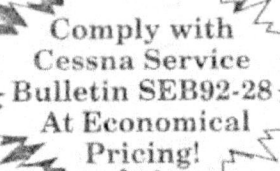

CESSNA

Typical Installation for Cessna Aircraft

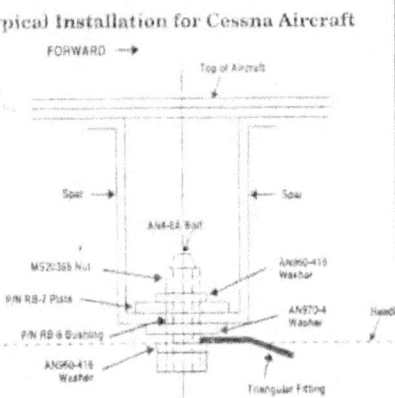
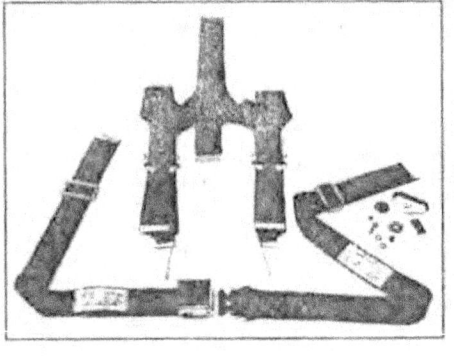
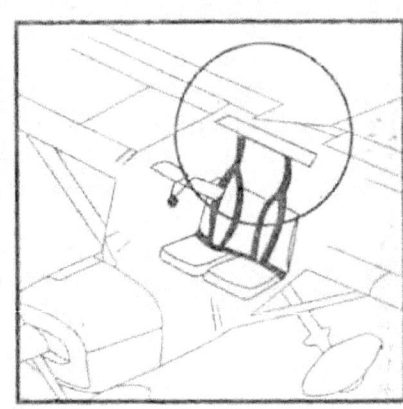

Only $138.50 each
All Code 4

Color	120/140	150/152	170	172A-I,K,L	175A,B,C	180	182, A-H J-N, P-Q T,TR182	185	206	207
Black	H600-000	H601-000	H602-000	H603-000	H615-000	H604-000	H605-000	H606-000	H612-000	H613-000
Brown	H600-100	H601-100	H602-100	H603-100	H615-100	H604-100	H605-100	H606-100	H612-100	H613-100
Red	H600-200	H601-200	H602-200	H603-200	H615-200	H604-200	H605-200	H606-200	H612-200	H613-200
Navy	H600-300	H601-300	H602-300	H603-300	H615-300	H604-300	H605-300	H606-300	H612-300	H613-300
Ryl Blue	H600-400	H601-400	H602-400	H603-400	H615-400	H604-400	H605-400	H606-400	H612-400	H613-400
Tan	H600-500	H601-500	H602-500	H603-500	H615-500	H604-500	H605-500	H606-500	H612-500	H613-500
Slv Gray	H600-600	H601-600	H602-600	H603-600	H615-600	H604-600	H605-600	H606-600	H612-600	H613-600
Olive Green	H600-700	H601-700	H602-700	H603-700	H615-700	H604-700	H605-700	H606-700	H612-700	H613-700

    For Customer Service Call 1-414-763-9586 For 24-Hr FAX Ordering Call 1-414-763-7595
Prices and Supply Subject To Change Without Notice

Catalog Section

GENERATORS

CHICAGO ELECTRIC PORTABLE & CONTRACTOR GENERATORS POWERED BY BRIGGS & STRATTON

5000 WATT MAX 4400 WATT RATED PORTABLE GENERATORS

Up to 14 hours of power to run lights, power tools, or even welders, anywhere. Heavy duty 8 HP Briggs & Stratton engines. 120V outlets: (2) 20 amp grounded receptacles; 240V outlets: (2) 15 amp grounded receptacles; Engine: 8 HP, 3600 RPM, 3" bore, 2-3/4" stroke, 60 Hz, single phase, 36.7/18.3 amp brushless alternator; Peak wattage: 5000; Fuel consumption: .8 gallons per hour @ full load; Fuel tank: 1 gallon capacity

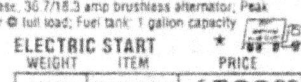

RECOIL START			ELECTRIC START		
WEIGHT	ITEM	PRICE	WEIGHT	ITEM	PRICE
119 lbs.	03123-7ANA	$479.99	114 lbs.	03357-0ANA	$589.99

8 HP, 125 AMP WELDER/4400 WATT MAX, 4000 WATT RATED GENERATOR

This unit delivers 125 amps of raw welding power and can also output AC power for tools, pumps, lights, etc. Excellent for emergency power and construction in remote locations.
- Powered by a 8 HP Briggs & Stratton engine
- 1 gallon fuel capacity
- 30% duty cycle for welding
- Continuous output welding control with a single dial • Full roll cage
- One year manufacturer warranty
- Two 120V receptacles and one 240V receptacle
- Use a wide range of electrodes up to 1/8" diameter • 21" x 20" x 30" dimensions
- 160 lbs. tool weight

Engine: 19.4 cu. in. displacement, recoil start, 3700 RPM operating speed, 1 cylinder, 4 cycle
Generator: 4000 watts rated, 4400 watts maximum, 120V/240V single phase, 100% duty cycle
Welder: 125 amp @ 30% duty cycle, 100 amps @ 50% duty cycle
ITEM 33163-2ANA **$1099.99**

11 KW MAXIMUM, 10 KW RATED DIESEL POWERED GENERATOR SET

High powered 10,000 watt generator is powered by a two-cylinder, air cooled, ACME diesel engine. Counter balance design for smooth running, low vibration, and reduced noise. Includes 12V key start with built-in 12V battery charger.
- Longer service life than gasoline powered generators; diesel fumes not volatile as are gasoline and natural gas; higher flashpoint than gasoline
- Manufacturer's warranty on generator
5 gallon fuel tank: 10.41 hrs. run time @ 50% load; 5.88 hrs @ 100%; Low oil shutdown; Includes 120V and 240V duplex receptacles, 240V twist lock and 50 amp range receptacle, and 12V battery charging capability; 37" x 24" x 26"; 274 lbs. shipping weight; Battery sold separately

ACME DIESEL ENGINE
ITEM 01665-2ANA **$3499.99**

5000 WATT PORTABLE GENERATORS

7.8 gallon fuel tanks provide long run times, up to 9 hours at full load, & more than 13 hours at 1/2 load. Features include 8 horsepower Briggs & Stratton standard and Industrial/Commercial engines, brushless design, revolving field construction, high surge capacities, all copper windings, circuit breakers, sealed ball bearings, waterproof enclosures, class F + H insulation, solid state engine ignition, rubber vibration mounts, and sturdy roll cage carrying frames. Fuel tanks are mounted over the generator, away from the hot engine for extra safety. Approx. run times @ 1/2 load: 13.3 hr; @ 3/4 load: 10.8 hr; @ full load: 9.2 hr.
Duplex receptacles: two 120V, 20 amp; two 240V, 15 amp; Fuel tank: 7.8 gallons; Dimensions: 23"W x 32"L x 28-3/4" H

5000 WATT MAX./4400 WATT RATED BRIGGS & STRATTON ENGINE
- 1 year manufacturer's warranty on engine and generator • Low oil shutdown
- Net weight: 139 lbs.
ITEM 04080-0ANA **$599.99**

5000 WATT MAX./4400 WATT RATED BRIGGS & STRATTON INDUSTRIAL/COMMERCIAL ENGINE
- 2 year manufacturer's warranty on engine, 1 year manufacturer's warranty on generator • Low oil shutdown
- Net weight: 148 lbs.
ITEM 04081-0ANA **$699.99**

5-1/2 HP, 2500 WATTS MAX., 2300 WATTS RATED GENERATOR

Overhead valve engine provides more electricity using less fuel to save you money. Cast iron cylinder sleeve and ball bearing supported crankshaft provide years of maintenance free operation when combined with electronic ignition and low Oil Alert™. Use AC and DC simultaneously without risk to tools, which are protected by circuit breakers on both currents. Automatic voltage regulator.
Includes Automatic Decompression for easy starts, USDA qualified spark arresting mufflers, and a protective full tubing frame.
120VAC outlets: two standard 3 pole
AC output @ 120VAC: 2500 watts (20.8 amps) maximum, 2300 watts (19.2 amps) rated
DC output @ 12VDC: 100 watts (8.3 amps)
Starter: Recoil; Fuel tank capacity: 1 gallon; Run time: 3 hours; Noise level: 70 db

ITEM 32533-1ANA **$939.99**

8 HP, 3500 WATTS MAXIMUM, 3000 WATTS RATED GENERATOR

Delivers 3500 watts maximum and 3000 continuous watts: 25 amps @ 120 volts and 12.5 amps at 240 volts. Long run 4.5 gallon tank provides over 8 hours of continuous operation. Features 8 HP engine, low oil alert for engine protection, and fuel saving idle control. Safe ground fault circuit interrupter for safety.
- Dimensions: 23-3/4" x 25-3/4" x 22-3/8"
- 193 lbs. shipping weight

8 HOURS OF POWER

ITEM 32531-2ANA **$1499.99**

11 HP, 5000 WATTS MAXIMUM 4500 WATTS RATED GENERATOR

5.6 HOURS RUN TIME

Large capacity 4.5 gallon tank lets you operate for 5.6 uninterrupted hours. Delivers 5000 maximum watts and 4500 continuous watts: 37.5 amps @ 120 volts, 18.8 amps @ 240 volts. Features Oil Alert™ to protect engine if oil levels drop too low. When power demands fall, idle control saves you fuel. Includes ground fault circuit interrupter and 11 HP Honda engine.

- Dimensions: 25-3/4" x 25-3/4" x 22-5/8"
- 228 lbs. shipping weight

ITEM 32537-3ANA **$1929.99**

SPECIAL SAVINGS!

This symbol can represent EXTRA SPECIAL SAVINGS FOR YOU OF UP TO $80.00 OR MORE on our free shipping policy for all orders over $50.00. Be sure to carefully compare our prices on these items with other companies who do not include free shipping in their purchase price.

TOLL FREE 1-800-423-2567

Catalog Section

Tailwheels by Scott — Page 43

Replacement Parts for Model 3200

Model 3200 8" Pneumatic

Designed for the most rugged conditions. Sealed bearings. Steerable and full swivel. The very best for tail dragger performance. 1³/₄" spring size.
List $905.80. Code 4
Cat. No. D-306-000 .... **$549.50**

Item	P/N	Cat. No.	Code	Price
1. Bracket Asssembly	3216	D-306-008	4	$197.95
2. Washer, Thrust	3207	D-306-009	4	2.99
3. Thrust Plate Assy.	3234	D-306-010	4	19.10
4. Compression Spring	3233	D-306-011	4	2.85
5. Dust Cap, Upper	3235-2	D-306-012	4	8.30
6. Washer, Thrust	3206	D-306-001	4	10.80
7. Arm Assembly	3214	D-306-002	4	99.95
8. Pawl	3219	D-306-003	4	15.80
9. Dust Cap	3235-1	D-306-004	4	8.30
10. Fork Assembly	3224	D-306-005	4	191.50
11. Cone Bearing	13286	D-306-013	4	35.95
12. Grease Retainer	1863	D-306-014	4	5.50
13. Spacer	2504-1	D-306-015	4	9.75
14. Washer	33451-020	D-306-016	4	1.80
15. Spacer	3258	D-306-017	4	1.80
16. Spring	3222	D-306-018	4	7.35
17. Pin	3257	D-306-019	4	5.65
18. Spacer	3258-1	D-306-020	4	2.40
Tire, not pictured		D-288-003	8	19.75
Tube, not pictured		D-288-004	8	9.77

Replacement Parts for Model 3400

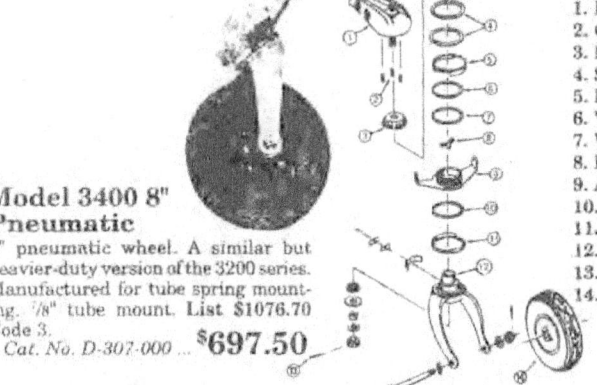

Model 3400 8" Pneumatic

8" pneumatic wheel. A similar but heavier-duty version of the 3200 series. Manufactured for tube spring mounting. ⁷/₈" tube mount. List $1076.70 Code 3.
Cat. No. D-307-000 ... **$697.50**

Item	P/N	Cat. No.	Code	Price
1. Bracket Assembly	3412-1	D-307-001	3	$238.50
2. Compression Spring	3233	D-306-011	4	2.85
3. Bearing "Timken"	08118	D-307-003	3	27.50
4. Spacer	3407	D-307-004	3	1.80
5. Dust Cap, LWR	3411-1	D-307-005	3	19.50
6. Washer	3408	D-307-006	3	5.65
7. Washer	3408-1	D-307-007	3	2.20
8. Pawl	3219	D-306-003	4	15.80
9. Arm Assembly	3410-3	D-307-009	3	112.00
10. Washer	3408-2	D-307-010	3	2.15
11. Dust Cap, Upper	3411-2	D-307-011	3	19.50
12. Fork Assembly	3423-1	D-307-012	3	199.95
13. Cotter Pin	18482	D-307-013	3	1.99
14. Tire & Wheel Assy.	2600	D-307-014	3	289.50

SEE TAILSPRING KITS ON PAGE 45

Replacement Parts for Model 2000

Model 2000 Lightweight

Utilizes 6 x 2 rubber tire. Features lightweight and rugged reliability. Steerable and full swivel. Usable on 1¹/₂" and 1¹/₄" spring size. 6 lbs. List $743.60. Code 4
Cat. No. D-305-000 .... **$509.50**

Item	P/N	Cat. No.	Code	Price
1. Self Locking Nut	MS20364-720	D-305-006	4	$ 2.20
2. Washer	2346	D-305-007	4	2.35
3. Shim	2085	D-305-008	4	2.45
4. Steering Arm Assy.	1709	D-305-002	4	166.50
5. Bracket Assembly	2077	D-305-004	4	133.50
6. Cap Bushing	1781	D-305-009	4	23.75
7. Bronze Center Bushing	1800	D-305-001	4	19.25
8. Fork Assembly	2078	D-305-003	4	199.50
9. Grease Retainer	1863	D-306-014	4	5.50
10. Spacer, Grease Rtnr	1862	D-305-011	4	6.75
11. Bearing Set (Includes Race)	1883	D-305-012	4	27.25
12. Hub Kit	1967	D-305-005	4	146.50
13. Hub Cap Assembly	1882	D-305-013	4	25.50
Tire, not pictured		D-288-001	8	34.75

Tailwheel Tires

All Code 8 Except Cat. No. D-288-002

Size	Tread Type	Price	Size	Tread Type	Price	Size	Tread Type	Price	Size	Tread Type	Price
6x2	Solid Tailwheel		2.80x2.50-4*	Tire		10x3.50-4	Tire		12.5x4.5-10	Tire	
Cat. No. D-288-001		$34.75	Cat. No. D-288-003		19.75	Cat. No. D-288-005		23.30	Cat. No. D-288-007		111.30
6.50x2.50	Solid Tailwheel		2.80x2.50	Tube		10x3.50-4	Tube		12.5x4.5-10	Tube	
Code 4											
Cat. No. D-288-002		41.50	Cat. No. D-288-004		9.77	Cat. No. D-288-006		8.99	Cat. No. D-288-008		32.07

*Replaces 6.00x3 Tire.

To Place Your Order By Catalog Number, Call Toll Free 1-800-558-6868

Prices and Supply Subject To Change Without Notice

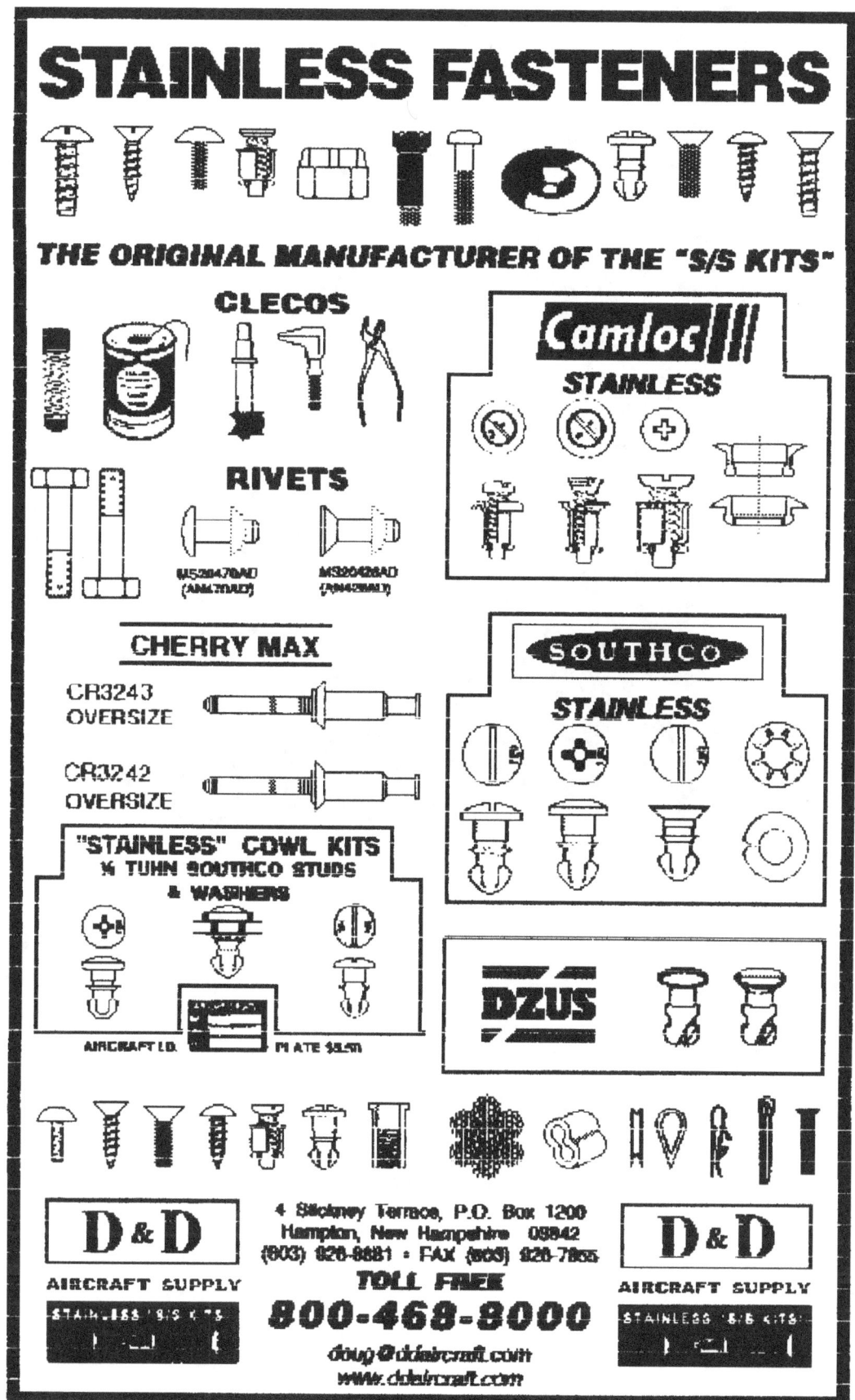

Catalog Section

It takes only seconds to discover why Bose® sets the standard.

But we'll give you a 30-day test flight anyway.

Thirty days will confirm your first impression. The Bose Aviation Headset sets the standard for the field. This is the only headset ever honored with the AOPA Special Citation. And since 1989, when we pioneered Acoustic Noise Cancelling headset technology in aviation, we've striven to set new standards for quality. Reputation. Customer service.

And most of all, performance — with clear differences in comfort and quiet you must experience for yourself.

To arrange your personal 30-day test flight or receive a free information booklet:

Call Bose Toll-Free Today
1-800-242-9008 Ext. H11

Outside USA: Call 508-879-7330 Ext. 2006

Better sound through research

AIRSHIP

TECHNICAL NOTES

Compiled 1980
by
Robert J. Rechs

AIRSHIP SHAPE
MATHEMATICAL
Possibilities

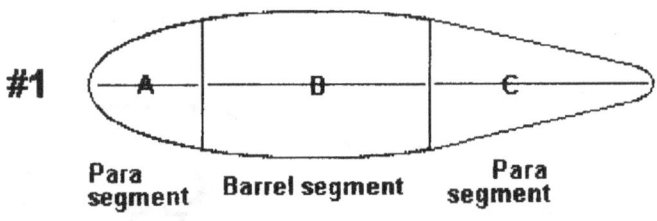

#1 — Para segment / Barrel segment / Para segment

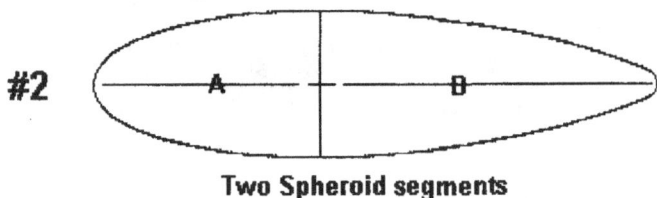

#2 — Two Spheroid segments

#3 — Continuous Radiuse of Curvature

Building Small Gas Blimps

1.0 Math_PBC
0.6 R2

Parabola - Barrel - Cone

#1

Shape

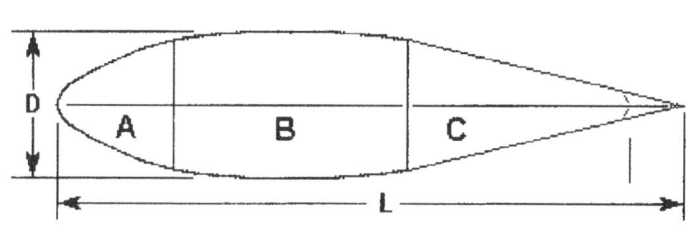

Definitions

L = Length = $D * AR$
D = Diameter (max.)
X = Distance from nose
R = Radius (max.)
h = $L/100$

Y = Radius at sta. X
V = Volume
AR = Aspect Ratio
VS = Segment Volume

For X (or h), L is divided into 100 incraments for ease of gore layout.

Paraboloid

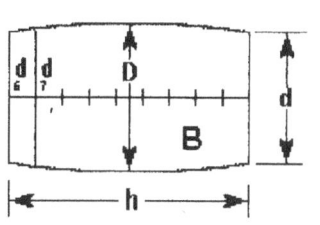

Formulas

Volume $VA = .5 \pi r^2 h = 0.3927 d^2 h$

Segment Volume $VS = 0.3927 h (d_5^2 + d_6^2) = 1.5087 h (r_5^2 + r_6^2)$

Area $PA = \frac{2\pi}{3p} \left| \sqrt{\left(\frac{d^2}{4} + p^2\right)^3} - p^3 \right|$ where $p = \frac{d^2}{8h}$

Barrel

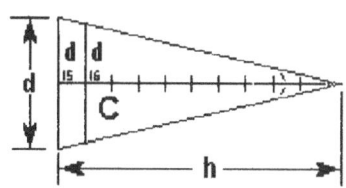

Formulas

Volume $VB = 0.262 h (2D^2 + d^2)$

Segment Volume $VS = 0.3927 h (d_5^2 + d_6^2) = 1.5087 h (r_5^2 + r_6^2)$

Cone

Formulas

Volume $VC = \frac{\pi r^2 h}{3} = 1.0472 r^2 h = 0.2618 d^2 h$

Segment Volume $VS = 0.3927 h (d_{15}^2 + d_{16}^2) = 1.5087 h (r_5^2 + r_6^2)$

Problems:

To many asumptions have to be made on the position of segments
Tail will droop if not trunicated.

NOTE: these formulas must be intagrated into the primary formula above.

Benifits:
None, May be the easiest way to make models.

Building Small Gas Blimps

Two Spheroid Segments

Shape

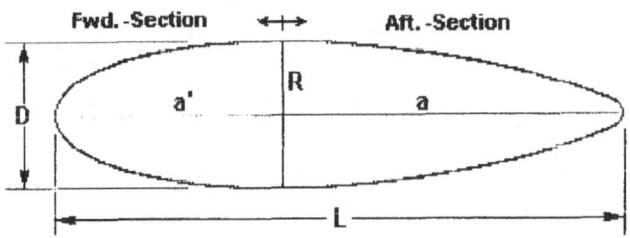

Definitions

- V = Volume
- L = Length
- VS = Segment Volume
- x = Distance from nose
- H = Segment length on the Longitudinal axis
- h = Segment length on the perimeter
- AR = Aspect Ratio
- D = Diameter (max.)
- R = Radius (max.)
- y = Hull radius at sta. x

Assumptions:

That the FWD(a') section and AFT(a) sections are always 1/3 & 2/3's the length (L) respectively.
That the max. diameter (D) always equals the length (L) divided by the aspect ratio (AR).
That the segment length on the Longitudinal axis (H) can be any number,
  but is best programed as a repetitive incrament (H1 = H2 etc.). L/50 works well.
V will not equal VT due to cumulative error, in which case V is considered the more accurate.
  V & VT errors are considered irrevelant compared to potiential stretch of single ply fabrics.

Formulas

Volume
$$V = 4.1888 * \left(\frac{L}{2}\right) * R^2$$

Gore Length
$$GL = \frac{\pi \sqrt{2\left(\frac{L}{2} + R^2\right)}}{2}$$

Total Surface Area
$$TA = \frac{2\pi}{3p}\left[\sqrt{\left(\frac{D^2}{4} + p^2\right)^3} - p^3\right] 2 \quad \text{Where} \quad p = \frac{D^2}{8 * L/2}$$

Paraboloidal Segments

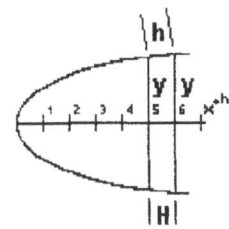

Segment Length
$$h = \sqrt{H^2 + g^2} \quad \text{Where } g = y_6 - y_5$$

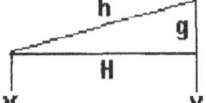

Gore Length
$$GL = \int_0^{50} h + h_1 + h_2 +$$

Segment Volume
$$VS = \frac{\pi}{2} H (y_5^2 + y_6^2) = 1.5087 * H * (y_5^2 + y_6^2)$$

Problems: It doesn't quite match the absolute minimum drag of the NPL formula.

Benefits: It easier to calculate. Easily programed in a spreadsheet.

Building Small Gas Blimps

mathematics of
AIRSHIP SHAPES
(as the mating of two different partial spheroids at their prime diameter)

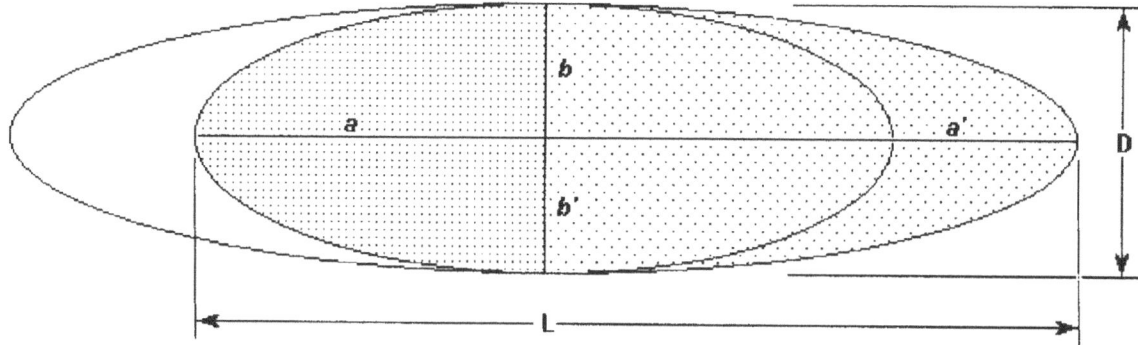

INPUTS (variables):
    A. That **V** is the volume (cu.ft. or cu.m.);
    B. That **AR** is the Aspect Ratio (usually between 3.8 to 4.3)
    C. That **N** is the Number of gores (any whole number).

PRIMARY ASSUMPTIONS:
    1. That *b* is always the radius of the envelope end view;
    2. That **D** is the Diameter, and **D = 2 *b***;
    3. That **C** is the Circumference, and **C = Pi** times **D**;
    4. That **L** is the summation of *a* and *a'*, or **L = *a* + *a'***;
        That **L** is the airship length, and **L = AR** times **D**;
    5. That **P** is the Perimeter of the envelope side view;
    6. That **G** is the Gore length, and **G = 0.5** times **P**, or **P/2**;
    7. That **GW** is Gore Width at a particular point on the Gore Length;
        GW = Circumference divided by the Number of gores, or **GW=C/N**.

SECONDARY POSSIBILITIES (for a reasonable aerodynamic shape):
    11. That *a* is equal to 2.5 times *b*, or **2.5*b***;
    12. That *a'* is equal to 4 times *b'*, or **4*b'***.
    13. That a fabric width (**FW**) may be substituted for the number of gores
        (to minimize fabric cutting loss);
    14. That the number of gores can be any number, odd or even;
        (optimum would be divisible by 4, or 2).

LIST (on screen)
    20. (L)   The inflated airship length
    21. (GL) The gore length apex to apex;
    22. (D)   The maximum circumference;
    23 (MW) The maximum gore width (MC/N);

PRINT (on command)
    24. (GA) The gore position, in 100 increments, in distance from the nose apex;
    25. (GW) The gore width at each of those 100 increments;
    26. (PL) The equivalent position on L at each of those 100 increments;
    27. (CM) The center of mass (distance on L from the nose);
    28. (CB) The center of buoyancy (distance on L from the nose);

R: 12/01/97

Building Small Gas Blimps

1.0 Math_CR
0.6 R3
Continuous Radius of curvature
#3

Shape
Developed by the Nat. Physical Lab. and used extensively by the U.S. Navy

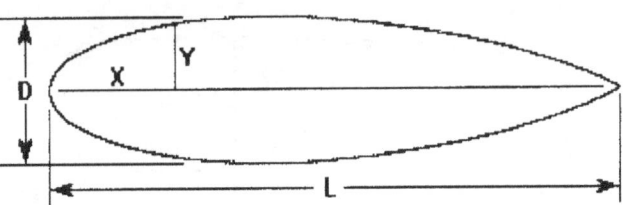

Definitions
L = Length = D * AR
D = Diameter (max.)
X = Distance from nose
R = Radius (max.)
h = L/100

Y = Radius at sta. X
V = Volume
AR = Aspect Ratio
VS = Segment Volume

Formula
(ONLY one given)

$$Y = D/2 * \sqrt{1-(\frac{X}{L/2}-1)^2} * (1.19072 - 0.21263 * \frac{X}{L/2})$$

$$1.19072 - 0.21263 = \underline{.97809}$$

Practice
For X (or h), L is divided into 100 incraments for ease of gore layout.

Best blimp Aspect Ratio for blimps is considered 4:1 (L=4D)

Paraboloidal Segments

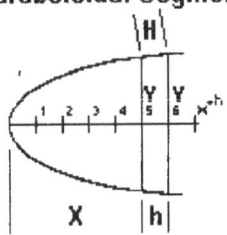

Segment Volume
$$VS = \frac{\pi}{2} h (Y_5^2 + Y_6^2) = 1.5087 h (Y_5^2 + Y_6^2)$$

Total Volume
$$VT = \int_0^{100} VS + VS_1 + VS_2 +$$

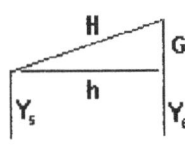

Curved Segment Length
$$H = \sqrt{h^2 + G^2} \qquad \text{where } G = Y_6 - Y_5$$

Gore Length
$$GL = \int_0^{100} H + H_1 + H_2 +$$

Problems:
The starting formula should be from a hull volume input.
H in the program doesn't know when to turn around
   when passing D. (h becomes a negative number).

NOTE: these formulas must be integrated into a loop program.

Benifits: Best method known.
Shape gives the optimum minimum drag regardless of volume ~~or Aspect Ratio~~.

Building Small Gas Blimps

1.0 Gore_CR
0.6

GORE DEVELOPMENT

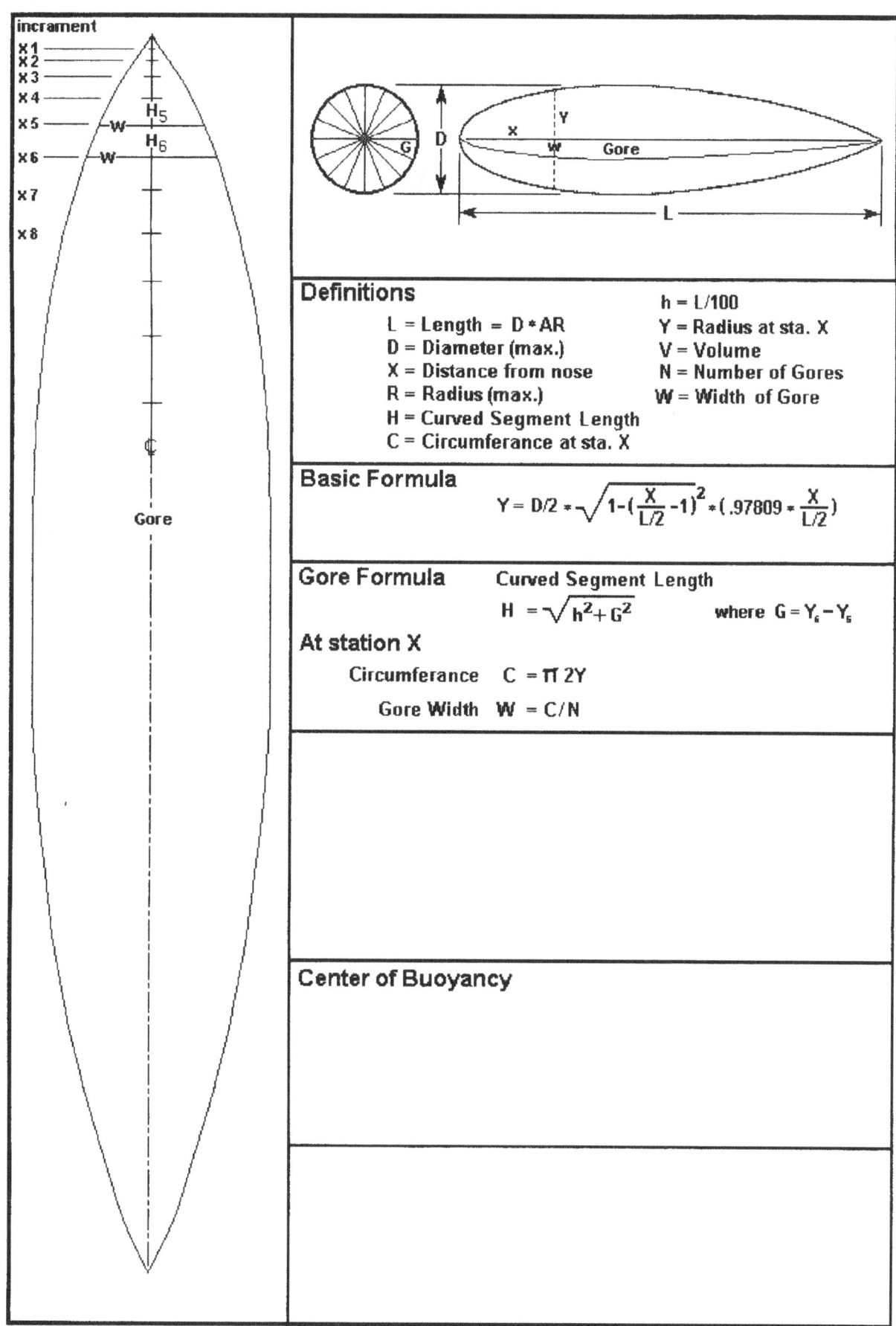

Definitions

- L = Length = D * AR
- D = Diameter (max.)
- X = Distance from nose
- R = Radius (max.)
- H = Curved Segment Length
- C = Circumferance at sta. X
- h = L/100
- Y = Radius at sta. X
- V = Volume
- N = Number of Gores
- W = Width of Gore

Basic Formula

$$Y = D/2 * \sqrt{1-\left(\frac{X}{L/2}-1\right)^2} * \left(.97809 * \frac{X}{L/2}\right)$$

Gore Formula

Curved Segment Length

$$H = \sqrt{h^2 + G^2} \qquad \text{where } G = Y_6 - Y_5$$

At station X

Circumferance $\quad C = \pi\, 2Y$

Gore Width $\quad W = C/N$

Center of Buoyancy

Building Small Gas Blimps

VOLUMES - Aspect Ratio 4:1

L	D	R	R x R	x 2.0944	Cu.Ft.	Cu.Mt.	Lbs.Lift
1	0.25	0.125	0.015625	0.032725	0.032725	0.0009267	0.0021926
2	0.5	0.25	0.0625	0.1309	0.2618	0.0074133	0.0175406
3	0.75	0.375	0.140625	0.294525	0.883575	0.0250201	0.0591995
4	1	0.5	0.25	0.5236	2.0944	0.0593068	0.1403248
5	1.25	0.625	0.390625	0.818125	4.090625	0.1158336	0.2740719
6	1.5	0.75	0.5625	1.1781	7.0686	0.2001604	0.4735962
7	1.75	0.875	0.765625	1.603525	11.224675	0.3178474	0.7520532
8	2	1	1	2.0944	16.7552	0.4744544	1.1225984
9	2.25	1.125	1.265625	2.650725	23.856525	0.6755415	1.5983872
10	2.5	1.25	1.5625	3.2725	32.725	0.9266687	2.192575
11	2.75	1.375	1.890625	3.959725	43.556975	1.2333961	2.9183173
12	3	1.5	2.25	4.7124	56.5488	1.6012835	3.7887696
13	3.25	1.625	2.640625	5.530525	71.896825	2.0358912	4.8170873
14	3.5	1.75	3.0625	6.4141	89.7974	2.542779	6.0164258
15	3.75	1.875	3.515625	7.363125	110.44688	3.1275069	7.3999406
16	4	2	4	8.3776	134.0416	3.7956351	8.9807872
17	4.25	2.125	4.515625	9.457525	160.77793	4.5527234	10.772121
18	4.5	2.25	5.0625	10.6029	190.8522	5.404332	12.787097
19	4.75	2.375	5.640625	11.813725	224.46078	6.3560207	15.038872
20	5	2.5	6.25	13.09	261.8	7.4133497	17.5406
21	5.25	2.625	6.890625	14.431725	303.06623	8.581879	20.305437
22	5.5	2.75	7.5625	15.8389	348.4558	9.8671685	23.346539
23	5.75	2.875	8.265625	17.311525	398.16508	11.274778	26.67706
24	6	3	9	18.8496	452.3904	12.810268	30.310157
25	6.25	3.125	9.765625	20.453125	511.32813	14.479199	34.258984
26	6.5	3.25	10.5625	22.1221	575.1746	16.287129	38.536698
27	6.75	3.375	11.390625	23.856525	644.12618	18.23962	43.156454
28	7	3.5	12.25	25.6564	718.3792	20.342232	48.131406
29	7.25	3.625	13.140625	27.521725	798.13003	22.600523	53.474712
30	7.5	3.75	14.0625	29.4525	883.575	25.020055	59.199525
31	7.75	3.875	15.015625	31.448725	974.91048	27.606388	65.319002
32	8	4	16	33.5104	1072.3328	30.365081	71.846298
33	8.25	4.125	17.015625	35.637525	1176.0383	33.301694	78.794568
34	8.5	4.25	18.0625	37.8301	1286.2234	36.421787	86.176968
35	8.75	4.375	19.140625	40.088125	1403.0844	39.730921	94.006653
36	9	4.5	20.25	42.4116	1526.8176	43.234656	102.29678
37	9.25	4.625	21.390625	44.800525	1657.6194	46.938551	111.0605
38	9.5	4.75	22.5625	47.2549	1795.6862	50.848166	120.31098
39	9.75	4.875	23.765625	49.774725	1941.2143	54.969062	130.06136
40	10	5	25	52.36	2094.4	59.306798	140.3248
41	10.25	5.125	26.265625	55.010725	2255.4397	63.866935	151.11446
42	10.5	5.25	27.5625	57.7269	2424.5298	68.655032	162.4435
43	10.75	5.375	28.890625	60.508525	2601.8666	73.67665	174.32506
44	11	5.5	30.25	63.3556	2787.6464	78.937348	186.77231
45	11.25	5.625	31.640625	66.268125	2982.0656	84.442687	199.7984
46	11.5	5.75	33.0625	69.2461	3185.3206	90.198226	213.41648
47	11.75	5.875	34.515625	72.289525	3397.6077	96.209526	227.63971
48	12	6	36	75.3984	3619.1232	102.48215	242.48125
49	12.25	6.125	37.515625	78.572725	3850.0635	109.02165	257.95426
50	12.5	6.25	39.0625	81.8125	4090.625	115.83359	274.07188

Building Small Gas Blimps

VOLUMES - Aspect Ratio 4:1

51	12.75	6.375	40.640625	85.117725	4341.004	122.92353	290.84727
52	13	6.5	42.25	88.4884	4601.3968	130.29704	308.29359
53	13.25	6.625	43.890625	91.924525	4871.9998	137.95966	326.42399
54	13.5	6.75	45.5625	95.4261	5153.0094	145.91696	345.25163
55	13.75	6.875	47.265625	98.993125	5444.6219	154.17451	364.78967
56	14	7	49	102.6256	5747.0336	162.73785	385.05125
57	14.25	7.125	50.765625	106.32353	6060.4409	171.61256	406.04954
58	14.5	7.25	52.5625	110.0869	6385.0402	180.80419	427.79769
59	14.75	7.375	54.390625	113.91573	6721.0278	190.31829	450.30886
60	15	7.5	56.25	117.81	7068.6	200.16044	473.5962
61	15.25	7.625	58.140625	121.76973	7427.9532	210.33619	497.67287
62	15.5	7.75	60.0625	125.7949	7799.2838	220.8511	522.55201
63	15.75	7.875	62.015625	129.88553	8182.7881	231.71073	548.2468
64	16	8	64	134.0416	8578.6624	242.92064	574.77038
65	16.25	8.125	66.015625	138.26313	8987.1031	254.4864	602.13591
66	16.5	8.25	68.0625	142.5501	9408.3066	266.41355	630.35654
67	16.75	8.375	70.140625	146.90253	9842.4692	278.70766	659.44543
68	17	8.5	72.25	151.3204	10289.787	291.3743	689.41574
69	17.25	8.625	74.390625	155.80373	10750.457	304.41901	720.28062
70	17.5	8.75	76.5625	160.3525	11224.675	317.84737	752.05323
71	17.75	8.875	78.765625	164.96673	11712.637	331.66493	784.74671
72	18	9	81	169.6464	12214.541	345.87725	818.37423
73	18.25	9.125	83.265625	174.39153	12730.581	360.48988	852.94895
74	18.5	9.25	85.5625	179.2021	13260.955	375.5084	888.48401
75	18.75	9.375	87.890625	184.07813	13805.859	390.93837	924.99258
76	19	9.5	90.25	189.0196	14365.49	406.78533	962.4878
77	19.25	9.625	92.640625	194.02653	14940.042	423.05485	1000.9828
78	19.5	9.75	95.0625	199.0989	15529.714	439.75249	1040.4909
79	19.75	9.875	97.515625	204.23673	16134.701	456.88382	1081.025
80	20	10	100	209.44	16755.2	474.45438	1122.5984
81	20.25	10.125	102.51563	214.70873	17391.407	492.46975	1165.2243
82	20.5	10.25	105.0625	220.0429	18043.518	510.93548	1208.9157
83	20.75	10.375	107.64063	225.44253	18711.73	529.85713	1253.6859
84	21	10.5	110.25	230.9076	19396.238	549.24026	1299.548
85	21.25	10.625	112.89063	236.43813	20097.241	569.09043	1346.5151
86	21.5	10.75	115.5625	242.0341	20814.933	589.4132	1394.6005
87	21.75	10.875	118.26563	247.69553	21549.511	610.21413	1443.8172
88	22	11	121	253.4224	22301.171	631.49879	1494.1785
89	22.25	11.125	123.76563	259.21473	23070.111	653.27272	1545.6974
90	22.5	11.25	126.5625	265.0725	23856.525	675.5415	1598.3872
91	22.75	11.375	129.39063	270.99573	24660.611	698.31067	1652.2609
92	23	11.5	132.25	276.9844	25482.565	721.58581	1707.3318
93	23.25	11.625	135.14063	283.03853	26322.583	745.37247	1763.613
94	23.5	11.75	138.0625	289.1581	27180.861	769.67621	1821.1177
95	23.75	11.875	141.01563	295.34313	28057.597	794.50259	1879.859
96	24	12	144	301.5936	28952.986	819.85718	1939.85
97	24.25	12.125	147.01563	307.90953	29867.224	845.74552	2001.104
98	24.5	12.25	150.0625	314.2909	30800.508	872.17318	2063.634
99	24.75	12.375	153.14063	320.73773	31753.035	899.14573	2127.4533
100	25	12.5	156.25	327.25	32725	926.66872	2192.575

Building Small Gas Blimps

VOLUMES - Aspect Ratio 4:1

101	25.25	12.625	159.39063	333.82773	33716.6	954.74771	2259.0122
102	25.5	12.75	162.5625	340.4709	34728.032	983.38826	2326.7781
103	25.75	12.875	165.76563	347.17953	35759.491	1012.5959	2395.8859
104	26	13	169	353.9536	36811.174	1042.3763	2466.3487
105	26.25	13.125	172.26563	360.79313	37883.278	1072.7349	2538.1796
106	26.5	13.25	175.5625	367.6981	38975.999	1103.6773	2611.3919
107	26.75	13.375	178.89063	374.66853	40089.532	1135.209	2685.9987
108	27	13.5	182.25	381.7044	41224.075	1167.3357	2762.013
109	27.25	13.625	185.64063	388.80573	42379.824	1200.0629	2839.4482
110	27.5	13.75	189.0625	395.9725	43556.975	1233.3961	2918.3173
111	27.75	13.875	192.51563	403.20473	44755.724	1267.3409	2998.6335
112	28	14	196	410.5024	45976.269	1301.9028	3080.41
113	28.25	14.125	199.51563	417.86553	47218.804	1337.0875	3163.6599
114	28.5	14.25	203.0625	425.2941	48483.527	1372.9005	3248.3963
115	28.75	14.375	206.64063	432.78813	[49770.634]	1409.3473	3334.6325
116	29	14.5	210.25	440.3476	51080.322	1446.4335	3422.3815
117	29.25	14.625	213.89063	447.97253	52412.785	1484.1647	3511.6566
118	29.5	14.75	217.5625	455.6629	53768.222	1522.5464	3602.4709
119	29.75	14.875	221.26563	463.41873	55146.828	1561.5841	3694.8375
120	30	15	225	471.24	56548.8	1601.2835	3788.7696
121	30.25	15.125	228.76563	479.12673	57974.334	1641.6502	3884.2804
122	30.5	15.25	232.5625	487.0789	[59423.626]	1682.6895	3981.3829
123	30.75	15.375	236.39063	495.09653	60896.873	1724.4072	4080.0905
124	31	15.5	240.25	503.1796	62394.27	1766.8088	4180.4161
125	31.25	15.625	244.14063	511.32813	63916.016	1809.8998	4282.373
126	31.5	15.75	248.0625	519.5421	65462.305	1853.6859	4385.9744
127	31.75	15.875	252.01563	527.82153	67033.334	1898.1725	4491.2334
128	32	16	256	536.1664	68629.299	1943.3652	4598.163
129	32.25	16.125	260.01563	544.57673	[70250.398]	1989.2695	4706.7766
130	32.5	16.25	264.0625	553.0525	71896.825	2035.8912	4817.0873
131	32.75	16.375	268.14063	561.59373	73568.778	2083.2356	4929.1081
132	33	16.5	272.25	570.2004	75266.453	2131.3084	5042.8523
133	33.25	16.625	276.39063	578.87253	76990.046	2180.1151	5158.3331
134	33.5	16.75	280.5625	587.6101	78739.753	2229.6613	5275.5635
135	33.75	16.875	284.76563	596.41313	[80515.772]	2279.9525	5394.5567
136	34	17	289	605.2816	82318.298	2330.9944	5515.3259
137	34.25	17.125	293.26563	614.21553	84147.527	2382.7924	5637.8843
138	34.5	17.25	297.5625	623.2149	86003.656	2435.3521	5762.245
139	34.75	17.375	301.89063	632.27973	87886.882	2488.6791	5888.4211
140	35	17.5	306.25	641.41	[89797.4]	2542.779	6016.4258
141	35.25	17.625	310.64063	650.60573	91735.407	2597.6572	6146.2723
142	35.5	17.75	315.0625	659.8669	93701.1	2653.3194	6277.9737
143	35.75	17.875	319.51563	669.19353	95694.674	2709.7712	6411.5432
144	36	18	324	678.5856	97716.326	2767.018	6546.9939
145	36.25	18.125	328.51563	688.04313	[99766.253]	2825.0654	6684.339
146	36.5	18.25	333.0625	697.5661	101844.65	2883.9191	6823.5916
147	36.75	18.375	337.64063	707.15453	103951.72	2943.5845	6964.7649
148	37	18.5	342.25	716.8084	106087.64	3004.0672	7107.8721
149	37.25	18.625	346.89063	726.52773	108252.63	3065.3729	7252.9263
150	37.5	18.75	351.5625	736.3125	[110446.88]	3127.5069	7399.9406

Building Small Gas Blimps

VOLUMES - Aspect Ratio 4:1

151	37.75	18.875	356.26563	746.16273	112670.57	3190.475	7548.9283
152	38	19	361	756.0784	114923.92	3254.2826	7699.9024
153	38.25	19.125	365.76563	766.05953	117207.11	3318.9354	7852.8762
154	38.5	19.25	370.5625	776.1061	119520.34	3384.4388	8007.8627
155	38.75	19.375	375.39063	786.21813	121863.81	3450.7985	8164.8752
156	39	19.5	380.25	796.3956	124237.71	3518.02	8323.9268
157	39.25	19.625	385.14063	806.63853	126642.25	3586.1088	8485.0306
158	39.5	19.75	390.0625	816.9469	129077.61	3655.0705	8648.1999
159	39.75	19.875	395.01563	827.32073	131544	3724.9108	8813.4477
160	40	20	400	837.76	134041.6	3795.6351	8980.7872
161	40.25	20.125	405.01563	848.26473	136570.62	3867.249	9150.2316
162	40.5	20.25	410.0625	858.8349	139131.25	3939.758	9321.794
163	40.75	20.375	415.14063	869.47053	141723.7	4013.1678	9495.4876
164	41	20.5	420.25	880.1716	144348.14	4087.4838	9671.3255
165	41.25	20.625	425.39063	890.93813	147004.79	4162.7117	9849.321
166	41.5	20.75	430.5625	901.7701	149693.84	4238.857	10029.487
167	41.75	20.875	435.76563	912.66753	152415.48	4315.9253	10211.837
168	42	21	441	923.6304	155169.91	4393.922	10396.384
169	42.25	21.125	446.26563	934.65873	157957.32	4472.8529	10583.141
170	42.5	21.25	451.5625	945.7525	160777.93	4552.7234	10772.121
171	42.75	21.375	456.89063	956.91173	163631.9	4633.5391	10963.338
172	43	21.5	462.25	968.1364	166519.46	4715.3056	11156.804
173	43.25	21.625	467.64063	979.42653	169440.79	4798.0284	11352.533
174	43.5	21.75	473.0625	990.7821	172396.09	4881.7131	11550.538
175	43.75	21.875	478.51563	1002.2031	175385.55	4966.3652	11750.832
176	44	22	484	1013.6896	178409.37	5051.9903	11953.428
177	44.25	22.125	489.51563	1025.2415	181467.75	5138.594	12158.339
178	44.5	22.25	495.0625	1036.8589	184560.88	5226.1818	12365.579
179	44.75	22.375	500.64063	1048.5417	187688.97	5314.7592	12575.161
180	45	22.5	506.25	1060.29	190852.2	5404.332	12787.097
181	45.25	22.625	511.89063	1072.1037	194050.77	5494.9055	13001.402
182	45.5	22.75	517.5625	1083.9829	197284.89	5586.4854	13218.087
183	45.75	22.875	523.26563	1095.9275	200554.74	5679.0772	13437.167
184	46	23	529	1107.9376	203860.52	5772.6865	13658.655
185	46.25	23.125	534.76563	1120.0131	207202.43	5867.3188	13882.563
186	46.5	23.25	540.5625	1132.1541	210580.66	5962.9798	14108.904
187	46.75	23.375	546.39063	1144.3605	213995.42	6059.6749	14337.693
188	47	23.5	552.25	1156.6324	217446.89	6157.4097	14568.942
189	47.25	23.625	558.14063	1168.9697	220935.28	6256.1898	14802.664
190	47.5	23.75	564.0625	1181.3725	224460.78	6356.0207	15038.872
191	47.75	23.875	570.01563	1193.8407	228023.58	6456.9081	15277.58
192	48	24	576	1206.3744	231623.88	6558.8574	15518.8
193	48.25	24.125	582.01563	1218.9735	235261.89	6661.8742	15762.547
194	48.5	24.25	588.0625	1231.6381	238937.79	6765.9642	16008.832
195	48.75	24.375	594.14063	1244.3681	242651.78	6871.1327	16257.67
196	49	24.5	600.25	1257.1636	246404.07	6977.3855	16509.072
197	49.25	24.625	606.39063	1270.0245	250194.83	7084.728	16763.054
198	49.5	24.75	612.5625	1282.9509	254024.28	7193.1658	17019.627
199	49.75	24.875	618.76563	1295.9427	257892.6	7302.7046	17278.804
200	50	25	625	1309	261800	7413.3497	17540.6

Building Small Gas Blimps

1.3 Formulas3
.75 R2

(original)

NASA CR-137692

Burgess formula
(continuous contour)

The equation for the contour of the airship is assumed to be that used on non-rigid airships:

$$Y = R\left(1.02062 - 0.21263 \times \frac{2X}{L}\right)\left[1 - \left(\frac{2X}{L} - 0.2\right)^2\right]^{1/2}$$

where

    X   is measured from maximum diameter
    R   is maximum radius, and
    L   is hull length

Employing the contour expressed by the above equation, one obtains for the volume:

* Editors notes:

$V = 0.64381 \; \pi R^2 L$ * (Where .64381 × π = 2.022587)

$V = 0.16095 \; \pi D^2 L$ * (Where .16095 × π = .5056389)

$V = 0.16095 \; \pi (D/L)^2 L^3$

or

$V = 0.16095 \; \pi (L/D) D^3 \sim (L/D) D^3$

which for L gives

$L = 1.25522 \; V^{1/3} (L/D)^{2/3} \sim V^{1/3} (D/L)^{2/3}$

For D

$D = 1.25522 \; V^{1/3} (D/L)^{1/3} \sim V^{1/3} (D/L)^{1/3}$

The surface area, S, derived from the contour equation is

$S = 3.80243 \; V^{2/3} (L/D)^{1/3} \sim V^{2/3} (L/D)^{1/3}$

Building Small Gas Blimps

HULL COORDINATES (Max.Dia.@ 40% Length) Page 1 of 2

INPUTS:	Length	L=	82.10954	feet	OUTPUTS	Diameter D=	18.246564 '
Fineness		FR=	4.5	to one	Max.circumference	C=	57.323407 '
Matl.width		MW=	5	feet	Volume (cu.ft)	V=	14000
# of Gores		N=	12	pieces		XX	YY
X-Sta-%	X-Distance	Y-Radius	Cum.Surf	Cum.Vol.		Gore Pos.	Gore Width
0	feet	feet	sq.ft.	cu.ft.		0	0
1	0.8210954	2.118	11	12		2.2715901	1.1089848
2	1.6421908	2.972	26	34		3.4562899	1.5561392
3	2.4632862	3.611	45	68		4.4967315	1.8907196
4	9.8531448	4.137	66	112		5.471859	2.1661332
5	4.105477	4.588	90	166		6.4086614	2.4022768
6	4.9265724	4.985	116	231		7.3206957	2.610146
7	5.7476678	5.34	143	304		8.2152474	2.796024
8	6.5687632	5.662	172	387		9.0972234	2.9646232
9	7.3898586	5.955	203	478		9.9690299	3.118038
10	8.210954	6.224	235	578		10.833066	3.2588864
11	9.0320494	6.471	269	686		11.690508	3.3882156
12	9.8531448	6.701	303	802		12.543208	3.5086436
13	10.67424	6.913	339	925		13.391231	3.6196468
14	11.495336	7.11	376	1056		14.235628	3.722796
15	12.316431	7.293	413	1193		15.076869	3.8186148
16	13.137526	7.464	452	1337		15.915581	3.9081504
17	13.958622	7.623	491	1487		16.75193	3.9914028
18	14.779717	7.771	531	1642		17.586257	4.0688956
19	15.600813	7.908	572	1804		18.418703	4.1406288
20	16.421908	8.036	613	1970		19.249715	4.2076496
21	17.243003	8.155	655	2142		20.079389	4.269958
22	18.064099	8.265	698	2318		20.90782	4.327554
23	18.885194	8.367	741	2499		21.735227	4.3809612
24	19.70629	8.462	785	2683		22.5618	4.4307032
25	20.527385	8.548	829	2872		23.387386	4.4757328
26	21.34848	8.628	873	3064		24.21237	4.5176208
27	22.169576	8.701	918	3259		25.036704	4.5558436
28	22.990671	8.767	964	3457		25.860448	4.5904012
29	23.811767	8.827	1009	3658		26.683732	4.6218172
30	24.632862	8.881	1055	3862		27.506601	4.6500916
31	25.453957	8.929	1101	4067		28.329099	4.6752244
32	26.275053	8.971	1147	4275		29.151267	4.6972156
33	27.096148	9.008	1194	4484		29.973196	4.7165888
34	27.917244	9.039	1240	4695		30.794876	4.7328204
35	28.738339	9.065	1287	4907		31.616383	4.746434
36	29.559434	9.087	1334	5120		32.437774	4.7579532
37	30.38053	9.103	1381	5334		33.259025	4.7663308
38	31.201625	9.114	1428	5548		34.080194	4.7720904
39	32.022721	9.121	1475	5763		34.901319	4.7757556
40	32.843816	9.123	1522	5977		35.722417	4.7768028
41	33.664911	9.121	1569	6192		36.543534	4.7757556
42	34.486007	9.115	1616	6406		37.364703	4.772614
43	35.307102	9.104	1663	6620		38.185936	4.7668544
44	36.128198	9.089	1710	6833		39.007251	4.7590004
45	36.949293	9.07	1757	7045		39.828668	4.749052
46	37.770388	9.047	1804	7257		40.650208	4.7370092
47	38.591484	9.02	1850	7466		41.471851	4.722872
48	39.412579	8.99	1896	7675		42.293692	4.707164
49	40.233675	8.955	1943	7882		43.115666	4.688838
50	41.05477	8.917	1989	8087		43.937785	4.6689412

R2

Building Small Gas Blimps

HULL COORDINATES (Max.Dia.@ 40% Length) Page 2 of 2

X-Sta-%	X-Distance	Y-Radius	Cum.Surf.	Cum.V	Gore Pos.	Gore Width
51	41.875865	8.876	2034	8290	44.760112	4.6474736
52	42.696961	8.831	2080	8491	45.582668	4.6239116
53	43.518056	8.782	2125	8690	46.405409	4.5982552
54	44.339152	8.73	2170	8887	47.228344	4.571028
55	45.160247	8.675	2215	9081	48.051485	4.54223
56	45.981342	8.617	2260	9272	48.874918	4.5118612
57	46.802438	8.555	2304	9461	49.698582	4.479398
58	47.623533	8.49	2348	9647	50.522489	4.445364
59	48.444629	8.422	2391	9830	51.346648	4.4097592
60	49.265724	8.351	2434	10010	52.171071	4.3725836
61	50.086819	8.277	2477	10187	52.995769	4.3338372
62	50.907915	8.2	2510	10360	53.820753	4.29352
63	51.72901	8.12	2561	10530	54.645932	4.251632
64	52.550106	8.038	2602	10697	55.471519	4.2086968
65	53.371201	7.952	2643	10860	56.297317	4.1636672
66	54.192296	7.864	2684	11020	57.12344	4.1175904
67	55.013392	7.773	2724	11176	57.949898	4.0699428
68	55.834487	7.679	2764	11328	58.776586	4.0207244
69	56.655583	7.583	2803	11476	59.603628	3.9704588
70	57.476678	7.484	2841	11620	60.430912	3.9186224
71	58.297773	7.383	2880	11761	61.258568	3.8657388
72	59.118869	7.279	2917	11898	62.086606	3.8112844
73	59.939964	7.172	2954	12030	62.914773	3.7552592
74	60.76106	7.064	2.991	12159	63.743472	3.6987104
75	61.582155	6.952	3.026	12284	64.572307	3.6400672
76	62.40325	6.839	3052	12404	65.401556	3.5809004
77	63.224346	6.723	3096	12521	66.231087	3.5201628
78	64.045441	6.605	3130	12634	67.06105	3.458378
79	64.866537	6.484	3164	12742	67.891307	3.3950224
80	65.687632	6.361	3197	12846	68.713106	3.3306196
81	66.508727	6.327	3229	12947	69.562392	3.3128172
82	67.329823	6.11	3260	13043	70.393559	3.199196
83	68.150918	5.981	3291	13135	71.225197	3.1316516
84	68.972014	5.849	3321	13224	72.056994	3.0625364
85	69.793109	5.716	3351	13308	72.889113	2.9928976
86	70.614204	5.581	3380	13388	73.722058	2.9222116
87	71.4353	5.441	3408	13465	74.555685	2.8489076
88	72.256395	5.297	3435	13537	75.390914	2.7735092
89	73.077491	5.144	3462	13605	76.228032	2.6933984
90	73.898586	4.981	3487	13669	77.067988	2.6080516
91	74.719681	4.804	3512	13729	77.911921	2.5153744
92	75.540777	4.609	3536	13783	78.761207	2.4132724
93	76.361872	4.392	3559	13833	79.618361	2.2996512
94	77.182968	4.146	3580	13878	80.487184	2.1708456
95	78.004063	3.862	3600	13916	81.373612	2.0221432
96	78.825158	3.528	3618	13948	82.2896	1.8472608
97	79.646254	3.122	3634	13973	83.258838	1.6346792
98	80.467349	2.607	3648	13991	84.351559	1.3650252
99	81.288445	1.886	3657	14000	86.408545	0.9875096
100	82.10954	0	3657	14000	87.229641	0

Gore width   GW= 4.7769506 feet   Gore L 87.229641 feet
Ctr.Buoyancy   CB= 36.54 feet   Solidity 0.6521 units

Building Small Gas Blimps

CON-R5   REV: 12/97

APPENDIX
CONVERSION UNITS
(Applicable to L.T.A.)

LENGTH: (conversions)   proven numbers   (equivalants)

INCHES	x	2.54	= centimeters	1 Inch	=	2.540	Cm.
"	/	39.37008	= Meters	1 Inch	=	.0254	Meters
FEET	x	.3048	= Meters	1 Foot	=	30.48	Cm.
YARDS	x	.9144	= Meters	1 Yard	=	.9144	Meters
"	/	1.09361	= Meters	1 Meter	=	3.280075	Feet
METERS	x	3.28084	= Feet	1 Meter	=	39.37008	Inches
"	/	.3048	= Feet	1 Meter	=	1.093613	Yards
METERS	x	1.09361	= Yards	1 Sq.Met.	=	10.7630	Sq.Ft.
"	/	.9144	= Yards	1 Sq.Met.	=	1.19589	Sq.Yd.

DISTANCE: (1 Nm= 6076.115 * Feet);   (1 Km=3280.8*Feet);   (1 Sm=5280')

S-MILES	x	.62137	= K-Miles	1 S-Mile	=	1.60934	K-Meter
S-MILES	x	.86898	= N-Miles	1 S-Mile	=	.86898	N-Miles
N-MILES	x	1.15079	= S-Miles	1 N-Mile	=	1.15078	S-Miles
N-MILES	x	1.85195	= K-Meters	1 N-Mile	=	1.85195	K-Miles
KILOMETER	x	.62137	= S-Miles	1 K-Meter	=	.53996	N-Miles

WEIGHT: (1 Pound = 453.6 * Grams)

POUNDS	x	2.2046229	= Kilograms	1 Kilogram	=	2.2046	Pounds
KILOGRAMS	/	.45359237	= Pounds	1 Pound	=	.45359	Kg.
GALLON		Gasolene = 6.0#; Methane = 6.62#; Kerozene = 6.75#; Oil =7.5#; Water = 8.33#.					

VOLUME: (1 Cu.Foot = 1728* Cu.In.)   (1 Cu.Yard = 46656. *Cu.Inches)

GALLONS	x	.133681	= Cu.Feet	1 Gallon	=	231.0163	Cu.In.
"	/	7.48052	= Cu.Feet	1 Gallon	=	.13369	Cu.Feet
LITERS	x	.26417	= Gallons	1 Gallon	=	3.78541	Liters
"	/	3.78541	= Gallons	1 Liter	=	.26418	Gallons
Cu.METERS	x	35.31467	= Cu.Feet	1 Cu.Foot	=	7.48052	Gallons
"	/	.0283168	= Cu.Feet	1 Cu.Foot	=	.02832	Cu.Mts
Cu.FEET	x	.0283168	= Cu.Meters	1 Cu.Yard	=	.76455	Cu.Mts
"	/	35.31467	= Cu.Meters	1 Cu.Meter	=	35.3147	Cu.Feet
Cu.YARDS	x	1.30795	= Cu.Meters	1 Cu.Meter	=	1.30795	Cu.Yds
				1 Cu.Meter	=	264.172	Gallons

CONSTANTS: (Pi = 3.141593)   (De = .785398)

TEMPERATURE: (1 Degree F =1.8C + 32)   (1 Degree C = (F-32)/(1.8)

CENTIGRADE =	5/9 (F - 32)	FAHRENHEIT =	9/5 (C + 32)

LIQUID PETROLEUM GAS: (LPG = 6.62 Lb/Gal. @ 59 degrees F)

PROPANE = 1925 BTU/Lb. and has a Molecular Wt. of 44.09
50/50   = 1975 BTU/Lb.   "   51.10
BUTANE   = 2025 BTU/Lb. and has a Molecular Wt. of 58.12

Propane (C3H8) boils at -44 degrees F (-6.6C)   reaches 60 PSI at 35 F, 100 PSI at 65 F;
Butane (C4H10) boils at 32 degrees F (0 C)   reaches 60 PSI at 120F, 120 PSI at 212F.

1 Gal. LPG @ 60 F = 1.07 Gal. @ 100 F = .92 Gal. @ 0 F.
10 Gal. BUTANE @ 60F loses 26% BTU at 212 F (if relief valve pressure stays constant at 250 PSI)

HELIUM: 65 lbs.lift/1000 C.F.   (Hydrogen = 69 lbs.lift/1000)

1 C.F. liquid = 7.6 Pounds (3.45 Kg.)	1000LL = 268.39 Pounds (121.7 Kg.)
1 C.F. liquid = 682 C.F. gas	1000LL = 768.231 C.M. Gas
1 C.F. liquid = 47 lbs. lift (21.3 Kg.)	1000LL = 1660 lbs.Lift (753 Kg)

Building Small Gas Blimps

1.5
1.5

Trailer   Cylinders

HELIUM

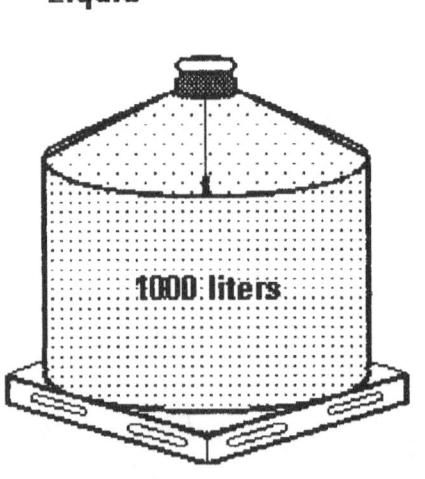

Liquid

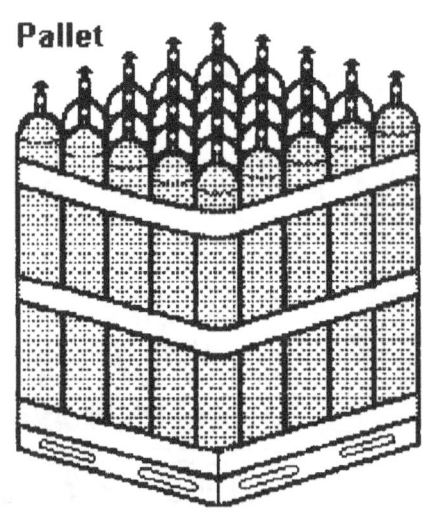

Pallet

194

Building Small Gas Blimps

HELIUM
R:12/02/97

Purity: It should be remembered that corrections are required for the purity of the gas. Check your purity anytime the lift-ballast required doesn't look right; but at least weekly. Purity meters are readily available, but in the 800 to $2500 range. The density, or chemical composition and purity of the lifting gas can measured by any of the following methods:

- (a) By measuring the gas pressure difference against air at different levels.
- (b) By taking gas samples in small containers of accurately known volume and weighing the containers on a sensitive analytical balance before and after the sample gas has been replaced by air or known pressure.
- (c) By the Bunsen method of timing the flow of a gas sample and an air sample through a small orifice.
- (e) By absorbing all the impurities over refrigerated charcoal, and noting the difference by weight or heat conductivity.
- (d) **The most common method of purity analysis is with a *Interferometer* which can compare gas and air densities by passing light rays through the samples.**

The chief impurity found in the lifting gas is air. If the purity falls as low as 97 percent (90% on large airship), the lift is seriously affected. The usual cause of purity loss are **bullet holes through the envelope, which are quite common** over certain sections of the country. Bullet holes are very hard to locate, and a very painstaking repair process at best, but especially if close to the nose battens or under the fin attachments. If one penetrates the ballonet, quick action will be directly related to cost of saving the ship.

1. Patch the BALLONET to helium hole(s) FIRST;
    - (a) get patches and glue;
    - (b) IMMEDIATELY open the ballonet access port;
    - (c) get inside, get the patch(es) on fast (time is critical for purity);
2. Patch the exit hole on the top SECOND;
    - (d) rent a Man-Lift (cherry-picker) ASAP;
    - (e) search the area that aligns with the bottom hole(s)
    - (f) while you are up there, take a quick look for other holes;
3. Bottom holes are the lowest priority, but don't ignore them.
4. Check your purity; if less than 90%, you lost the race.
    - (g) deflate and refill with new gas (dump & pump).

NOTE: The atmospheric humidity in warm climate also has a decided influence on the gas purity. Atmospheric humidity decreases buoyancy of the airship in two ways:
- (a) The moist air has less carrying capacity than the heavier dry air.
- (b) The moisture penetrates through the fabric of the gas cells into the gas, and also absorbed by the fabric.
- (c) Do not confuse *Superheat* with purity.

Building Small Gas Blimps

AIRCO ▬▬▬▬▬▬▬▬▬▬▬▬▬▬▬▬▬▬▬▬▬▬▬▬▬▬▬▬▬ **DATA**

Helium (He)
A chemically inert, colorless gas. Also available in bulk quantities, tube trailers and aluminum cylinders.

	Purity	Cyl. Size	Contents @ NTP	(psig) @70 F	Valve Outlet	Equipment Recommendation
Grade 6	99.9999%	300	291 cf	2640	CGA580	
		200	219 cf	2200	"	
Research Grade 5.5	99.9995%	300	8241 it	2640	"	
		200	6202 it	2200	"	
		5	100 it	1000	"	
		2	50 it	1742	"	
		2	25 it	842	"	
		2	10 it	322	"	
–+Ultra High Purity Grade 5	99.999%	300	291 cf	2640	"	
		200	219 cf	2200	"	
		80	80 cf	2200	"	
		30	30 cf	2200	"	
		2	2 cf	1800	"	
High Purity Grade 4.7	99.997%	300	291 cf	2640	"	
		200	219 cf	2200	"	
		80	80 cf	2200	"	
		30	30 cf	2200	"	
		1.8	2 cf	1800	CGA-180	
Zero Grade 0.5	99.997%	300	291 cf	2600	CGA-580	
		200	219 cf	2200	"	
High Press.	99.995%	500	524 cf	6000	CGA-877	

Purity Specifications (ppm unless noted)	Minimum Impurities Purity	ppm Max.	H_2	Ne	N_2	O_2	CO_2	H_2O
Grade 6	99.9999%	1	-	-	-	-	-	-
5.5	99.9995%	5	1	2	1	0.5	0.5	1
5	99.9990%	10	-	-	4	1.0	0.5	1
4.7	99.9970%	30	-	-	-	3.0	-	3
0.5	99.9970%	30	-	-	-	-	-	-
H.P.	99.9950%	50	-	-	-	-	-	-

Technical Data:
Mol. Wt.	4.000
Boiling Point	-452.1 F
Sp. Volume	96.7 cf/lb
Critica Temp.	-450.3 F
Critica Press.	33.2 psia
Flamable Limits	Non Flamible
Toxicity	Simple Asphyxiant
Compatability	Noncorrosive

Shipping Information:
DOT Name	Helium, Compressed
Hazard Class	2.2
ID Number	UN 1046
DOT Label	Nonflamable Gas
Gas Number	7440-59-7
MSDS No.	G-5

F.O.B. Point: Grades 6 and 5; Carol Stream, IL; City of Industry, CA; Research Triangle, NC; Riverton, NJ.
Other Grades: CA, IL, TN, LA, NJ.

Building Small Gas Blimps

AEROSTAT RULES
R: 12/05/97

Constants: (with all other factors remaining equal)
1. The lift of an aerostat varies with a change in volume.
2. The lift of an aerostat varies with a change in barometric pressure.
3. The lift of an aerostat varies with a change in temperature.
4. The lift of an aerostat varies with a change in humidity.
5. The barometric pressure decreases approximately 1-inch for each 1000' of altitude.
6. The temperature will decrease approximately 1-degree of Fahrenheit for each 300' ascent.

Changes:
7. The lift of gas increases as barometric pressure increases,
    and decreases if the pressure decreases.
8. The lift of a fixed volume of gas decreases if the atmospheric temperature increases,
    and increases if the temperature decreases.
9. The lift does not change due to a change in barometric pressure,
    if the gas is free to expand.
10. The lift decreases as the Atmospheric Humidity increases,
    for a fixed volume of gas.
11. The lift does not change in lift when air and gas temperature change an equal amount,
    if the gas is free to expand.
12. An aerostat in equilibrium at any altitude will be in equilibrium at the surface,
    providing there is no superheating of the gas.
13. An aerostat rising from the surface in equilibrium will be in equilibrium at any altitude
    below pressure height, providing no weight is lost and there is no superheating.

Numbers: (if the gas is free to expand)
14. The gas volume will increase approximately 1% for each 375 feet of altitude.
15. The gas volume will increase 1% for every 5 F-degrees of temperature.
16. The gas volume will increase 1% for every 5 F-degrees of superheat.

Formula for Lift:

$$L = \frac{A_s \times P_p \times (1-S_g) \times T_s}{P_p \times P_s}$$

S_g = Specific Gravity of He gas*	A_s = Standard Air Density (.0765)
P_p = Present baro. Pressure	P_s = Standard Pressure (29.92)
T_p = Present Temperature	T_s = Standard Temperature (59)

NOTE: "Present" means true readings at a specific altitude above Mean Sea Level (MSL)
    "Sg" of helium is normally *0.1381, but varies slightly by source.

WARNING: "Balloon Grade" gas is only **85%** Helium. <u>DO NOT ACCEPT THIS GAS</u>

Building Small Gas Blimps

HELIUM LIFT / ALTITUDE TABLE
(international standard atmosphere)

ALTITUDE in 100's feet	PRESSURE in Inches	PRESSURE in Milabars	TEMPERATURE -0.3566 F-degrees	TEMPERATURE in R-Degrees	TEMPERATURE C-Degrees	HELIUM Lift/Exansion cubic feet	HELIUM coefficient
0	29.92	1013.25	59.01	518.7	15.01	0.065949	1
100	29.81	1009.85	58.64	518.3	14.79	0.065768	1.0028
200	29.71	1006.44	58.29	517.6	14.59	0.065585	1.0036
300	29.59	1003.04	57.93	517.2	14.41	0.065403	1.0074
400	29.49	999.64	57.57	516.9	14.21	0.065221	1.0112
500	29.39	996.23	57.22	516.5	14.01	0.065038	1.0141
600	29.28	992.83	56.86	516.2	13.81	0.068556	1.0168
700	29.18	989.42	56.51	515.8	13.61	0.064673	1.0196
800	29.07	986.02	56.14	515.4	13.41	0.064491	1.0224
900	28.97	982.62	55.79	515.4	13.22	0.064308	1.0252
1000	28.86	979.21	55.43	515.1	13.02	0.064043	1.0298
100	28.76	975.81	55.07	514.7	12.82	0.063944	1.0313
200	28.65	972.41	54.72	514.4	12.62	0.063761	1.0346
300	28.55	969.01	54.36	513.9	12.42	0.063579	1.0379
400	28.45	965.59	54.01	513.6	12.23	0.063396	1.0412
500	28.35	962.19	53.65	513.3	12.03	0.063214	1.0445
600	28.24	958.79	53.29	512.9	11.83	0.063032	1.0478
700	28.14	955.39	52.93	512.6	11.63	0.062849	1.0511
800	28.04	951.98	52.57	512.2	11.43	0.062667	1.0544
900	27.93	948.58	52.22	511.8	11.23	0.062484	1.0577
2000	27.82	945.18	51.87	511.5	11.04	0.062177	1.0611
100	27.72	941.77	51.51	511.1	10.84	0.062119	1.0642
200	27.62	938.37	51.15	510.8	10.64	0.061937	1.0674
300	27.52	934.97	50.79	510.4	10.44	0.061755	1.0706
400	27.42	931.56	50.44	510.1	10.24	0.061572	1.0738
500	27.32	928.16	50.09	509.7	10.04	0.061391	1.0771
600	27.22	924.76	49.72	509.3	9.85	0.061208	1.0802
700	27.12	921.35	49.37	508.9	9.65	0.061025	1.0834
800	27.02	917.95	49.02	508.6	9.45	0.060843	1.0866
900	26.92	914.55	48.66	508.2	9.25	0.060661	1.0898
3000	26.82	911.14	48.31	507.8	9.05	0.060478	1.0931
100	26.72	907.74	47.94	507.5	8.86	0.060296	1.0963
200	26.63	904.33	47.58	507.2	8.66	0.060113	1.0996
300	26.53	900.93	47.23	506.8	8.46	0.059931	1.1029
400	26.43	897.52	46.88	506.5	8.26	0.059748	1.1062
500	26.34	894.12	46.52	506.1	8.06	0.059566	1.1095
600	26.24	890.72	46.16	505.7	7.86	0.059384	1.1128
700	26.14	887.32	45.81	505.4	7.67	0.059201	1.1161
800	26.04	883.91	45.45	505.1	7.47	0.059019	1.1194
900	25.95	880.51	45.09	504.7	7.27	0.058836	1.1227
4000	25.84	875.11	44.74	504.3	7.07	0.058654	1.126
100	25.75	873.69	44.38	503.9	6.87	0.058471	1.129
200	25.65	870.29	44.02	503.6	6.68	0.058289	1.133
300	25.56	866.89	43.67	503.2	6.48	0.058107	1.137
400	25.46	863.49	43.31	502.9	6.28	0.057924	1.141
500	25.38	860.09	42.95	502.5	6.08	0.057742	1.144
600	25.29	856.68	4261	502.1	5.88	0.057559	1.147
700	25.19	853.28	42.24	501.8	5.68	0.057377	1.151
800	25.11	849.88	41.88	501.4	5.49	0.057194	1.154
900	25.01	846.47	41.53	501.1	5.29	0.057012	1.158
5000	24.92	843.07	41.17	500.7	5.09	0.056828	1.161

Building Small Gas Blimps

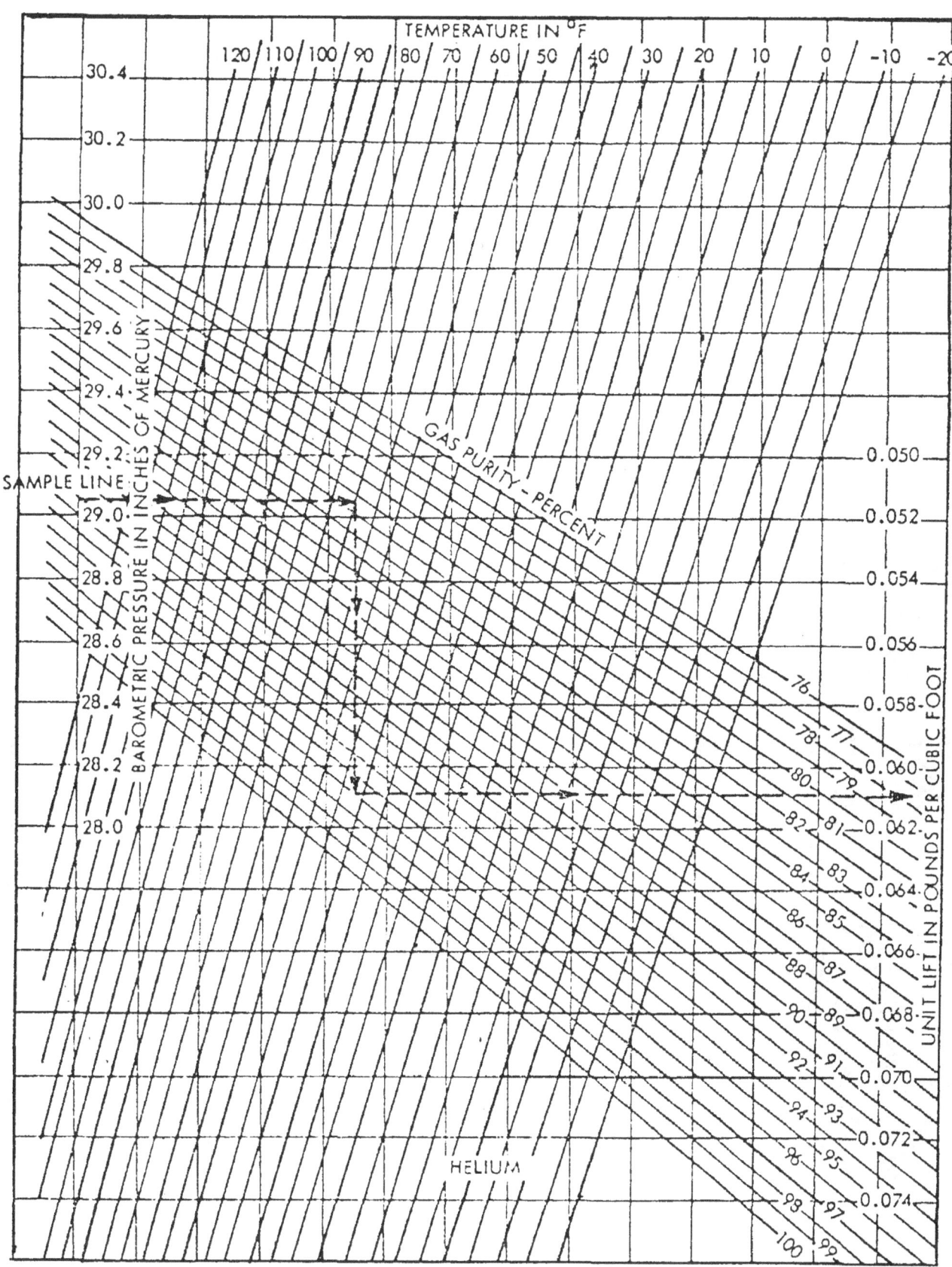

Lift of Dry Helium Versus Temperature,
Pressure, and Purity

Building Small Gas Blimps

Precision instruments designed to locate leaks in gas or volatile liquid filled systems in laboratory and industrial locations.

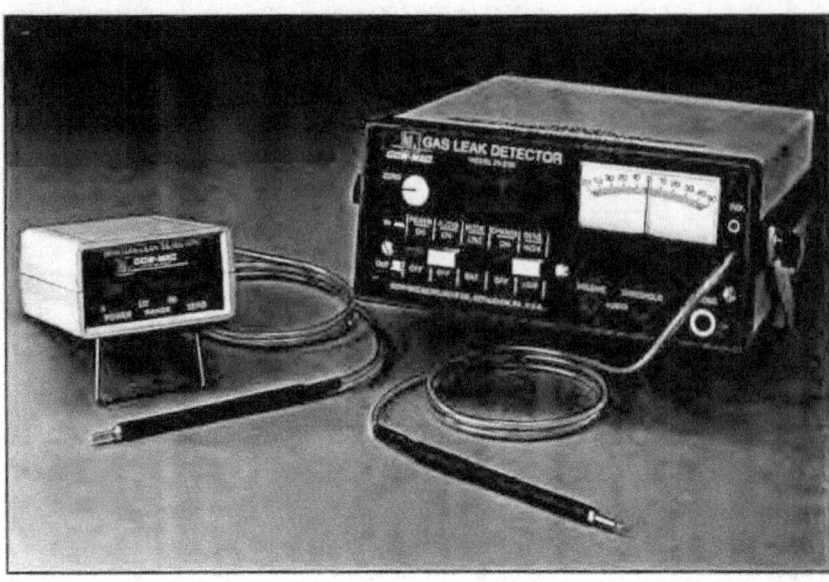

Model 21-050 and Model 21-250 Gas Leak Detectors

The **GOW-MAC Gas Leak Detectors** easily and quickly pinpoint gas leaks emitting from pressurized systems. Utilizing a thermal conductivity detector with signal amplification, the instruments are zeroed in ambient air and respond to any gas mixture with a thermal conductivity different from that of air.

The **GOW-MAC Gas Leak Detectors** are highly sensitive. Having an intrinsically high signal to noise ratio, amplification provides maximum usable sensitivity. Helium leaks of 1×10^{-5} cc/sec. are easily detected as are refrigerant leaks of 1.1×10^{-4} cc/sec.

Easy to Operate

The Model 21-050 or Model 21-250 Gas Leak Detector can be operated with little or no training. Turn it on, zero, and probe for leaks. As the instrument probe passes over the leak, a sample is drawn into the thermal conductivity cell. When a leak is discovered a signal is registered on the meter. The larger Model 21-250 will emit an audible signal as well. No messy soap solutions. No system contamination.

The probe is designed to reach difficult and confined locations. In close areas where several connections are grouped together the probe is small enough to pinpoint the exact problem location. Both units incorporate a high/low switch for controlling the instrument's sensitivity when sniffing for very small leaks.

Industrial Applications

Flow Meters
Welds
Seams
Joints
Valves
Pressure Regulators
Compression Fittings
High Pressure Vessels
Refrigeration Equipment
Exposed High Pressure Compressed Gas Lines

Laboratory Applications

Gas Chromatographs
Mass Spectrometers
Purge & Trap Systems
Injection Ports/Septa
Hot/Cold Column Fittings
Capillary Columns
Packed Columns
Gas Purifier Connections
Mass Flow Controllers
Cylinder Connections
Regulator Diaphragms
Inlet Tube Fittings

GOW-MAC
INSTRUMENT CO.

SB-GLD

Building Small Gas Blimps

PORTABLE THERMAL CONDUCTIVITY GAS ANALYZER
MODELS 20-600

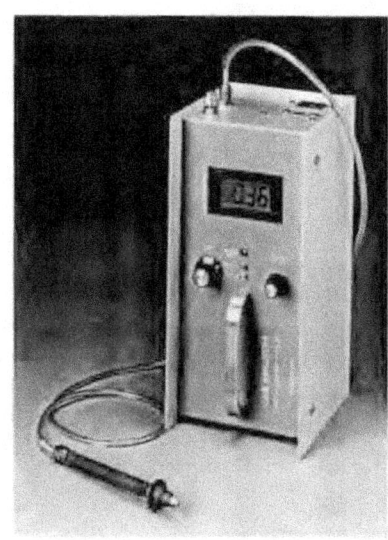

RESPONSE TIME LESS THAN 30 SECONDS

BATTERY OPERATED WITH RECHARGE CAPABILITY

CALIBRATION: 2 RANGES

SENSITIVITY: CAN BE CALIBRATED FOR AS LOW AS 0.5% CO_2 IN AIR.

COMMON APPLICATIONS:
- ARBON - HELIUM (S/R)
- METHANE - AIR
- CARBON DIOXIDE - HELIUM (S/R)
- HELIUM - AIR
- HYDROGEN - METHANE (S/R)
- REFRIGERANT 12 - AIR
- HALON 1301 - AIR

*S/R = SEALED REFERENCE

The GOW-MAC Model 20-600 Portable Gas Analyzer is designed for the quantitative measurements of binary mixtures. The primary gas being air, the secondary gas having sufficient difference in thermal conductivity value to be able to produce a signal. The Model 20-600 is portable, rugged, and suitable for field use.

The need for an instrument such as the 20-600 was first presented by the engineering staff of a major chemical manufacturer. They required a unit to measure Halon™ 1301 Fire Extinguishant in the field. The instrument described is a result of meeting their requirements and extending its use to other gases.

The GOW-MAC Model 20-600 Portable Gas Analyzer operates on the principle of thermal conductivity, i.e. gases vary in the amount of heat that they can carry away from a given source. Four rhenium-tungsten filaments are incorporated in the thermal conductivity cell used in this analyzer. A built-in sample pump draws the sample gas into the cell and the sample concentration, relative to air or a sealed reference (S/R) gas, is displayed on either a 4½" analog meter or on a 3½ digit digital led meter. The meters can be calibrated to read directly in terms of percent gas of the sample. The method is simple, reliable, and relatively fast.

The lead/acid gel battery will provide up to 31 hours continuous operation without a recharge. Recharging takes 15 hours from a completely discharged state and is easily accomplished by connecting the instrument to a source of 115 volts, 60 Hz. Should the battery become discharged in the field, the instrument may be operated from a source of 115 volt AC power.

Calibration is easily accomplished by sampling from a known mixture and adjusting the proper reading on the scale. Two separate adjustments are provided, one for each range.

The Model 20-600 comes with operating instructions and sample probe.

TRADEMARK E.I. DuPont De Nemours

GOW-MAC
INSTRUMENT CO.
The Basics of Good Science

U.S.A./International Office
GOW-MAC Instrument Co.
P.O. Box 25444
Lehigh Valley, PA 18002-5444
Tel: (610) 954-9000
Fax: (610) 954-0599

European Office
GOW-MAC Instrument Co.
Bay K, 14a, Industrial Estate
Shannon, Co. Clare, Ireland
Tel: +353-61-471632
Fax: +353-61-471042

U.K. Office
GOW-MAC Instrument Co.
6 Livingston Circus
Gillingham, Kent, U.K. ME7 4HA
Tel: Medway (0634) 575661
Fax: Medway (0634) 2800531

Canadian Office
GOW-MAC Instrument Co.
P.O. Box 6220
Ottawa, Ontario, Canada K2A
Tel: (613) 725-3621
Fax: (613) 747-9001

SB-206

Building Small Gas Blimps

- Detects Any Gas
- Digital Display
- NIST/NAMAS Traceable

LEAK DETECTION

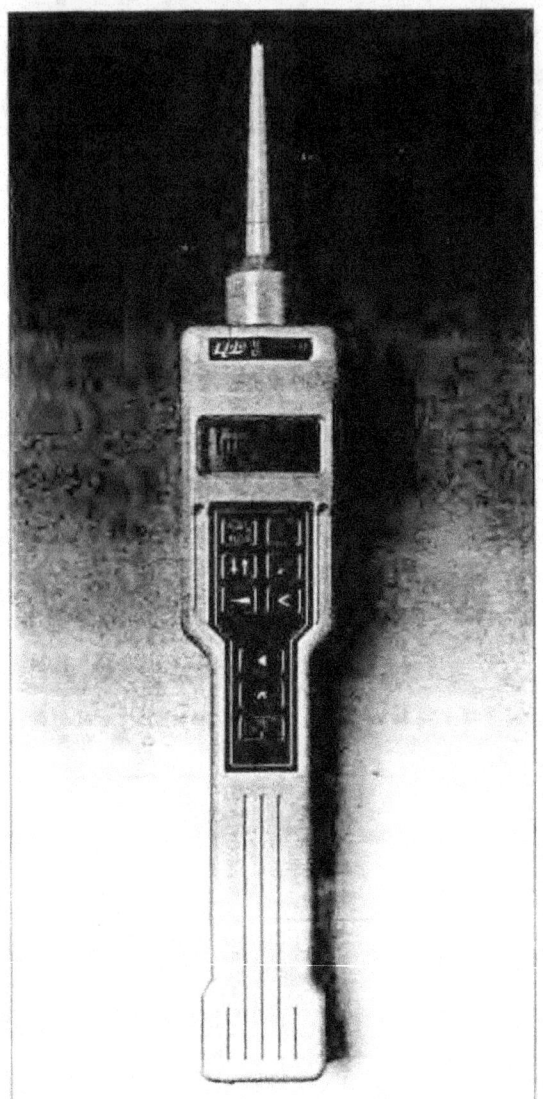

The **Digital Gas Leak Detector** is a microprocessor controlled multigas monitor that has been miniaturized, producing a handheld instrument capable of detecting very small gas leaks quickly, safely and without risk of contamination.

The heart of the Gas Leak Detector is a tiny sensor that can detect minute changes in thermal conductivity. When the detector is switched on, it automatically zeroes on the ambient air around it.

A small fan draws a continuous sample of air into the nozzle and over the sensor. When a change in the sample is detected, the reading is quantified and directly displayed on the screen, accompanied by an audible signal.

Once the general area of the leak has been determined the nozzle is removed, exposing a thin probe. This reduces the sample area and increases sensitivity further, allowing the leak's location to be pinpointed. The detector is also equipped with a long flexible probe for leak detection on hard-to-reach areas.

The Gas Leak Detector is powered by 4 conventional AA batteries, good for 40 hours of operation. The display includes a warning of low battery power, and the instrument automatically shuts itself off to conserve energy after five (5) minutes without input.

There are two models of the detector available. The **EPD-B3** model is designed for applications using **helium**. The instrument's display is calibrated to quantify Helium leaks in ml/sec.

The **EPD-B4** general purpose model offers the user a choice of units and Gas Groups, allowing accurate readings of all gases in ml/sec, ppm or cu.ft/yr. Advanced software features allow the user to program the unit to auto-range, to hold maximum readings, and to screen out interference from background gases. The **EPD-B4** is used if accurate readings are needed of multiple gases, or when working in industrial environments.

The use of rugged polymers for the casing and surface mount hybrid technology for the electronics has produced a lightweight, highly durable instrument. The detector requires minimum servicing, **it auto-zeroes and auto-calibrates**, and the sensor is not subject to deterioration so does not require periodic replacement. A one year parts/labor warranty is standard.

All instruments come complete with a carrying case, instruction manual, and a full range of accessories.

An Intrinsically Safe version of the EPD-B4 is also available, for use in hazardous areas or when working with flammable gases.

Helium Gas Leak Detector	EPD-B3	$1695.00
Digital Multigas Leak Detector	EPD-B4	$2195.00
Intrinsically Safe Digital Multi-Gas Leak Detector	EPD-B4IS	$2995.00
Deluxe Headset	EPD-DHC-4	$ 99.00
NIST Calibration Certificate	EPD-NIB4	$ 75.00
Use Portable Reference Leak (For description visit our web site or call customer care.)	EPD-PRL	$ 995.00

Extended Warranty Available

*  GSA Listed

EPD • 14 HAYES STREET, ELMSFORD, NEW YORK 10523 USA

TOLL FREE 1-800-892-8926 USA & Canada • INTL: 1-914-592-1234 • FAX: 914-347-2181 • E-MAIL: epd@epdtech.com

Building Small Gas Blimps

AFTERTHOUGHTS
on Small Gas Airships

By Brian Hval, P. Eng.
ZARD Aerospace Corp. Calgary, Canada

In recent years there has been great Interest in small airships sometimes described as "mini-blimps", "sport" or "light utility" airships. The interest stems from the perception of a market for small gas airships f or advertising or surveillance duties. Furthermore such airships should be low in cost to develop. Investors are more inclined to risk thousands of dollars on a baby blimp than millions on a heavy lift airship! We still hear a lot about small airships, but see very little evidence of flying hardware. The reasons for this are:

1. Many projects are still on the drawing board. For every project that is financed to construction, there must be at least 10 sitting. in the design stage .

2. Few organizations have the skill necessary to build such airships ... and the financiers know it! That is why most designs will stay where they belong. .. on a sheet of paper. It is very hard to convince someone to lend you money to build an airship if you have never done it before.

3. Financial scams abound. For every legitimate project there must be at least two that are not . I am personally aware of several schemes in Canada that bilked at least $400,000 in 1986 from unsuspecting shareholders. This is most unfortunate, especially for the bonifide projects which seem to live on shoe-string budgets.

4. Small airships may not be viable. 'Baby' blimps may be cute, but can they be supported by their intended markets? There definitely does seem to be a market for the Goodyear size airships as witness the growth of the competitive Airship Industries. But what about small airships under 50,000 cu.ft.? Can you prove there really is a market for these?

5. Several small airships are under construction but have yet to be flown. Some have run out of money. There should be an interesting crop of attempts in 1988-89.

6. Several small airships have flown in the last year, but all had varying degrees of success, and none were completely successful. It is this last category that is of the greatest interest. What has happened to those projects which cleared the financial, marketing and design hurdles? What happened during the first flights? As a consultant on airship design problems, I have become aware of the outcome of several of these projects. The lessons (some VERY COSTLY) will be of Interest to those either building or contemplating small airships. Here then, is a nuts and bolts account of what can happen once you have it ready to fly:

Caution: This is a composite account of the experiences of several small airships recently built. It did not happen to one project entirely, Thank God!

A. Engine Reliability - Most mini-blimps are planned around two-cycle engines. If you must use 2-cycles, use two of them! Most of these engines are plagued with carburetor problems, spark plug fouling,, and crankshaft failures. A "Reliable" 2-cycle engine used in ultralights, or experimental, is one that runs 300 hours without falling apart. Plan for engine failure unless you can manage a certified 4-cycle aircraft engine on your project. You will probably make several free balloon landings unless you have a spare engine to keep you going. Any engine used should have a proper break-in period followed by a test run of at least 10 hours with the engine mounted in the car. An engine that. will not run continuously on the ground for more 10 minutes without problems certainly cannot be expected to run more sweetly when it takes to the air!

B. Fuel Contamination - Pay strict attention to providing reliable means of fuel filtration and water removal of debris in the tiny carburetor ports can quickly stop you flying.

C. Engine Vibration - anything that can vibrate loose, soon will. Check your design thoroughly to make sure all nuts, etc., are securely safety wired or *Nylocked*. Watch for fatigue in parts. Avoid unsupported fuel lines or electrical wiring. The engine test run period previously described is an excellent opportunity to identify problem areas before taking to the air on the first flight.

Source: *AEROSTATION*; **Summer 1989 p12-16**

Building Small Gas Blimps

D. Emergency Landings - Free balloon landings without power will occur sooner or later. Make sure the flight crew has some means of making a landing without assistance of the ground crew. A lightweight folding grapnel anchor is an essential means for temporarily mooring a small airship. Remember airships are always flying even when the engine stops! They continue to fly even when on the ground so appropriate precautions must be taken to avoid embarrassment or disaster.

E. Knife - in the past, you could always recognize an airshipman by the large knife he carried. Nothing has changed. A large sharp knife will find plenty of uses in many situations such as cutting the envelope fabric in an unexpected emergency.

F. Airship Design - It is far more difficult than it looks! Nothing could be simpler than hanging a car under a gasbag and starting the engine. Although there are some excellent technical design references available, designing a successful airship is still a very tricky business. It is a mistake to rely 100% on the design formulas you find in some texts. Remember nature will do what it has to do regardless of what the book says. Provide lots of flexibility in your design to change things once the airship is built, particularly items related to trim control.

G. Fabric Selection - This can be and usually is a major problem area. Fabric selection is more critical in small airships. Remember, a small airship has a higher surface area to volume ratio . Hence the daily lift loss due to helium permeability can be serious. The material properties supplied by textile producers are usually for fabric fresh off the roll. Major changes in fabric strength can occur on exposure to sunlight.

H. Fabric Handling - Great care was taken by past airship operators in minimizing the folds in envelopes. Gas airship envelopes cannot be treated like hot air balloon envelopes. They must be gently packed with minimum *creasing or crushing*. Otherwise the folding considerably increases the permeability. This applies to modern fabrics/coatings/films as it did to the old varnished cotton envelopes.

I. Communications - During flight trials a lot of surprises can happen, some of which the ground crew can spot and the pilot cannot. Make sure you have reliable radio contact between ground and air at all times.

J. Inspection - During flight trials and during the construction of the airship the Airworthiness Inspectors can be very helpful. Airship cars and airplanes have many points in common. Involve the appropriate authorities in the project as early as possible so you can benefit from their advice.

K. Drag - Inflatable control surfaces are found in some of the more recent designs. They are, however, a significant source of drag. Early airships such as the French Clement-Bayard-1 in 1908 used inflatable fins, but they were abandoned because of the high drag penalty. High drain of course, means reduced speed pilots who have flown thermal airships often remark that their operations largely consist of f lying backwards or traveling a zig-sag course down wind. Such airships have proven useful only under ideal no wind conditions. Similar performance in a gas airship is unacceptable because of the need to return to a place where it can be safely moored. Otherwise the airship may have to be deflated with the loss of its expensive helium. Small gas airships therefore must have sufficient speed under *al] Normal] flight condi*tions to return to base. Improved speed can likely be achieved only by using an envelope with a fineness ratio (Length to Diameter) of 4 or above and replacing "soft" fins with 8 hard ones. These improvements are not without penalty. A higher fineness ratio can lead to a requirement for a 24 hour pressure watch to prevent buckling of the envelopes. Rigid fins cannot be packed easily, are more difficult to rig, and can be damaged very easily in a "crunch". Trade-offs in airship design can be more complex than first visualized.

L. Rate of climb/descent - climbing ability is essential to avoid obstacles and overcome downdrafts. Descent ability is important to be able to penetrate thermal updrafts with out being carried above pressure height, the point where the expanding gas must escape before the envelope bursts. Generally aim for 300+ fpm plus in climb & descent.

Source: *AEROSTATION*; Summer 1989 p12-16

Building Small Gas Blimps

M. Preflight Tests - Try out or check every operating piece of hardware on the ground before it takes to the air. Do not take to the air until you are certain everything is working properly. *Test, test, test! Check, check,* check! Then fly.

N. Data Collection - The excitement of seeing a new airship fly can cause one to neglect collection of basic data. Such data may include items such as weighing off the ship on a daily basis to determine lift losses. Flight trials should begin with an agreed upon plan (in writing) to determine flight characteristics and other essential data. The plan should then be followed in a disciplined manner.

O. Ground Crew - Claims are made by some developers that very small ground crews can handle small airships. Not true. A small airship may indeed have a small volume compared to its larger cousins But airships follow the volume cube/dimensions squared relationship. So although an airship is much smaller in volume, the length and diameter shrink only slightly. So the surface area exposed to wind and gusts is only slightly smaller in "small" airships as compared to the Goodyear blimp. From personal experience I nave found a crew of at least 6 necessary to handle a 25,000 cu.ft. airship in a 15 knot wind if no mechanical assistance is available.

P. Hangars - Some hopeful airship enthusiasts talk about conducting regular operations from a large conventional aircraft hangars In theory after flying you can tuck your baby blimp in a hangar. In practice, crosswinds can prevent you getting out of the hangar or into it. you need a hangar that has sufficient height and width to turn the ship into the hangar regardless of cross winds. Also watch out for tail kiting slamming your fins into the hangar door sill. The ideal solution is to have a hangar which rotates to align with the wind. The Germans did have these in World War I ... but can your project budget stand It?

The above provides some appreciation of the problems associated with small airships, whether homebuilt or otherwise. Do not venture lightly into these projects unless you:

1. Have adequate financial resources to build the project. A 2-man airship can cost well over $100,000 to build. I know of one developer who spent nearly $300,000 in 1985. He gave up after 4 flights and 4 crashes involving injuries.

2. Obtain qualified help in the design. Why re-invent the airship? Many starry eyed inventors arc shocked to discover their pet ideas were applied, tried, and discarded by airship builders over 50 years agog.

3. Have qualified pilots, either balloon or airship. Your airship will often end up flying balloon so be ready for it!

4. Have adequate financial resources to sustain the monstrous appetite that new-born baby blimps have for time, money, helium, money and more money!

Biography:

Brian Hyal is a professional engineer who has been involved with LTA technology for since 1967. He is a licensed balloon pilot and was founder of the two largest balloon clubs in Canada. he is the author of "Practical Ballooning", a training manual for balloon pilots, in use worldwide and in its third edition. In 1982 he was the first recipient of the Joan Martin Award presented by the Canadian Balloon Association in recognition of outstanding contributions to the sport of ballooning in Canada. Currently, Brian is associated with ZARD Aerospace of Calgary, which provides engineering services in innovative and advanced technologies including airship design. He resides in Calgary with his wife and 3 cats.

Editors Note:

This article was selected for incorporation because it was appropriate when it was published, AND is even more appropriate today. Brian has been a friend of sport aviation in general, and a credit to the airship fraternity in particular.

Source: *AEROSTATION*; Summer 1989 p12-16

Building Small Gas Blimps

AIRSHIP RECORDS
CLASS 'B' - GAS AIRSHIPS

SUBCLASS BA-1 (less than 400 cubic meters)

DISTANCE (UK)
Donald Cameron   94.86 km
Cameron DG/14   58.94 mi
10/12/90

DURATION (USA)
Bryan Allen   8 hrs/50 min/12 sec
Raven Helium "White dwarf"
Thermal to Brawley, CA   2/12/85

SUBCLASS BA-2 (401 - 900 cubic meters)

ALTITUDE (CANADA)
Hokan Colting   1,898 m
21st Century SPAS   6,227 ft
Holt, ONT   11/8/92

DISTANCE (UK)
Donald Cameron   94.86 km
Cameron DG/14   58.94 mi
10/12/90

DURATION (USA)
Bryan Allen   8 hrs/50 min/12 sec
Raven Helium "White Dwarf"
Thermal to Brawley, CA   2/12/85

SUBCLASS BA-3 (900 - 1,600 cubic meters)

ALTITUDE (CANADA)
Hokan Colting   1,898 m
21st Century SPAS   6,227 ft
Holt, ONT   11/8/92

DISTANCE (USA)
Paul Woessner, Pilot   374.71 km
Dr. Coy Foster, Copilot   232.83 mi
Thunder and Colt GA-42
Dallas, TX to Marion, TX   10/25/90

DURATION (USA)
Bryan Allen 8 hrs/50 min/12 sec
Raven Helium "White Dwarf"
Thermal to Brawley, CA   2/12/85

SPEED OVER A 3 KILOMETER COURSE (USA)
Paul Woessner, Pilot   77.50 kmh 48.16 mph
Dr. Coy Foster, Copilot
Thunder and Colt GA-42
Red Oak, TX   10/24/90

SUBCLASS BA-4 (1,600 - 3,000 cubic meters)

ALTITUDE (CANADA)
Hokan Colting   1,898 m
21st Century SPAS   6,227 ft
Holt, ON   11/8/92

DISTANCE (USA)
Paul Woessner, Pilot   374.71 km/232.83 mi
Dr. Coy Foster, Copilot
Thunder and Colt GA-42
Dallas, TX to Marion, TX   10/25/90

DURATION (USA)
Bryan Allen   8 hrs/50 min/12 sec
Raven Helium "White Dwarf"
Thermal to Brawley, CA   2/12/85

SPEED OVER A 3 KILOMETER COURSE (USA)
Paul Woessner, Pilot   77.50 kmh
Dr. COY Foster, Copilot   48.16 mph
Thunder and Colt GA-42
Red Oak, TX   10/24/90

Building Small Gas Blimps

AIRSHIP RECORDS
CLASS 'B' - GAS AIRSHIPS

SUBCLASS BA-5 (3,000 - 6,000 cubic meters)

ALTITUDE (CANADA)
Hokan Colting
21st Century SPAS
Holt, ONT

1,898 m
6,227 ft
11/8/92

DISTANCE (USA)
Paul Woessner, Pilot
Dr. Coy Foster, Copilot
Thunder and Colt GA-42
Dallas, TX to Marion, TX

374.71 km
232.83 mi

10/25/90

DURATION (USA)
Bryan Allen
Raven Helium "White Dwarf"
Thermal to Brawley, CA

8 hrs/50 min/12 sec

2/12/85

SPEED OVER A 3 KILOMETER COURSE (USA)
Paul Woessner, Pilot
Dr. Coy Foster, Copilot
Thunder and Colt GA-42
Red Oak, TX

77.50 kph
48.16 mph

10/24/90

SUBCLASS BA-6 (6,000 - 12,000 cubic meters)

ALTITUDE (CANADA)
Hokan Colting
21st Century SPAS
Halt, ONT

1,898 m
6,227 ft
11/8/92

DISTANCE (USA)
Paul Woessner, Pilot
Dr.Coy Foster, Copilot
Thunder and Colt GA-42
Dallas, TX to Marion, TX

374.71 km
232.83 mi

10/24/90

DURATION (USA)
Brian Allen
Raven, Helium "White Dwarf"
Thermal to Brawley, CA

8 hrs/50 min/12 sec

2/12/85

SPEED OVER A 3 KILOMETER COURSE (USA)
Paul Woessner, Pilot
Dr.Coy Foster, Copilot
Thunder and Colt GA-42
Red Oak, TX

77.50 kmh
48.16 mph

10/24/90

SUB CLASS BA-7 (12,000 - 25,000 cubic meters)

ALTITUDE (CANADA)
Hokan Colting
21st Century SPAS
Host, ON

1,898 m
6,227 ft
11/8/92

DISTANCE (USA)
Paul Woessner, Pilot
Dr.Coy Foster, Copilot
Thunder and Colt GA-42
Dallas, TX to Marion, TX

374.71 km
232.83 mi

10/25/90

DURATION (USA)
Bryan Allen
Raven Helium "White Dwarf"
Thermal to Brawley, CA

8 hrs/50 min/12 sec

2/12/85

SPEED OVER A 3 KILOMETER COURSE (USA)
Paul Woessner, Pilot
Dr.Coy Foster, Copilot
Thunder and Colt GA-42
Red Oak, TX

77.50 kmh
48.16 mph

10/24/90

Building Small Gas Blimps

AIRSHIP RECORDS
CLASS 'B' - GAS AIRSHIPS

SUBCLASS BA-8 (25,000 - 50,000 cubic meters)

ALTITUDE (CANADA)
Hokan Colting　　　　　　　　1,898 m
21st Century SPAS　　　　　　6,227 ft
Holt, ON　　　　　　　　　　　11/8/92

DISTANCE (USA)
Paul Woessner, Pilot　　　　　374.71 km
Dr. Coy Foster, Copilot　　　　232.83 mi
Thunder and Colt GA-42
Dallas, TX to Marion, TX　　　10/25/90

DURATION (USA)
Bryan Allen　　　　　　　　　　8 hrs/50 min/12 sec
Raven Helium "White Dwarf"
Thermal to Brawley, CA　　　　2/12/85

SPEED OVER A 3 KILOMETER COURSE (USA)
Paul Woessner, Pilot　　　　　77.50 kmh
Dr. Coy Foster, Copilot　　　　48.16 mph
Thunder and Colt GA-42
Red Oak, TX　　　　　　　　　10/24/90

SUBCLASS BA-9 (50,000 - 100,000 cubic meters)

ALTITUDE (CANADA)
Hokan Colting　　　　　　　　1,898 m
21 Century SPAS　　　　　　　6,227 ft
Holt, ONT　　　　　　　　　　11/8/92

DI ACE (USA)
Paul Woessner, Pilot　　　　　374.71 km
Dr sy Foster, Copilot　　　　　232.83 mi
Th er and Colt GA-42
Dallas, TX to Marion, TX　　　10/24/90

DURATION (USA)
Brian L.Allen　　　　　　　　　8 hrs/50 min/12 sec.
Ra -I Helium "White Dwarf"
Thermal to Brawley, CA　　　　2/12/85

SPEED OVER A 3 KILOMETER COURSE (USA)
Pal.l Woessner, Pilot　　　　　77.50 kmh
Dr. Coy Foster, Copilot　　　　48.16 mph
Thunder and Colt GA-42
Red Oak, TX　　　　　　　　　10/24/90

SUBCLASS BA-10 (more than 100,000 cubic meters)

ALTITUDE (CANADA)
Hokan Colting　　　　　　　　1,898 m
21st Century SPAS　　　　　　6,227 ft
Halt, ONT　　　　　　　　　　11/8/92

DISTANCE (USA)
Paul Woessner, Pilot　　　　　374.71 km
Dr. Coy Foster, Copilot　　　　232.83 mi
Thunder and Colt GA-42
Dallas, TX to Marion, TX　　　10/25/90

DURATION (USA)
Bryan Allen　　　　　　　　　　8 hrs/ 50 min/ 12 sec
Raven Helium "White Dwarf"
Thermal to Brawley, CA　　　　2/12/85

SPEED OVER A 3 KILOMETER COURSE (USA)
Paul Woessner, Pilot　　　　　77.50 kmh
Dr. Coy Foster, Copilot　　　　48.16 mph
Thunder and Coit GA-42
Red Oak, TX　　　　　　　　　10/24/90

Building Small Gas Blimps

In Flight USA, June 1998

FAA Clarifies Policy on Amateur Experimental Flights

WASHINGTON—The FAA has issued a bulletin clarifying limitations for Experimental amateur-built aircraft flying over populated areas, which will hopefully end confusion over the issue that emerged last month in the Western United States.

The EAA and its members in the region had requested the clarification after confusion over operating limitations arose among local pilots and FAA officials in the Los Angeles area. The confusion over proper operating procedures threatened to slow or halt flying activity, as pilots were reluctant to fly their Experimental "homebuilt" aircraft for fear of being found in violation of regulations or FAA policy.

"EAA always maintained that FAA had established its policy on amateur-built over flights of populated areas more than 25 years ago", EAA President Tom Poberezny says. "That policy was based on homebuilt aircraft's excellent safety record after the test period. Although this situation grew out of a singular case regarding an Experimental exhibition aircraft, it's important to reiterate current policy for amateur-built operations."

The FAA bulletin states that once flight testing is completed in a homebuilt, an Experimental amateur-built aircraft may "operate over densely populated areas, both en route and during takeoffs and landings, and operate within congested airways of the National Airspace System (NAS)."

The EAA also asked FAA headquarters to reiterate to local and regional offices that FAA headquarters should establish policy for U.S., airspace to prevent a patchwork of regulations that might vary throughout the country.

The situation began when FAA's Flight Standards District Office (FSDO) in Long Beach, Calif., contacted Experimental aircraft owners operating at John Wayne Airport in Orange County, Calif. Representatives from that FSDO informed those airplane owners that they may be in violation of FAR Part 91, which prohibits Experimental category aircraft over populated areas "unless otherwise approved."

That was in conflict with an FAA policy established in 1972, which stated that amateur-built aircraft could operate over populated areas, once certain conditions were met. The policy was created because such operations are not deemed a safety concern, but rather a means of limiting risk during the initial test phase of such aircraft. The clarification by FAA headquarters reinforces that policy nationwide. The FAA requested the EAA's input on the issue, particularly in drafting a clarification for operating limitations, after initial test flights. The EAA is continuing to work with FAA headquarters and its Western Region offices to solve the particular situations regarding operations at John Wayne Airport. The Long Beach FSDO is also involved in the issue, so a reasonable solution to the current situation can be implemented quickly. The EAA has also asked the FAA to clarify the same issue for Experimental-exhibition aircraft.

Building Small Gas Blimps

U.S. Department of Transportation
Federal Aviation Administration

General Aviation Airworthiness Alerts

AC No. 43-16

ALERT NO. 230
SEPTEMBER 1997

NOTE: This was included to show you a good source of information on where problems in ultralights are. This is unbiased and free from the FAA on a subscription basis.

**Improve Reliability-
Interchange Service
Experience**

210

AMATEUR, EXPERIMENTAL, AND SPORT AIRCRAFT

All aircraft hoses should be closely examined at every opportunity.

Part total time not reported.

AVID

Avid Flyer   **Tailwheel Failure**
3222

After approximately 30 landings, the tailwheel leaf spring broke while taxiing.

The manufacturer stated a "bad batch of springs" had been received from the supplier. These springs were too thin, and the heat treatment was not correct.

The submitter suggested that anyone having a kit from early 1989 to 1990 should have the springs checked.

Part total time not reported.

RANS

Rans   **Engine Power Loss**
Model S-12   7322
Engine Rotax
Model 582

During flight, the engine lost power and eventually failed. A landing was made without personal injury or aircraft damage.

An inspection revealed the carburetor socket had vibrated loose from the intake manifold. The manufacturer recommends safety wiring the attachment clamps. In this case, safety wire was not used. The submitter stated: "While the preflight inspection was being completed, the carburetors felt secure."

Part time since overhaul-100 hours.

SKYSTAR

Skystar   **Engine Seizure**
Model IV-1200   8520
Engine Rotax
Model 582

During a ground operational test, the engine failed.

An engine teardown and inspection disclosed that the crankshaft connecting rod bearings failed, causing the connecting rods to fail. The engine seized and would not rotate. Before the failure, all of the engine operating parameters (such as temperature, oil pressure, etc.) were normal. The submitter did not offer a cause for this problem.

Part total time-110 hours.

KITFOX

Kitfox   **Engine Starter Ring**
Model II   **Gear Failure**
Engine Rotax   8011
Model 582

During a preflight inspection, metal filings were found in the engine starter area.

Further inspection revealed a large crack near a bolt hole on the starter ring gear (P/N 834-050). The crack was approximately .125 inch. (Refer to the following illustration.) Additional smaller cracks were found emanating from the vicinity of the other bolt holes. Also, the ring gear displayed evidence of heat damage. This is not an uncommon occurrence, and Rotax has issued Service Bulletin (SB) 4UL90-E which gives specific instructions for inspection and/or replacement criteria. This defect could result in a catastrophic failure during flight and deserves prompt attention.

It was recommended that all operators using the Model 532 or Model 582 engines should check the starter ring gear in accordance with SB 4UL90-E.

Part total time-185 hours.

Building Small Gas Blimps

INDEX

Actuators, Lin.
Actuators, Rot.
Airship parts
Altimeter
Altitude
Anchor bolts
Anchor stakes
Anchoring
Aspect ratio
Ballonets
Banners
Base plate
Battens
Batton tunnel
Bibliography
Binders, load
Blind nuts
Blowers
Books
Burnout
Cable, control
Cable, wire
Lacing channel
Cam rings
Catalogs
Clamshells
Composits
Cone ties
Control surfaces
Controllers
Controls
Cover, valve
Crews
Cruciform
Cutting
Cutting fabric
Definitions
Deflating
Design, basic
Disclaimer
Docking cone
Drilling
Drip fence
Dual controls
Ducted Fans
Ducted Fans
Ellipsoid
Engines
Envelopes
Envelopes

Expertese
Eyes
FAA publicate.
FAA limits
Fabric
Fin guys
Finger patches
Fins
Fittings
Flange, Mast
Flange, valve
Generators
Glueing
Glues
Gondola
Gore panels
Gores
Ground support
Formulas
Gussets
Handles
Hangars
Hardware, std.
Hardware, MS
Helium
Hinges
Inflation
Instruments
Inverted-Y
Kits
Landing gear
Lights
Lock pins
Machining
Magazines
Maintainability
Manuals
Mast Heads
Masts
Materials
Money
Mooring locks
Nets
Nicopress
Nose cone
Nose details
Offset hinges
Pigtails
Plumbing
Pole barns

Poles, mast
Ports, View
Ports, Access
Ports, Gas
Pressure
Program, gores
Program, shape
Propellers
Releases
Reports
Rigging
Rings, Dee
Rings, triangle
Rip panels
Safety
Safety factors
Shackles
Skids
Sleeves
Snaps
Spindles
Spread sheets
Stabilizer fins
Standards
Strength
Subcontracting
Suppliers
Eyebolts
Supplies
T-Box
Thimbles
Tools
Trailers
Training
Turbines
Turnbuckles
Ultralight
Useful load
Valves
Vectored thrust
Vehicles
Volumes
Webbing
Weight control
Welding
Windows
Workmanship
Workplace
Wheels
Work platforms

Building Small Gas Blimps

ISBN LOG BOOK

OTHER PUBLICATIONS by Robert J. Recks

ISBN	Title	Date
0-937568 - 00 - 7	PROBLEMS OF HELICOPTER USE	6/1961
0-937568 - 01 - 5	HELICOPTER WEIGHT AND BALANCE	8/1961
0-937568 - 02 - 3	BALLOONING EXAMINATION GUIDE	9/1969
0-937568 - 03 - 1	BALLOONING TRAINING MANUAL	10/69
0-937568 - 04 - X	BALLOONING CLUB INTRODUCTION	11/69
0-937568 - 05 - 8	BALLOONING FLIGHT CIRRICULUM	12/69
0-937568 - 06 - 6	BALLOONING PROPOSAL, FAR-AMMENDMENTS	1/1970
0-937568 - 07 - 4	BALLOON COMPETITION & SANCTION GUIDE	2/1970
0-937568 - 08 - 2	BALLOON PILOT SKILL RATING	4/1970
0-937568 - 09 - 0	TEACHING BALLOONING TO ADULTS	5/1970
0-937568 - 10 - 4	BRIEFING BALLOON CREWMEMBERS	6/1970
0-937568 - 11 - 2	BALLOONS & PILOTS CURRENTLY ACTIVE	7/1970
0-937568 - 12 - 0	BALLOON RECORDS, NATIONAL & IN'TL	8/1970
0-937568 - 13 - 9	KITE BALLOONING, HISTORICAL NOTES	1/1971
0-937568 - 14 - 7	FREE BALLOONING, HISTORICAL NOTES	2/1971
0-937568 - 15 - 5	BUILDING YOUR OWN SPORT BALLOON, VOL-1	1/1975
0-937568 - 16 - 3	BUILDING YOUR OWN SPORT BALLOON, VOL-2	8/1981
0-937568 - 17 - 1	BUILDING YOUR OWN SPORT BALLOON, VOL-3	3/1975
0-937568 - 18 - X	BALLOONS ON STAMPS, 1783-1983	12/80
0-937568 - 19 - 8	HELICOPTER EXTERNAL LOADS, VOL-1	3/1981
0 937568 - 20 - 1	AIRSHIPS ON STAMPS	6/1981
0-937568 - 21 - X	HELICOPTER EXTERNAL LOADS, VOL-II	8/1981
0-937568 - 22 - 8	AIRSHIPS & BALLOONS ON STAMPS	10/81
0-937568 - 23 - 6	BALLOON STAMP HISTORY ALBUM	12/81
0 937568 - 24 - 4	IT BEATS WORKING, GYPSY PILOT AUTOBIO.	Res.
0-937568 - 25 - 2	AIRCRAFT TERMS, ENGLISH-PORTUGUESE	12/82
0-937568 - 26 - 0	WHO'S WHO OF BALLOONING, 1783-1983	11/83
0-937568 - 27 - 9	BALLOONING WITH SMOKE	Res.
0-937568 - 28 - 7	BUILDING GAS BLIMPS	12/97
0-937568 - 29 - 5	BUILDING THERMAL BLIMPS	1998
0-937568 - 30 - 2	MUSCLE POWERED BLIMPS	01/99
0-937568 - 31 - 0	BUILDING & FLYING GAS BALLOONS	Res.

ABOUT THE AUTHOR

Born and raised in Southern California; Served as a mechanic in the U.S.Army; Engineering major at Compton College; Started flight training 1957; Graduated from the Univ.of Miami (FL); Served as a flight instructor; Flew worldwide as an Airline Engineer/Pilot, Holder of pilots licenses in 6 countries. Active Balloon & Airship pilot since the 1960's; Served as a FAA Designated Airworthiness Rep. specializing in Lighter-Than-Air aircraft; Holder of a Master Riggers License; Serves as an Aviation engineering consultant specializing in Inflatable Technology.

Building Small Gas Blimps

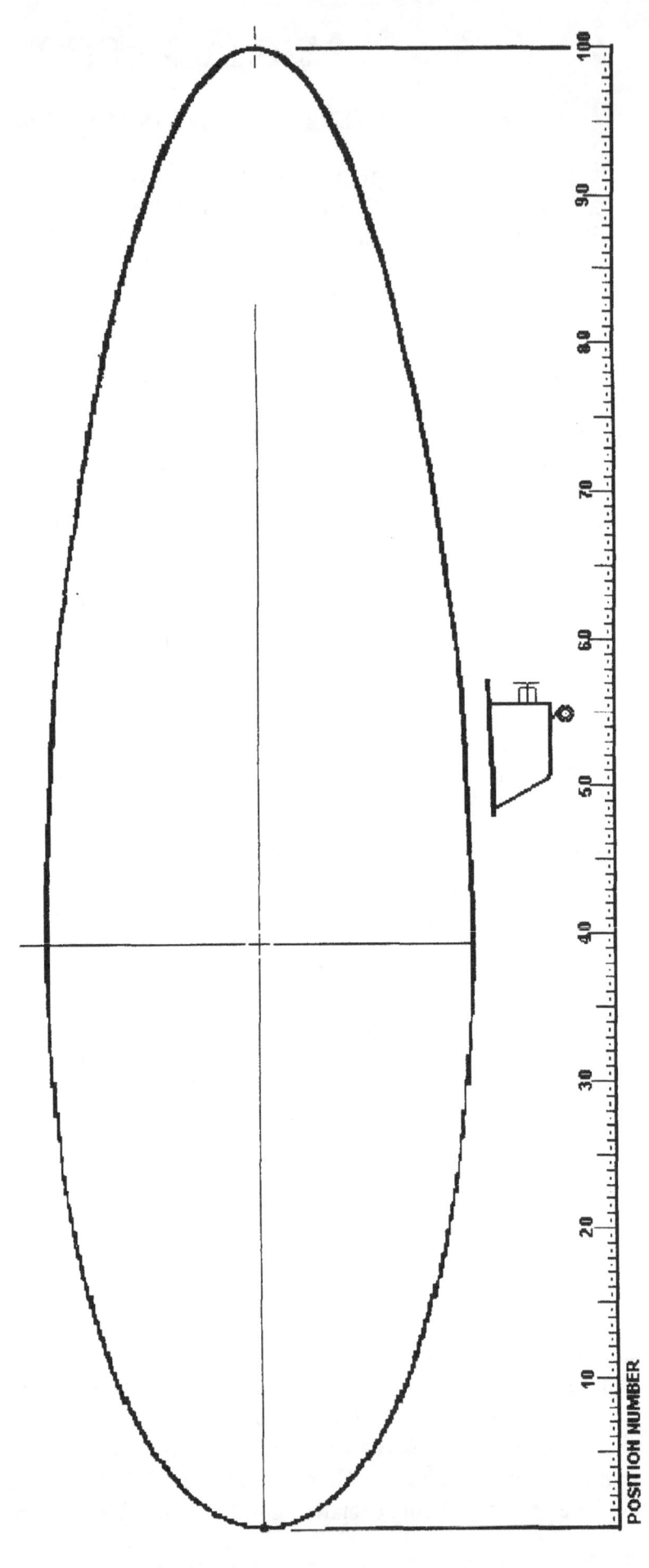

AIRSHIP LAYOUT
WORK SHEET

1.5 C_DOCUM
1.0

DOCUMENTATION
(Making peace with the FAA)

Building Small Gas Blimps

APPENDIX
DOCUMENTATION

The day you start cutting fabric, *not* the day you start building the metal stuff, is *THE* day you become fully committed. Hopefully, you've already gotten the FAA involved; but now you have to start the paper trail. You need to get two forms from them and start filling in the blanks.

FIRST, Apply for a "N" registration number **(costs $15)**:
- A. FAA Form 8050-1, Application for Registration;
- B. A formal letter of Request for Registration (asking for a specific set of numbers/letters, see sample);
- C. Copies of major receipts for purchase of materials, as Evidence of (YOUR) Ownership;

(Make sure YOUR NAME is on each receipt as *purchaser*) at least FABRIC, ENGINE, and PROP.

NOTE: It normally takes 3-6 weeks to get a reply.
Some "N1xx" numbers are still available, incl. N1RR (or your initials)

SECOND, When you are ready for your FAA inspection **(no cost*)**:
- C. FAA Form 8130-6, Application for Certification;
- D. A formal letter requesting inspection (see sample).

NOTE: Phone your local FSDO for his best dates before writing letter.
Don't try to dazzle him with how far, fast, and high you expect to fly, or he may not be available until after next Christmas.

* unless you go to a DAR (you still need the 8130-6).

AIRCRAFT OWNER RESPONSIBILITIES

You, as an aircraft owner, will be assuming responsibilities similar to those you have if you own an automobile. Owning an automobile usually means that you must register it in your State of residence and obtain license plates. As the registered owner of an aircraft, you will be responsible for:

1—Having a current Airworthiness Certificate appropriately displayed in your aircraft.
2—Maintaining your aircraft in an airworthy condition.
3—Assuring that maintenance is properly recorded.
4—Keeping abreast of current regulations concerning the operation and maintenance of your aircraft.
5—Notifying the FAA Aircraft Registry immediately of any change of permanent mailing address **or of the sale or export** of your aircraft.

Some States require that your car be inspected periodically (most States every 12 calendar months) to assure that it is in a safe operating condition. Your aircraft will have to be inspected Within every 12 calendar months in order to maintain a current Airworthiness Certificate. As With your automobile, accidents involving your aircraft must also re ported. Some similarities between automobile and aircraft responsibilities are shown in the following chart;

Building Small Gas Blimps

YOUR LETTERHEAD
Your contact phone number

(todays date)

Registration Branch, AC-250
Federal Aviation Administration
Monroney Aeronautical Center
Post Office Box 25082
Oklahoma City, OK 73125

                                                                Re: Registration
                                                                 Number Request

Dear Sir:

Please issue a FIVE symbol, historically significant, registration number for my HOMEBUILT aircraft now under construction. As per AC-20-21, I request "N" plus 2-digits and two last letters as follows:

        Preferred Registration number is: N_____; if available
        Preferred second choice is: N_____; if available
        OR   N x x _____ with any two available first numbers
              N_____ x x with any two available last numbers

        The Aircraft Category is: **LIGHTER-THAN-AIR**
                Class is: **AIRSHIP**
                Type is: **HELIUM GAS**
                Make is: (your last name)
                Model is: **BA-3** (or other FAI class)
                Serial #: **01** (assuming it is your first)

    I hereby certify that this aircraft has never been previously registered in any country, and is being constructed of parts and materials in conformance with FAR 47.33(b). Please call me at the above phone number if there are any conflicts or suggestions.

Very truly yours;

_____
builder / owner / operator

encl: $15 Money Order
    Copy, FAA Form 8050-1 (completed)
    Material Receipt copies (largest expenditures)
    Identification drawing (showing any unique features or colors)

Building Small Gas Blimps

APPENDIX
DOCUMENTATION

Automobile / Airplane Comparison Chart

Responsibility	Automobile	Aircraft
Registration	Yes	Yes
Annual Inspection	Yes	Yes
Compulsory Insurance	Yes	No
Reporting of Accidents	Yes	Yes
Required maintenance records	No	Yes
Maximum speed restrictions	Yes	Yes
Controlled maintenance	No	Yes

CERTIFICATE OF REGISTRATION

If you build or purchase an aircraft, you must apply for a Certificate of Registration before you fly it. An Aircraft is eligible for registration only if it is owned by a citizen of the United States, and it is not registered under the laws of any other foreign country.

The Application for Aircraft Registration, FAA Form 8050-1 (see the example) consists of an original (white) and two duplicate copies (green and pink). Instructions for preparing and submitting the form are attached to the top of the form.

When a homebuilder is applying for an ORIGINAL Registration Certificate, he must also submit satisfactory evidence that he purchased (or sources of) the components therein. Receipts, and a letter certifying his purchase of the major components, may be considered as satisfactory.

When applying for a new Registration Certificate on a manufactured aircraft, an aircraft bill of sale or other evidence of ownership must be submitted. A Bill of Sale (FAA Form 8050-2, not shown) meets the recording requirements.

It must be emphasized that before the aircraft can be legally flown, you must have sent a duly executed Application for Aircraft Registration and the proper fees to the FAA Aircraft Registry. The Temporary or "pink" copy of the application should be placed in your aircraft until you receive the permanent Certificate of Registration from the Federal Aviation Administration. When received, the Certificate of Aircraft Registration, FAA Form 8050-3 (see the examples), should replace the "pink" copy of FAA Form 8050-1 in the aircraft. Aircraft last registered in a foreign country may not be operated until the PERMANENT registration certificate is displayed in the aircraft.

The Certificate of Registration expires when:
    (a) the aircraft is registered under the laws of another foreign country;
    (b) the registration of the aircraft is canceled at the written request of the owner;
    (c) the aircraft is totally destroyed or scrapped; or
    (d) the ownership of the aircraft is transferred.

When the aircraft is sold, the previous owner (seller) must notify the FAA by filling in the back of his Certificate of Registration, and mailing it to the FAA Aircraft Registry. The FAA does not issue any certificate of ownership or endorse any information with respect to ownership on a certificate of aircraft Registration.

Building Small Gas Blimps

FORM APPROVED OMB NO. 04-R0076

UNITED STATES OF AMERICA DEPARTMENT OF TRANSPORTATION
FEDERAL AVIATION ADMINISTRATION-MIKE MONRONEY AERONAUTICAL CENTER
AIRCRAFT REGISTRATION APPLICATION

CERT. ISSUE DATE

UNITED STATES REGISTRATION NUMBER	N

AIRCRAFT MANUFACTURER & MODEL

AIRCRAFT SERIAL No.

FOR FAA USE ONLY

TYPE OF REGISTRATION (Check one box)

☐ 1. Individual ☐ 2. Partnership ☐ 3. Corporation ☐ 4. Co-Owner ☐ 5. Gov't.

NAME OF APPLICANT (Person(s) shown on evidence of ownership. If individual, give last name, first name, and middle initial.)

ADDRESS (Permanent mailing address for first applicant listed.)

Number and street: _____

Rural Route: _____ P.O. Box: _____

CITY	STATE	ZIP CODE

☐ **CHECK HERE IF YOU ARE ONLY REPORTING A CHANGE OF ADDRESS**

ATTENTION! Read the following statement before signing this application.

A false or dishonest answer to any question in this application may be grounds for punishment by fine and/or imprisonment (U.S. Code, Title 18, Sec. 1001).

CERTIFICATION

I/WE CERTIFY:

(1) That the above aircraft is owned by the undersigned applicant, who is: Check one as appropriate

   a. ☐ A citizen of the United States;

   b. ☐ A resident alien, with alien registration (Form 1-151 or Form 1-551) No. _____

   c. ☐ A foreign-owned corporation organized and doing business under the laws of (state or possession) _____, and said aircraft is based and primarily used in the United States. Records of flight hours are available for inspection at _____

(2) That the aircraft is not registered under the laws of any foreign country; and
(3) That legal evidence of ownership is attached or has been filed with the Federal Aviation Administration.

NOTE: If executed for co-ownership all applicants must sign. Use reverse side if necessary.

	SIGNATURE	TITLE	DATE
EACH PART OF THIS APPLICATION MUST BE SIGNED IN INK.	SIGNATURE	TITLE	DATE
	SIGNATURE	TITLE	DATE

NOTE: Pending receipt of the Certificate of Aircraft Registration, the aircraft may be operated for a period not in excess of 90 days, during which time the PINK copy of this application must be carried in the aircraft.

AC FORM 8050-1 (11-79) (0052-00-628-9005) SUPERSEDES PREVIOUS EDITION

APPENDIX
DOCUMENTATION

CERTIFICATE OF AIRWORTHINESS

An Airworthiness Certificate is issued by a representative of the Federal Aviation Administration after the aircraft has been inspected and it is found that it meets the requirements of the Federal Aviation Regulations FARs). and is found to be in a condition for safe operation. The Certificate must be displayed in the aircraft according to applicable Regulations.

When issued, it indicates that the aircraft must be operated in accordance with the limitations specified for the category. It also Indicates that the aircraft is considered in a condition for safe operation at the time of inspection and issuance of the Certificate. Any exemptions from the applicable airworthiness standards are briefly noted or the word NONE will be indicated if no exemption exists. The Airworthiness Certificate is in effect indefinitely if the aircraft is maintained in accordance with FAR Part 01, General Operating and Flight Rules, which requires inspections and maintenance as necessary to keep the aircraft in a condition for safe operation.

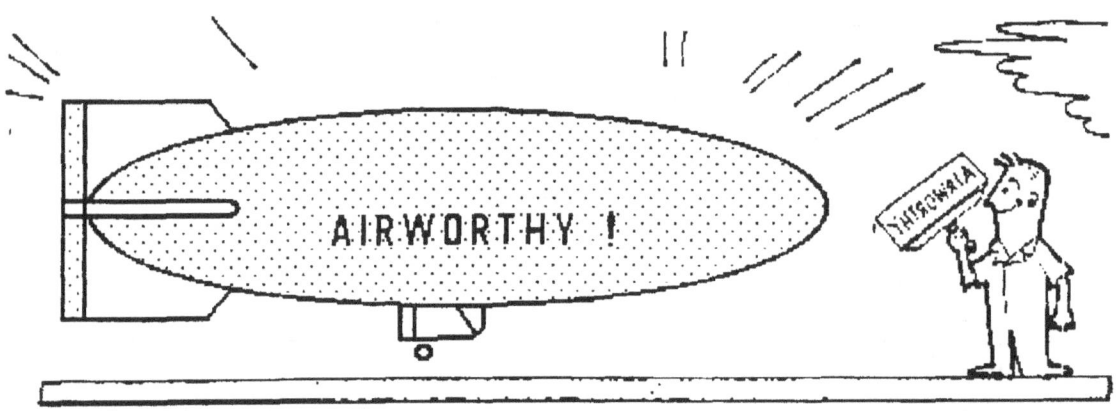

OPERATING LIMITATIONS

When the FAA Airworthiness Specialist invites you over to pick up your certificate, he will also hand you another piece of paper. It will tell you just who can fly (designated pilot - you only), where you can fly (within 25 nautical miles of point "X"), where you can't fly (over any people, places, or things), and when (between sunrise and sunset). Just be aware that they don't like catastrophic surprises.

I DON'T EITHER

Building Small Gas Blimps

UNITED STATES OF AMERICA
DEPARTMENT OF TRANSPORTATION — FEDERAL AVIATION ADMINISTRATION
CERTIFICATE OF AIRCRAFT REGISTRATION

This certificate must be in the aircraft when operated

NATIONALITY AND REGISTRATION MARKS	AIRCRAFT SERIAL NO.
N 0000	6969

MANUFACTURER AND MANUFACTURER'S DESIGNATION OF AIRCRAFT

ISSUED TO:
SMITH, MOE M.
102 E. 9th Street
Kansas City, Missouri
12903

This certificate is issued for registration purposes only and is not a certificate of title. The Federal Aviation Administration does not determine rights of ownership as between private persons.

It is certified that the above described aircraft has been entered on the register of the Federal Aviation Administration, United States of America, in accordance with the Convention on International Civil Aviation dated December 7, 1944, and with the Federal Aviation Act of 1958, and regulations issued thereunder.

DATE OF ISSUE: 1/20/70
Administrator

DURATION - See reverse side

AC Form 8050-3 (4-69) SUPERSEDES PREVIOUS EDITION

DEPARTMENT OF TRANSPORTATION
FEDERAL AVIATION ADMINISTRATION
AERONAUTICAL CENTER
P.O. BOX 25082
OKLAHOMA CITY, OKLAHOMA 73125

POSTAGE AND FEES PAID
FEDERAL AVIATION ADMIN.

OFFICIAL BUSINESS

TO:
SMITH, MOE M.
102 E. 9th Street
Kansas City, Missouri
12903

CHANGE OF ADDRESS

Federal Aviation Regulations require that you report within 30 days any change in permanent mailing address. A revised Certificate of Aircraft Registration will be issued without charge.

REPLACEMENT OF CERTIFICATE

If this certificate is lost, destroyed, or mutilated, a duplicate may be obtained at the written request of the holder. Send your request and $2.00 (check or money order made payable to Federal Aviation Administration) to:

FAA Aircraft Registry
P.O. Box 25082
Oklahoma City, Oklahoma 73125

NOTE: All correspondence should describe the registration "N" number, make, model, and serial number of the aircraft.

EFFECT OF REGISTRATION

Section 501 (f) of the Federal Aviation Act of 1958 (49 U.S.C. 1401) provides: "...... Registration shall not be evidence of ownership of aircraft in any proceeding in which such ownership by a particular person is, or may be, in issue."

NOTICE

It is the responsibility of persons needing information as to recorded instruments affecting the aircraft identified in this certificate to make a personal search of the records or avail themselves of the services of an agent or attorney. Instruments affecting ownership and encumbrances are recorded by the Federal Aviation Administration for recordation. Such records are public records open for inspection at the FAA Aviation Records Building, Aeronautical Center, Oklahoma City, Oklahoma.

DURATION OF REGISTRATION

Each certificate of registration issued by the FAA is effective, unless suspended or revoked, until the date upon which:

a. ☐ The registration is cancelled at the written request of the registered owner.
b. ☒ The aircraft is totally destroyed or scrapped.
c. ☐ The registered owner loses his U.S. citizenship.
d. ☐ Thirty days have elapsed since the death of the registered owner.
e. ☐ The aircraft is registered under the laws of a foreign country:

(NAME OF FOREIGN COUNTRY) _____

f. ☐ The ownership of the aircraft is transferred to:

NAME _____
ADDRESS _____
CITY _____ STATE _____

UPON EXPIRATION FOR ANY OF THE FOREGOING REASONS, CHECK THE APPROPRIATE BOX ABOVE, SIGN IN INK BELOW, AND RETURN THIS CERTIFICATE TO:
FAA AIRCRAFT REGISTRY, P.O. BOX 25082, OKLAHOMA CITY, OKLA. 73125

SIGNATURE: *John Q. Public*   TITLE: N/A   DATE: 2/29/71

Aircraft Certification Standardization

Special Airworthiness Certificates

- Issued for aircraft that do not meet "standard requirements.
- ✓ Restricted, limited, and provisional airworthiness certificates.
  - ✓ Special flight permits.
  - ✓ experimental certificates.
- Lower level of airworthiness compensated by operating limitations.
- Do not meet requirements for international operations.
- Do not permit operation for compensation or hire.

Experimental Certificates

- Flight testing.
  - ✓ Research and development.
  - ✓ Show compliance with FAR's.
- Crew Training (applicant's).
- Exhibition.
- Air racing.
- Market surveys.
  - ✓ Sales demonstrations.
  - ✓ Customer crew training.
  - ✓ Operating amateur built aircraft.

Building Small Gas Blimps

SPECIAL AIRWORTHINESS CERTIFICATES
FAA FORM 8130-7

	UNITED STATES OF AMERICA DEPARTMENT OF TRANSPORTATION - FEDERAL AVIATION ADMINISTRATION **SPECIAL AIRWORTHINESS CERTIFICATE**	
A	CATEGORY/DESIGNATION Experimental	
	PURPOSE Operating Amateur-built Aircraft	
B	MANU-FACTURER NAME N/A	
	ADDRESS N/A	
C	FLIGHT FROM N/A	
	TO N/A	
D	N— 12345	SERIAL NO. AB1
	BUILDER Arthur E. Black	MODEL Black Special
E	DATE OF ISSUANCE November 1, 1968	EXPIRY November 1, 1969
	OPERATING LIMITATIONS DATED 11/1/68	ARE A PART OF THIS CERTIFICATE
	SIGNATURE OF FAA REPRESENTATIVE Edward E. Smith	DESIGNATION OR OFFICE NO. ASW EMIDO 20 4341

Any alteration, reproduction or misuse of this certificate may be punishable by a fine not exceeding $1,000 or imprisonment not exceeding 3 years, or both. THIS CERTIFICATE MUST BE DISPLAYED IN THE AIRCRAFT IN ACCORDANCE WITH APPLICABLE FEDERAL AVIATION REGULATIONS

FAA FORM 8130-7 (10/82)   SEE REVERSE SIDE

EXPERIMENTAL CATEGORY

	UNITED STATES OF AMERICA DEPARTMENT OF TRANSPORTATION - FEDERAL AVIATION ADMINISTRATION **SPECIAL AIRWORTHINESS CERTIFICATE**	
A	CATEGORY/DESIGNATION Experimental	
	PURPOSE Showing Compliance With Regulations	
B	MANU-FACTURER NAME N/A	
	ADDRESS N/A	
C	FLIGHT FROM N/A	
	TO N/A	
D	N— 12345	SERIAL NO. 100
	BUILDER Piper	MODEL PA-700
E	DATE OF ISSUANCE January 8, 1969	EXPIRY July 30, 1969
	OPERATING LIMITATIONS DATED 1/8/69	ARE A PART OF THIS CERTIFICATE
	SIGNATURE OF FAA REPRESENTATIVE John R. Doe	DESIGNATION OR OFFICE NO. DOA PC No.7

Any alteration, reproduction or misuse of this certificate may be punishable by a fine not exceeding $1,000 or imprisonment not exceeding 3 years, or both. THIS CERTIFICATE MUST BE DISPLAYED IN THE AIRCRAFT IN ACCORDANCE WITH APPLICABLE FEDERAL AVIATION REGULATIONS

FAA FORM 8130-7 (10/82)   SEE REVERSE SIDE

Building Small Gas Blimps

	Form Approved
	O.M.B. No. 2120-0018

APPLICATION FOR AIRWORTHINESS CERTIFICATE

US Department of Transportation
Federal Aviation Administration

INSTRUCTIONS — Print or type. Do not write in shaded areas; these are for FAA use only. Submit original only to an authorized FAA Representative. If additional space is required, use an attachment. For special flight permits complete Sections II and VI or VII as applicable.

I. AIRCRAFT DESCRIPTION

1. REGISTRATION MARK	2. AIRCRAFT BUILDER'S NAME (Make)	3. AIRCRAFT MODEL DESIGNATION	4. YR. MFR	FAA CODING
5. AIRCRAFT SERIAL NO	6. ENGINE BUILDER'S NAME (Make)	7. ENGINE MODEL DESIGNATION		
8. NUMBER OF ENGINES	9. PROPELLER BUILDER'S NAME (Make)	10. PROPELLER MODEL DESIGNATION	11. AIRCRAFT IS (Check if applicable) IMPORT	

II. CERTIFICATION REQUESTED

APPLICATION IS HEREBY MADE FOR: (Check applicable items)

A	1	STANDARD AIRWORTHINESS CERTIFICATE (Indicate category)			NORMAL	UTILITY	ACROBATIC	TRANSPORT	GLIDER	BALLOON
B		SPECIAL AIRWORTHINESS CERTIFICATE (Check appropriate items)								
	2	LIMITED								
	5	PROVISIONAL (Indicate class)	1	CLASS I						
			2	CLASS II						
	3	RESTRICTED (Indicate operation(s) to be conducted)	1	AGRICULTURE AND PEST CONTROL	2	AERIAL SURVEYING	3	AERIAL ADVERTISING		
			4	FOREST (Wildlife conservation)	5	PATROLLING	6	WEATHER CONTROL		
			7	CARRIAGE OF CARGO	0	OTHER (Specify)				
	4	EXPERIMENTAL (Indicate operation(s) to be conducted)	1	RESEARCH AND DEVELOPMENT	2	AMATEUR BUILT	3	EXHIBITION		
			4	RACING	5	CREW TRAINING		MKT SURVEY		
			0	TO SHOW COMPLIANCE WITH FAR						
	8	SPECIAL FLIGHT PERMIT (Indicate operation to be conducted, then complete Section VI or VII as applicable on reverse side)	1	FERRY FLIGHT FOR REPAIRS, ALTERATIONS, MAINTENANCE OR STORAGE						
			2	EVACUATE FROM AREA OF IMPENDING DANGER						
			3	OPERATION IN EXCESS OF MAXIMUM CERTIFICATED TAKE-OFF WEIGHT						
			4	DELIVERING OR EXPORT	5	PRODUCTION FLIGHT TESTING				
			6	CUSTOMER DEMONSTRATION FLIGHTS						
C	6	MULTIPLE AIRWORTHINESS CERTIFICATE (Check ABOVE "Restricted Operation" and "Standard" or "Limited" as applicable)								

III. OWNER'S CERTIFICATION

A. REGISTERED OWNER (As shown on certificate of aircraft registration)   IF DEALER, CHECK HERE ⟶

NAME	ADDRESS

B. AIRCRAFT CERTIFICATION BASIS (Check applicable blocks and complete items as indicated)

AIRCRAFT SPECIFICATION OR TYPE CERTIFICATE DATA SHEET (Give No. and Revision No.)	AIRWORTHINESS DIRECTIVES (Check if all applicable AD's complied with and give latest AD No.)
AIRCRAFT LISTING (Give page number(s))	SUPPLEMENTAL TYPE CERTIFICATE (List number of each STC incorporated)

C. AIRCRAFT OPERATION AND MAINTENANCE RECORDS

CHECK IF RECORDS IN COMPLIANCE WITH FAR 91.173	TOTAL AIRFRAME HOURS		3	EXPERIMENTAL ONLY (Enter hours flown since last certificate issued or renewed)

D. CERTIFICATION — I hereby certify that I am the registered owner (or his agent) of the aircraft described above, that the aircraft is registered with the Federal Aviation Administration in accordance with Section 501 of the Federal Aviation Act of 1958, and applicable Federal Aviation Regulations, and that the aircraft has been inspected and is airworthy and eligible for the airworthiness certificate requested.

DATE OF APPLICATION	NAME AND TITLE (Print or type)	SIGNATURE

IV. INSPECTION AGENCY VERIFICATION

A. THE AIRCRAFT DESCRIBED ABOVE HAS BEEN INSPECTED AND FOUND AIRWORTHY BY (Complete this section only if FAR 21.183(d) applies)

2	FAR PART 121 OR 127 CERTIFICATE HOLDER (Give Certificate No.)	3	CERTIFICATED MECHANIC (Give Certificate No.)	6	CERTIFICATED REPAIR STATION (Give Certificate No.)
5	AIRCRAFT MANUFACTURER (Give name of firm)				

DATE	TITLE	SIGNATURE

V. FAA REPRESENTATIVE CERTIFICATION

(Check ALL applicable blocks in items A and B)

A. I find that the aircraft described in Section I or VII meets requirements for

B. Inspection for a special flight permit under Section VII was conducted by:

				THE CERTIFICATE REQUESTED			
			4	AMENDMENT OR MODIFICATION OF CURRENT AIRWORTHINESS CERTIFICATE			
			FAA INSPECTOR	FAA DESIGNEE			
			CERTIFICATE HOLDER UNDER	FAR 65	FAR 121, 127 or 135	FAR 145

DATE	DISTRICT OFFICE	DESIGNEE'S SIGNATURE AND NO	FAA INSPECTOR'S SIGNATURE

FAA Form 8130-6 (11-88) SUPERSEDES PREVIOUS EDITION

4-35

???
Small SPORT Airships

LICENSING
Regulations

Which way to go?

*If your blimp weighs **MORE** than 496 pounds empty* ↙   ↘ *If your blimp weighs **LESS** than 496 pounds empty*

FAR 91 — (aircraft) — **FAR 103**
FAR 61 — (pilots) — Waiver qualifications

see your local FAA 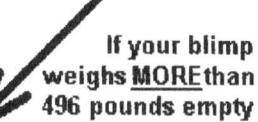   contact the

Building Small Gas Blimps

POB 589
Marshall, Mi.
49068-0589

PHONE 616-781-4021
FAX   616-781-7400

PRESS RELEASE   May 11, 1995

Aero Sports Connection Granted Two-place Training Exemption

On May 9th, the Federal Aviation Administration granted ASC a Two-place Training Exemption for the purposes of improving ultralight instruction and safety. Aero Sports Connection is the nonprofit organization designed to support the ultralight flying community, for the good of the Ultralight Community. ASC supports a club network and safety programs such as vehicle registration, pilot registration and competition recording systems. This support is designed to cross the boundaries of vehicle type.

The two-place training exemption allows ASC to administer a program of dual instruction in exempt vehicles which meet specific requirements. The exempt vehicles may have two seats but must have, an empty weight not to exceed 496 pounds, a tank capacity not to exceed 10 gallons, a stall speed not to exceed 35 knots, and a maximum speed not to exceed 75 knots. In addition, the instructor must be a person recognized by ASC as qualified to give instruction. This exemption joins with the ASC Basic Flight Instructor (BFI) and Advanced Flight instructor (AFI) Programs to support the "open structure" ASC Pilot Registration program. These safety programs are designed to allow qualified instructors to teach ultralight pilots through dual instruction in two-place vehicles. The exemption will allow ASC to help in expanding the availability of ultralight instruction in the ultralight flying community.

ASC's "open structure" programs are designed to improve participation by making them open to both ASC members and non-members. The ASC programs of "Vehicle Registration", "Pilot Registration", "Informal Data System" (competition), and "National Club Cross Country Awards" are all "open structured" programs. These programs are now significantly enhanced by the addition of two-place instruction and the associated BFI and AFI programs under the exemption.

ASC is supporting the ultralight flying community, for the good of the ultralight community. For more information, write:

ASC
POB 589
Marshall, MI 49068-0589 or, E-mail KIMOjim@AOL.COM
Jim Stephenson
CEO, ASE

2plpr04

Building Small Gas Blimps

POB 589
Marshall, Mi.
49068-0589

PHONE 616-781-4021
FAX 616-781-7400

AERO SPORTS CONNECTION
SUPPORTING THE ULTRALIGHT FLYING COMMUNITY
FOR THE GOOD OF THE ULTRALIGHT COMMUNITY

Annual Membership includes:
- Ultraflight Magazine
- National Wing organizations
- Unified National representation
- Club and Airpark Network
- Vehicle registration program
- Pilot registration program
- Competition recording
- 2 Place Training Exemption

You will have to select a Wing of your interest: (This does not limit you, it lets ASC know where to fund.)
 __ **NAPPF** (Para Wing) - Supporting the Powered Parachute Community.
 __ **Designer Wing** - Supporting the Ultralight Designers.
 __ **Training Wing** - Supporting flight training in the Ultralight Community.
 __ **Powered Para Glider Wing** - Supporting the foot launched powered chute community.
 __ **Trike Wing** - Supporting the trike and flex wing community.
 __ **Fixed Wing** - Supporting the fixed wing community.
 __ **General Wing** - Temporary selection until a perferred Wing is defined.

If a new Wing were to form, what Wing type would you support? _____
If you perfer your name may be withheld from outside mailing lists, if so, initial here. _____
When you renew you will be required to vote for ASC officers. The renewal form will provide the details.

Membership is $40 annually:

Name _____

Address _____

City _____ State ___ Zip _____

Phone _____ Date of Birth _____

Signiture _____

Make checks payable to: ASC, POB 589, Marshall, MI 49068-0589
    phone 616-781-4021 (after 1 PM EST)   ASC is a registered service mark. all rights reserved.

Ascappl7

Building Small Gas Blimps

optional program

POB 589
Marshall, Mi
49068-0589

PHONE 616-781-4021
FAX   616-781-7400

AERO SPORTS CONNECTION
SUPPORTING THE ULTRALIGHT FLYING COMMUNITY
FOR THE GOOD OF THE ULTRALIGHT COMMUNITY

Vehicle Registration Request Form

ASCVRRF1
10/94

This is the form used to request an ultralight vehicle registration number from ASC. The form should be completed and returned to ASC at the above address with a check for $25. Vehicle registration applies for ultralight vehicles and exempt vehicles used for training purposes.

At completion of registration you will recieve:
   1) a set of vehicle registraion numbers unique to your vehicle. These will consist of an "A", followed by two numbers and three letters. Where possible the letters will be your initials. Custom numbers may be requested for an additional $10 if they are not already taken.
   2) two sets of vinyl numbers in black vinyl. The owner may prepare his own letters if replacement or alternate color is required. These are 3" high letters to be placed on the vertical tail, if possible. For example, on powered parachutes with only tube fuselages 1 1/2 inch letters are acceptable.
   3) an ASC card showing the valid vehicle registration and identification of the owner. You do not have to be an ASC member to register your vehicle, the card will simply show "non-member". Members will recieve updated membership cards with the registartion shown.

Required information:

owners name(include middle initial)_____
owners address_____
owners phone number_____ spec. type *_____
vehicle maker _____ vehicle model _____ yr ____
engine type and HP_____
serial no. or unique VIN _____

Mail the data and check payable to ASC to:

ASC
POB 589
Marshall, Michigan 49068-0589
phone 616-781-4021   *pfw= powered fixed wing, ppw= powered para wing, rtw= rotor wing,
                      *pws= powered weight shift, ppg= powered para glider foot launch,
                       *pla= powered lighter than air, prb= powered rocket belt,
                      (substitute u for p for unpowered)

www.ingramcontent.com/pod-product-compliance
Lightning Source LLC
Chambersburg PA
CBHW080653190526
45169CB00006B/2094